Insects of the Great Lakes Region

**DO NOT REMOVE
CARDS FROM POCKET**

GREAT LAKES ENVIRONMENT

Matthew M. Douglas, Series Editor

A Focus on Peatlands and Peat Mosses
Howard Crum (Warren H. Wagner, Jr., Series Editor)

Mammals of the Great Lakes Region: Revised Edition
Allen Kurta

Ancient Life of the Great Lakes Basin: Precambrian to Pleistocene
J. Alan Holman

Insects of the Great Lakes Region
Gary A. Dunn

Insects of the Great Lakes Region

Gary A. Dunn

Matthew M. Douglas
Series Editor

Ann Arbor
THE UNIVERSITY OF MICHIGAN PRESS

Copyright © by the University of Michigan 1996
All rights reserved
Published in the United States of America by
The University of Michigan Press
Manufactured in the United States of America
⊛ Printed on acid-free paper

1999   1998   1997   1996   4   3   2   1

A CIP catalog record for this book is available from the British Library.

Library of Congress Cataloging-in-Publication Data

Dunn, Gary A.
    Insects of the Great Lakes Region / Gary A. Dunn.
        p.   cm. — (Great Lakes environment)
    Includes bibliographical references and index.
    ISBN 0-472-09515-3 (hc : alk. paper). — ISBN 0-472-06515-7 (pbk :
alk. paper)
    1. Insects—Great Lakes Region.   I. Title.   II. Series.
QL473.D85   1996
595.70977—dc20   95-50882
                                                                        CIP

Preface

Insects are by far the largest and most diverse group of animals on our planet. Most people have had at least one memorable encounter with insects—good or bad—that sticks out in their mind; but despite such encounters, the average person knows very little about the amazing world of insects.

I was oblivious to these wonderful creatures for most of the early part of my life. Like many other people, I made an insect collection for a Boy Scout merit badge and a high school class assignment, but for the most part I rarely noticed insects. When I was a freshman in college I decided, on the spur of the moment, to take an introductory entomology course to help fill my class schedule. That decision changed my life forever. After completing several more insect courses I knew that a career in entomology was my calling.

I obtained an advanced degree in entomology from the University of New Hampshire and went on to become a professional entomologist at Michigan State University. My enthusiasm for insects exceeded the boundaries of my employment, and I found myself collecting and studying insects even in my spare time. I spent weekends and vacation time searching for my favorite insects—tiger beetles and ground beetles. My search took me to remote places throughout the United States and the Great Lakes region. These trips gave me an opportunity to see for myself the magnificent grandeur of the United States. I have many fond memories of the places I visited, and no one can ever take these memories away from me.

More recently, I have developed an interest in entomology youth education. I have come to enjoy sharing my knowledge and enthusiasm about insects with young people. These experiences have their own unique rewards. My role as director of education for the Young Entomologists' Society and editor of *Young Entomologists' Society Quarterly* and *Insect World* have brought me into contact and friendship with hundreds of youth and adult "bug" enthusiasts around the world. I have learned a great deal from these people, and I hope they have learned a lot from me.

The purpose of this book is to assist you in recognizing some common insects of the Great Lakes region and to summarize information on their interesting habits, unusual lifestyles, changing distributions and populations, and economic importance. One of my primary goals in preparing this book was to deemphasize (as much as possible) technical entomological jargon, such as scientific names and anatomical terms, so that the book will be appealing and informative to beginning entomophiles. I also wanted to supply enough useful information (bibliography) that this book would serve advanced entomophiles and professionals in their studies. I hope I have succeeded in this difficult task!

Every book on insects has its limitations. This book is no exception. In the Great Lakes region alone there are at least 20,000 different kinds of insects. Quite simply there are just too many of these six-legged creatures to discuss each and every one. Therefore, I have been selective and present only the most significant insects of this region. Many of these are the common insects that you will most likely see any time you investigate the insects in your "neck of the woods", but others are rare and unique to the Great Lakes region. If and when you desire more detailed information on any particular insect or insect group that occurs in the Great Lakes states, I have provided you with lists of additional identification resources.

You might not plan a whole vacation just so you can search for a rare beetle, but you can certainly find a pleasurable activity in watching butterflies or bees at backyard flowers on a sunny afternoon. You do not need professional training to appreciate insects. I hope that this book will awaken you to the world of insects and encourage you to carry out your own entomological investigations.

Contents

Chapter 1

Geological and Biological History of the Great Lakes Region

The Great Lakes region lies one-third of the way between the Atlantic and Pacific coasts of North America, in the temperate region between 50° and 40° N. latitude. The boundaries of the region can be defined in two different ways. Using the political borders of those states and provinces that have parts of the five Great Lakes within their boundaries, the region includes New York, Pennsylvania, Ohio, Indiana, Illinois, Wisconsin, Michigan, Minnesota, and Ontario (see map 1). Of course insects are notorious for their complete disregard of political boundaries, and these imaginary boundaries in no way limit the distribution of insects in the Great Lakes region.

Using naturally established drainage patterns defines the region in a different way, by including only the land where water drains directly into one of the five Great Lakes. This puts major portions of Pennsylvania, Indiana, Illinois, and Minnesota outside the drainage basin. New York, Ohio, Ontario, and Wisconsin are represented in the region by moderately large areas (30 or more counties). Michigan is the only state that lies entirely within the Great Lakes drainage basin.

Biologically speaking, it perhaps makes more sense to use the natural drainage basin to delineate the boundaries of the Great Lakes region, because the distributions of insects have been, and continue to be, influenced by geological processes, climate, and the evolving bioscape. However, since collection of data on the insects of the Great Lakes region is definitely controlled by political considerations, and since insect distribution is not, I found it necessary to compromise a little bit. Therefore, when making references in this book about the distribution of an insect or insect group in the Great Lakes region, it was often necessary to include information from areas in Great Lakes states that are not located in the Great Lakes drainage basin. As our knowledge about the distribution of the region's insect fauna improves, it may be possible someday to abide by the stricter definition of the Great Lakes region.

The focal point of the region's uniqueness lies in the five Great Lakes (Superior, Michigan, Huron, Erie, and Ontario) and the St. Lawrence River. The lakes are in large part responsible for many of the current conditions.

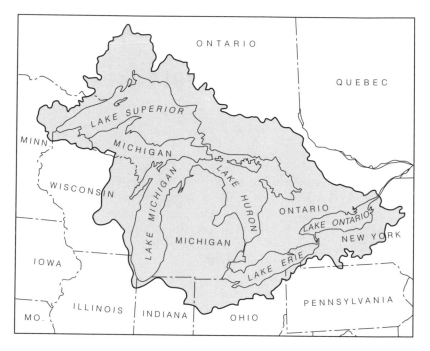

The drainage basin of the Great Lakes. (From Kurta 1995.)

They have influenced geological processes, the weather and climate, and the establishment of plant and animal communities. It is generally accepted that the Great Lakes did not exist at all in preglacial times. Therefore, the Great Lakes as we know them today came into existence very recently (geologically speaking), with the final development taking place less than 3,300 years before present (B.P.).

The Great Lakes are so vast that our astronauts could actually see them from the moon. Interconnected by a series of short rivers, they span 2,400 miles from the interior of the continent all the way to the Atlantic Ocean. They hold 20 percent of the earth's supply of fresh water—about 6 quadrillion gallons. If this amount of water were uniformly spread across the entire continental United States, it would cover our nation with 9.5 feet of water! The entire Great Lakes watershed, the area where all the rivers and streams drain into the lakes, includes 295,000 square miles (767,000 km²).

Lake Superior is the northern most, and largest, of the Great Lakes. It is so large that it could hold all the water from the other Great Lakes and still have room to hold the water from three additional lakes the size of Lake Erie. It also has the distinction of being the largest freshwater lake in the world (by surface area). The lake is 350 miles (563 km) in length, about 160 miles (259 km) wide, and has a surface area of 2,934 square miles (82,100 km²). It is a cold, deep lake, with an average depth of 489 feet (149 m); the greatest depth is 1,333 feet. Water flows into the lake from a variety of small streams and rivers in its drainage basin, which covers 49,300 square miles (127,700 km²) and includes parts of Ontario, Minnesota, Wisconsin, and the Upper Peninsula of Michigan. The shoreline, including the islands, is 2,726 miles (4,385 km) long. There are nine major islands, including world-famous Isle Royale (Keweenaw Co., Michigan).

Lake Huron is the second largest of the Great Lakes and has the longest shoreline of any of the lakes (3,827 mi [6,157 km]). It is 206 miles (331 km) long and 183 miles (294 km) wide with a surface area of 22,973 square miles (59,500 km²). Georgian Bay, the northeasternmost part of the lake, is almost a sixth great lake, because it is nearly separated from the rest of the lake by Manitoulin Island and the Bruce Peninsula. Lake Huron has an average depth of 194 feet (59 m) and a maximum depth of 748 feet (229 m). Eleven major rivers and many small streams flow into the lake from a drainage area that covers 50,700 square miles (131,300 km²) and includes parts of Ontario and Michigan. The lake has more than 30,000 islands, at least ten of them of significant size to be inhabited at one time or another. Manitoulin Island, the largest of the Lake Huron islands, is the largest freshwater island in the world.

Lake Michigan is the third largest of the Great Lakes and the sixth largest freshwater lake in the world. The lake is 307 miles (494 km) long and 118 miles (190 km) wide, and it has a surface area of 22,278 square miles (57,750 km²). The lake has an average depth of 279 feet (85 m) and reaches a maximum depth of 925 feet (282 m). The lake actually has a documented lunar tidal movement of minimal proportions. Ten major rivers and many streams flow into the lake from a drainage area that covers 45,598 square miles (118,100 km²) and includes parts of Wisconsin, Illinois, Indiana, and Michigan. Lake Michigan is the only one of the Great Lakes situated entirely within the boundaries of the United States. The shoreline, including the shoreline of the islands, is 1,659 miles (2,670 km) long. There are three major island groups: the Beaver Islands (Charlevoix Co., Michigan), the Fox Islands (Leelanau Co., Michigan), and the Manitou Islands (Leelanau Co., Michigan).

Lake Erie is the southernmost and fourth largest of the Great Lakes. It is the eleventh largest lake in the world. The lake is 210 miles (338 km) long and 57 miles (92 km) wide, with a surface area of 9,906 square miles (25,657 km²). Lake Erie only has an average depth of 62 feet (19 m), with a maximum depth of 210 feet (64 m), so it has the smallest volume of any of the Great Lakes. It is the second smallest of the Great Lakes (Lake Ontario is slightly smaller), but it is the shallowest and warmest, and it lies on rich soil. The lake averages 95 percent ice cover in the winter. The prevailing westerly wind pushes the water from one end of the lake to the other, and short-term differences in water levels between the western and eastern ends of the lake may be as much as 16 feet. Ninety-five percent of Lake Erie's water comes from the "upper" lakes via the St. Clair and Detroit Rivers, but some also comes from rivers and streams in a drainage area that covers 22,720 square miles (58,800 km²) and includes parts of Ontario, New York, Ohio, and Michigan. The shoreline, including the islands, is 871 miles (1,400 km) long. Major islands include Pelee Island (Essex Co., Ontario), Kellys Island (Huron Co., Ohio), and the Bass Islands (Ottawa Co., Ohio).

Lake Ontario is the smallest of the Great Lakes and is the fourteenth largest freshwater lake in the world. The lake is 193 miles (311 km) long and 53 miles (85 km) wide, with a surface area of 7,340 square miles (18,960 km²). The lake ranks fourth in its maximum depth (804 ft [245 m]), but its average depth of 282 feet (86 m) places it second after Lake Superior. Lake Ontario is 325 feet (95 m) below the level of Lake Erie to the west, with the incoming waters dropping over world-famous Niagara Falls. Most of the water (about 80 percent) comes from the "upper" Gr'eat Lakes and from rivers and streams that drain an area that covers 23,400 square miles (60,600 km²) and includes parts of Ontario and New York. The shoreline, including the islands, is 726 miles (1,168 km) long. There are only a few islands of significant size; the largest group, the Thousand Islands, is located at the northern end of the lake at the beginning of the St. Lawrence River.

Other lakes of significant size that are a part of the Great Lakes system are Lake Nipigon (1,870 mi^2 [4,825 km^2]), Lake St. Clair (460 mi^2 [1,175 km^2]), Lake Nipissing (400 mi^2 [1,010 km^2]), and Lake Simcoe (345 mi^2 [900 km^2]).

Waters from the Great Lakes reach the Atlantic Ocean via the St. Lawrence River which is 870 miles (1,200 m) long. Six tributaries feed into the St. Lawrence and double its flow from the volume that leaves Lake Ontario. Some of the water that enters the western or northern portion of Lake Superior will be retained in the Great Lakes system for up to 230 years before it reaches the Atlantic Ocean; water entering the southern end of Lake Michigan will be retained for 130 years.

FORMATION OF THE GREAT LAKES

The formation of the Great Lakes is a rather recent geological event and is the result of many changes in the geological landscape of the region. The existence of these lakes was made possible over time by several significant agents of change—namely, shifting bedrock and the movements of ice, water, and wind. The process that eventually led to the formation of the Great Lakes actually began about 3.5 billion years before the present (in a period of time known as the Precambrian era) with a very long period of rock formation and evolution. Over billions of years the region was subject to explosive volcanic activity, followed by long periods of erosion and shifting of the rock substrates. Eventually the region was inundated periodically by shallow seas. About 300 million years B.P., conditions stabilized somewhat and the land was eroded into northern highlands and southern highlands with a series of vast lowlands between them. The lowlands were transformed into a system of vast valleys with northward flowing rivers (the Michigan Basin). This development set the stage for the most significant geological event in the history of the Great Lakes region.

During the Pleistocene epoch, starting about 2 million B.P., a great ice age began. The world's climate cooled somewhat and the winter period became much longer. The frequent snowfalls began to accumulate into ice, because there was no period of melting during the short summers. A great ice cap (continental glacier) expanded southward, eventually covering the eastern half of North America as far south as the Ohio River and Missouri River valleys. The ice sheet did not advance like an ice cube sliding down a sloped surface; rather, it expanded like a mass of pancake batter that spreads out from its center because of the weight of the mass. In northern regions the ice was 3 to 6 miles thick; in southern regions, such as in the vicinity of Chicago, it was up to 1 mile thick. Glaciers advanced and retreated into the Michigan

Basin four times. Each time, they followed the preexisting "valleys" within the basin, gouging them deeper and deeper.

The last, or Wisconsinan, glaciation was the most important in the development of the modern-day Great Lakes and lasted from 75,000 to 10,000 B.P. When the glacier advanced across the Michigan Basin it moved most rapidly in the lowlands, where the ice was thicker and heavier. Thus the forward edge of the glacier advanced in three tonguelike lobes: one in the "Lake" Michigan basin, one in the Saginaw basin, and one in the "Lake" Erie/Huron basin. The three lobes of ice worked like independent glaciers, and the ice moved not only south but laterally to the east and west. As the glacier advanced past the entire Michigan Basin it reunited in the areas that are now Ohio, Indiana, and Illinois. The glacier reached its southernmost limit about 20,000 B.P.

When the glacier retreated northward, which occurred in a series of back-and-forth seesaw movements beginning in 14,800 B.P., the ice front once again assumed a distinctly lobate form. Large amounts of meltwater flowed from the glacier and became trapped between the glacier and the glacial debris (terminal moraines) to the south. As a result, crescent-shaped glacial lakes developed between the ice mass and these "dams."

The first postglacial lakes were followed by a long and complex succession of transitional lakes that ultimately led to the current configuration of the Great Lakes. The many glacial lakes that formed left telltale clues about their existence. Like the modern-day Great Lakes, they created an assortment of beaches and dunes, sand bars, and lake beds that we can still see today— even though they are now far "inland."

Over the thousands of years of lake development and evolution, dozens of lakes were formed and drained. It is generally accepted that during this process the Erie basin was occupied by four glacial lakes while the Huron basin was occupied by seven lakes. The combined Erie and Huron basins were also occupied by four different lakes. The Ontario basin was occupied by two intermediate glacial lakes. The Superior basin was occupied by five lakes while the Michigan basin had two lakes. The combined Superior, Michigan, and Huron basins were occupied by two glacial lakes.

At their peak the early glacial Great Lakes covered an area considerably greater than the current 94,000 square miles (244,000 km²). One other great glacial lake, Lake Agassiz, formed to the northwest of the present-day Great Lakes, and it once covered 110,000 square miles (280,000 km²) in the area that is now Minnesota, North Dakota, Manitoba, and Saskatchewan. Lake Winnipeg is a modern-day remnant of this once-great lake. Another large lake, Lake Barlow-Ojibway, was located 90 miles (145 km) to the northeast of Lake Superior in the Ontario clay belt.

Readers who would like a more complete account of the formation of the

Great Lakes are encouraged to read *Outline of the Geological History of Michigan* (Martin 1952), *Canada as an Environment for Insects* (Munroe 1956), *Glacial History of the Great Lakes* (Zoltai 1968), *Geology of Michigan* (Dorr and Eschman 1970), and *Atlas of Michigan* (Sommers 1977).

LANDFORMS OF THE GREAT LAKES REGION

The Wisconsinan glacier left its imprints on the landscape of the Great Lakes region. The glaciers provided the sediments that form today's landscape: millions of tons of rock and sediment that were deposited across the landscape. These deposits, varying in depth from a few inches up to thousands of feet, now cover the bedrock. The glacial meltwaters sorted many of these sediments and formed broad outwash plains. Forward thrusts of the glacier formed parallel sets of moraine. Chunks of ice that broke off of the main glacier and became buried in the sediments eventually melted, leaving large holes that filled with water and formed "pothole" lakes.

The Great Lakes region can be characterized as an area of low relief composed of more or less flat lacustrine (lake) plains with scattered hill lands and highlands composed of glacial debris or remnants of bedrock (often covered with glacial debris as well). The elevation in the Great Lakes region ranges from less than 250 feet (77 m) above sea level up to 2,300 feet (700 m). The highest points in the region are Eagle Mountain, Cook County, Minnesota: 2,301 feet (701 m) elevation; Mt. Arvon, Baraga County, Michigan: 1,979 feet (604 m) elevation; and Timms Hill, Price County, Wisconsin: 1,951 feet (595 m) elevation.

The area west and northwest of Lake Superior (the former lake bed of Lake Agassiz) and the area east and northeast of Lake Huron (former Lake Barlow-Ojibway) are broad lacustrine plains and lowlands filled with many lakes and marshes. North and northeast of Lake Superior there are some remnants of the Canadian Shield (Precambrian bedrock), and these higher areas separate the watersheds of Lake Superior from those of James Bay. South of Lake Superior, in parts of northwestern Wisconsin, the western Upper Peninsula of Michigan, the north central portion of the Lower Peninsula, the Irish Hills of southeastern Michigan, and parts of Ontario (between Lake Simcoe and the Ottawa River), there are hilly uplands and highlands ranging from 1,000 to 1,800 feet (550 m) in elevation. The eastern and southeastern part of Wisconsin, Illinois, northern Indiana, northwestern Ohio, and southern Ontario (including the Ottawa River and St. Lawrence River valleys) are relatively flat till plains, lacustrine plains, and lowlands. The highest spot in Indiana, 1,257 feet (375 m), is located on the till plains of the east central part of the state. Central and northeastern Ohio, western Pennsylvania, and western New York (except for the lake border plains) are highlands and mountain remnants of the once mighty Allegheny and Adirondack ranges of the Appalachian Mountains.

SOILS OF THE GREAT LAKES REGION

Glacial drift (moraine, till, outwash, and lake sediment) is the parent material for virtually all the present-day soils in the Great Lakes region. These soils developed over time through changes caused by weather (moisture and temperature), topography (elevation, slope, and aspect), and biological organisms (especially plants). Areas in the southwestern part of the region have soils that developed from primitive soils with annual incorporation of prairie grasses. Those soils in the central and eastern part of the region developed under deciduous broadleaf forests, while those of the northern parts of the region developed under conifers and mixed hardwoods. Soils range from well-drained sands and gravels of hill lands and highlands to poorly drained mucks of lowlands and bottomlands.

The Great Lakes region has the largest collection of freshwater sand dunes in the world. In Michigan alone there are 275,000 acres of coastal dunes. Great Lakes sand dunes range from under 4 feet (2 m) to almost 100 feet (31 m) in height. In the Great Lakes region there are also many examples of "perched dunes" (sand dunes built high atop glacial moraine). Famous examples of perched dunes include Sleeping Bear Dunes of northern Lake Michigan, the Manitou and Fox Island dunes of northern Lake Michigan, and the Grand Sable dunes of southern Lake Superior.

As the dunes grow, the sand supply is restocked by the never-ending process of erosion, sorting, and transport by wind and water. In time, however, growth and migration of most dunes is slowed and the dunes are stabilized by dune grasses and other plants, leading to the development of a rich dune community. The diverse plant and animal life associated with these dunes is unique to the Great Lakes region. These dunes are fertile because of the moderate climate and the availability of moisture (in the form of rainfall and a water table close to the surface).

Sand dune formations are not equally distributed throughout the Great Lakes region. The most highly developed dunes occur on the eastern shore of Lake Michigan, the southern shore of Lake Superior, the eastern shore of Lake Huron below Georgian Bay, and the southeastern shore of Lake Erie. The pattern of lake currents and prevailing winds in these areas have been the most favorable for dune formation. Examples of large dune formations include the Whitefish Park and Kohler Park dunes of Wisconsin (western shore of Lake Michigan); the Illinois Beach dunes of northeastern Illinois (western shore of Lake Michigan); Indiana Dunes of northwestern Indiana (southern shore of Lake Michigan); the dunes of Holland, Muskegon, Silver Lake Parks, and the Warren, Saugatuck, and Sleeping Bear Dunes in Michigan (eastern shore of Lake Michigan); the Grand Sable dunes of Michigan's eastern Upper Peninsula (southern shore of Lake Superior); the Ipperwash Park dunes of southwestern Ontario (eastern shore of Lake Huron); and the Headlands Beach dunes of Ohio (southeastern shore of Lake Erie).

The "ancient" dunes of the many glacial Great Lakes can also still be found in the Great Lakes region. These dunes are now located many miles inland from the current lakeshores, at sites where the old shoreline was when the water levels were higher. Remnants of these dune systems can be seen in Tuscola, Saginaw, Midland, and Cheboygan Counties in lower Michigan; in Schoolcraft, Luce, and Mackinac Counties in upper Michigan; in Kittson County (Norway Park dunes) and Marshall County (Old Mill Park dunes) in Minnesota; in the Homewood and south Chicago areas of northern Illinois; and in northwestern Ohio (north of Findlay). In contrast to the current coastal dunes, which are only 4,500 years old, some of the dunes associated with these earlier glacial lakes were formed up to 13,000 years ago.

CLIMATIC FACTORS

Climate, especially minimum winter temperature and seasonal precipitation, directly and indirectly influences the distribution of many insect species (and the plants they are often dependent on) in the Great Lakes region. The exact nature of the climate during the glacial period is not fully understood, but most scientists agree that an ice age of this magnitude had to have had some influence on the world's climate. The prevailing opinion is that the ice mass had a predominately local impact on climate, with its influence decreasing with increasing southerly distance. For example, at the height of the Pleistocene Ice Age, central Florida probably had an average July temperature of 65°F (roughly equivalent to the present-day July mean temperature in Marquette, Michigan), with a winter average of around 35°F (Howden 1969). In the tropical areas of Central and South America there was apparently no cooling.

The post-Wisconsinan climate has gone from slightly cooler (and wetter) to present-day conditions (warmer and drier). After the glacier retreated from the Great Lakes region between 13,000 and 8,000 B.P. the climate was slightly cooler than today. During the period between 7,600 and 3,000 B.P. there was a so-called climatic optimum when the temperatures were (on average) about 4° warmer than today. There was another global cooling trend during the period from 2,500 to 1,000 B.P., followed by an exceptionally cold period known as the "Little Ice Age," which occurred from 1650 to 1750 C.E. The warming trend that started in the mid-nineteenth century seems to have reached a peak in the first half of the twentieth century (Thomas 1968).

Presently the climate in the Great Lakes region can be described as more or less continental, with many parts of the region's temperatures modified by the Great Lakes (areas near the shoreline are warmer in winter and cooler in summer and have more frost-free days than the interior). The mean annual temperature ranges from less than 35°F (2°C) in the northern part of the Great Lakes region to more than 50°F (10°C) in the southern part of the region. The mean January temperature ranges from less than 10°F (-12°C) in the north to greater than 27°F (-4°C) in the south. Average July temperatures range from less than 64°F (18°C) in the northern part of the region to greater than 74°F (22°C) in the southern part of the region. The length of the growing season, the period of time between last spring frost and first fall frost, ranges from less than 60 days in the northern part of the region to greater than 180 days in the southern part (190 days in northwestern Indiana in the vicinity of Lake Michigan).

The Great Lakes region is situated in the westerly wind belt of North America, so the westerly wind brings weather fronts and air masses into the region from the west. The lack of a significant barrier (mountains) permits the frequent and rapid intrusions of hot, humid weather from the Gulf of Mexico; cool, dry air from the Midwest; and cold, Arctic air from the north. The mixing of these contrasting air masses creates the potential for extremes in weather conditions.

Precipitation is for the most part uniformly distributed during the year, although the warmer, moister air of spring and summer tends to produce slightly more precipitation than the cooler, drier air of fall and winter. As a result there is a slight tendency toward being wettest in the spring. The region receives sufficient moisture, on average 10 to 15 inches (250–380 mm) per year (in some parts of Indiana up to 35 in), which is adequate to support vigorous plant and tree growth throughout the region. In the northern part of the region the bulk of the annual precipitation comes in the form of snow. The average annual snowfall increases from south to north and reaches its peak in the highlands of the western Upper Peninsula of Michigan, northern Wisconsin, and Minnesota. Other areas on the eastern edges of the Great Lakes also receive larger amounts of snowfall than inland areas because of lake-effect snowfall.

THE BIOSCAPE

Preglacial Status

So far we have discussed only the abiotic influences in the Great Lakes region—the geological processes (and the resultant regional landscape) and the general climatic conditions. While geological features do not directly determine the distribution of plant and animal species in the region per se, they have a strong influence on the availability of habitats by setting up repeating patterns of topographic features throughout the region. These topographic features (lakes, rivers, wetlands, dunes, uplands, caves, etc.), along with climate constraints, have more directly determined which species can occur within the region, provided the proper macro- and microhabitats are available.

What then do we know about the living creatures that have inhabited the Great Lakes basin? During the earliest parts of geological history the region was uninhabitable over long periods of time. But eventually, with the advent of warm inland seas during the late Cambrian period (500 million years B.P.), living creatures, including primitive arthropods, began to inhabit the Great Lakes basin. The fossil record shows that two groups of arthropods had species that occurred in the Michigan Basin during the Paleozoic era. Trilobites (members of the class Trilobita), marine bottom crawlers and swimmers, were once common in the region but are now extinct. Trilobite fossils have been found in Ontario, Ohio, Michigan (Dorr and Eschman 1970), and southeastern Minnesota (Tufford and Hogberg 1965). Ostracods (class Crustacea) also inhabited the Paleozoic seas, and their fossils have been found in Michigan (Dorr and Eschman 1970) and southeastern Minnesota (Tufford and Hogberg 1965).

We know very little about the preglacial insect inhabitants of the Great Lakes region, because the glaciers of the Pleistocene erased most records and evidence from earlier periods. Thanks to fossils found in other areas, however, we know that some primitive arthropods left the seas and took up residence in moist terrestrial habitats about 400 million B.P., that the first primitive insect species made their appearance about 350 million B.P., and that winged insects had taken to the air as early as 300 million B.P. We also know that during the Pennsylvanian period ("coal swamp age") the Michigan Basin was a vast swampland that was home to many unusual plants and animals, including giant dragonflies with wingspans in excess of 2 feet. There are many fossils of mayflies from as early as the Upper Carboniferous period, and they were especially common during the Permian period (250 million B.P.), when aquatic habitats were in great abundance. The appearance of flowering plants about 100 million B.P. led to a great diversification of insect species, especially among the butterflies and bees, which developed an intimate interaction with these plants.

Between 65 and 25 million years B.P., North America was covered by tropical and subtropical forests (undoubtedly inhabited by lots of insects). About 25 million B.P. the climate started to cool, and this cooling trend continued up until 500,000 years B.P. With the change in climate the tropical forests of North America were replaced by tundra, dry coniferous forests, and then temperate forests. Although the fossil record for insects is poor worldwide (and virtually nonexistent in the Great Lakes region), the small amount of available evidence does suggest that by the Oligocene epoch (mid-Tertiary period) most North American insect genera were present. In the time since, insects have become the most abundant and diverse group of living organisms on earth.

The Pleistocene or Great Ice Age

There is little doubt that in preglacial times the Great Lakes region had a rich and varied insect fauna, but the advance of the glaciers changed all that. The glaciers made it impossible for arthropods, and for all other animals and plants for that matter, to inhabit the Great Lakes region. For a period of 2.3 million years the area was alternately covered and uncovered in glacial ice. Periods of stability alternated with advances of 10 or more feet a day. Total movement of the ice front was on the order of not more than 175 to 200 feet per year (Howden 1969). Most plants and animals living in the northern regions were blanketed by snow and ice pack and smothered. Plants and animals in other parts of the region that could not move away from the advancing ice front were also covered and obliterated. Some species, however, may have been able to move to more favorable areas to the south, but it appears that migration out of the glaciated areas was very minimal, and that those plants and animals that survived south of the glacier were already residents of those unglaciated areas and were not affected by glacial ice. When we look at the distribution of many insects currently inhabiting these formerly glaciated areas, we find that at least a part of their range extends outside the glaciated areas.

When the glaciers occupied the Great Lakes basin the climate was undoubtedly cooler than it is today. For a distance of 10 to 50 miles ahead of the glacier there existed a treeless tundra, and beyond this tundra was a forest of spruce, pine, and other cold-hardy plants. These forests probably had many wet, marshy openings and grasslands. It is generally believed that the great eastern hardwood forests of North America located south of the glaciated areas remained mostly unchanged, except in the immediate vicinity of the ice front. In areas 10 to 50 miles distant from the ice front, life was pretty much the same as it had been for many thousands of years. During the summer large amounts of meltwater and sediment flowed away from the glaciers. In the winter the flow decreased to a trickle. Large cold-climate animals, such as mammoths, mastodons, giant beavers, elk, musk ox, ground sloths, and others, inhabited the land in the vicinity of the ice front. A few cold-hardy aquatic and terrestrial insects probably inhabited these areas as well.

From an evolutionary standpoint, the longer a major geographical area remains climatically stable, the more diversified are the flora and fauna that ultimately evolve there. Since the tropical areas of the world were not as directly affected by the Pleistocene glaciers, they have had a longer history of unaltered climate than the Great Lakes region. The climatic instability of the Great Lakes region explains at least in part why the insect fauna is so much smaller and less diverse than that of the New World tropics.

Postglacial Plant Dispersal into the Great Lakes Region

As the climate warmed and the ice retreated northward over a period of thousands of years, the barren land was gradually colonized by plants and animals. The recession rate of the glacier was relatively fast, probably melting back at a rate of 400 to 500 feet per year. The warming trend was frequently reversed, and at times the ice stopped retreating or even advanced again. The retreat of the ice was not uniform along its front, and areas in the western part of the Great Lakes region were uncovered thousands of years before some of the areas in the Lake Michigan basin, the Lake Huron basin, the Upper Peninsula of Michigan, and northern Ontario. It is safe to say that all the plant and animal species that exist in the Great Lakes basin today either established or reestablished themselves in the region during the past 10,000 years, after the glaciers left the area.

Plants, without which most insect species cannot survive, had to colonize the Great Lakes region after the retreat of the Wisconsinan glacier. While plants are not as mobile as animals, their spread back into the previously glaciated areas was assisted by water and wind. Before long the once barren landscape was repopulated with plants.

The areas uncovered by the glacial ice were more or less level and had poor natural drainage. They were therefore well suited to wet prairies and bog-loving trees (such as spruce and balsam fir). Studies of postglacial pollen deposits in southern Michigan show that boreal species (such as spruce, balsam fir, larch, and northern white cedar) were at first favored by the cool, damp conditions south of the retreating ice between 13,200 and 11,000 B.P. As the glacier retreated completely from the area, conditions in the southern part of the region became warmer and drier, and the boreal species were at first mixed with and then replaced by pines. For example, the boreal forest of southern Ontario (the area east and northeast of Lake Huron and east of Lake Superior) was replaced by deciduous trees by at least 8,000 B.P. In many areas of Michigan and Wisconsin jack pine became dominant by 8,000 B.P., and by 6,000 B.P. white pine was mixed in with the other pines. Between 5,000 and 3,500 B.P. great prairies, oak savannas, and oak-chestnut forests existed in southwestern Michigan (as well as in parts of Illinois, Indiana, and Ohio). Walnut, butternut, sycamore, elm, ash, beech, and basswood were abundant in southeastern Michigan and parts of Ohio and Ontario. Beginning around 4,000 B.P. cooler and moister conditions began to favor oak, hemlock, and other deciduous broadleaves, and the prairies receded. By 2,500 B.P. the increasing dryness in southwestern Michigan favored a pine-oak mix.

The forests of 500 years ago were different from the forests that occur in the Great Lakes region today. The forests of northwestern Pennsylvania, northwestern and western Ohio, southern Michigan, extreme northeastern Illinois, and southwestern Wisconsin were beech-maple forests. To the immediate north there was a broad zone between the deciduous forests of the eastern United States and the boreal forests of the northern United States and Canada. This zone was dominated by forests containing a mix of hemlock, white pine, maple, beech, and birch trees (so-called northern hardwood forests) and was located in Ontario (Huron region), northern Lower Michigan, Upper Michigan, northern Wisconsin, and northeastern Minnesota. The only area of true boreal forest (spruce-fir) existed in parts of northern Wisconsin, northern Minnesota, northern Michigan (extreme northern Lower Peninsula and eastern Upper Peninsula), and Ontario north of Lakes Huron and Superior. The boundaries between these different forest types were generally broad zones of transition with significant overlapping of tree species.

Other forest types are represented within these large forest regions. In parts of the northern Lower Peninsula of Michigan, where the soils are sandy and well drained, there were large areas of pine-oak forests. In the drier areas of southern Michigan, northwestern Ohio, and northeastern Indiana, oak-hickory forests were common. In some areas of southwestern Michigan and northwestern Indiana the upland oak forest gave way to oak savannah and prairie. Within the expansive beech-maple forests the beech and sugar maple occupied only the better sites; in depressions and low-lying areas elm-ash-red maple-cottonwood forests dominated, forming deciduous swamps, such as the Great Black Swamp of northwestern Ohio.

For readers wishing to know more about the woody plants of the Great

Lakes region, both past and present, *Geology of Michigan* (Dorr and Eschman 1970), *Atlas of Michigan* (Sommers 1977), *Michigan Trees* (Barnes and Wagner 1981), and *The Woody Plants of Ohio* (Braun 1989) are recommended.

Postglacial Insect Dispersal into the Great Lakes Region

As the glaciers began to retreat, animals also colonized uncovered areas. Insects undoubtedly migrated into these areas as well, and many of them had the advantage of winged flight, which allowed them to move long distances and bypass areas of unsuitable habitat. Very little evolution at the species level has occurred within the last 10,000 years, so nearly all insect species found in the Great Lakes region today already existed elsewhere at the time of deglaciation. Generally speaking there were three major pathways into the Great Lakes region. Some species moved into the region from the great Appalachian forest of eastern North America. Some traveled up a prairie peninsula that extended from central North America into northern Minnesota and western Ontario. Other species dispersed to the east along the transcontinental boreal forests of upper Canada, many of them originating from the glacial refugium known as Beringia, a rather large portion of Alaska and Siberia that remained unglaciated during the Pleistocene. Furthermore, the lower water levels of the oceans exposed the continental shelf and created a land bridge between Siberia and Alaska. Thus many of our northern boreal insects also occur in Siberia or have closely related species there.

The dispersal of insects into the Great Lakes region was relatively simple. There were no significant geological barriers, other than the lakes. At times insect dispersal into the newly exposed Great Lakes region was assisted by the developing Great Lakes, and at other times it was impeded by the lakes. Many of the transitional Great Lakes lasted for relatively short periods of time. Recolonization was frequently disturbed, because some areas that are now dry were at times under water, while other areas that are now under water were once dry land. The changing water levels exposed large islands and archipelagos, as well as large peninsulas and land bridges, only to cover all or part of them again with rising water levels. Eventually, as the entire basin was uncovered, insects took their place among the animal communities of the recolonized region. Stability returned to the Great Lakes region about 7,000 B.P., and in the time since then insects (and other organisms) have had a period of relatively stable conditions under which to continue colonizing the region.

Of course some of these insects are still in the process of expanding their ranges and have not yet occupied their potential range. Some of them are just slower than others. There are also many insects that have distributional ranges just south of the maximum extent of Wisconsinan glaciation and that to date have not moved back into the Great Lakes region in the 8,000 to 10,000 years that the ice has been gone. This phenomenon is observable in many insect groups, and the trend is unmistakable. Perhaps there are other limiting factors that are preventing some insects from inhabiting the region north of the terminal moraines.

Some modern-day insects are temporary and seasonal inhabitants of the Great Lakes region, migrating into the area on an annual basis during the warm season, only to be wiped out by the cold winters. Leafhoppers, milkweed bugs, gray hairstreak butterflies, and painted lady butterflies are examples of insects that migrate seasonally into the region. The monarch butterfly also migrates in and out of the region on a regular basis, leaving the area in the fall to avoid the winter conditions, and dispersing back into the region the following spring and summer from roosts in Mexico.

THE CURRENT GREAT LAKES ENVIRONMENT

The first postglacial human inhabitants moved into the region between 10,000 and 11,000 B.P. These Paleo-Indians were at first nomadic (hunters and gatherers) but later developed woodland settlements and raised some crops (corn, beans, and squash). These early native Americans had little impact on the environment (except for the occasional clearing of small areas with fire) (Cleland 1975).

The discovery and colonization of the Great Lakes region by explorers from the east led to many dramatic changes in the region. Starting about 350 years ago, European explorers, first the French and later the British, roamed the area and eventually established permanent settlements and forts. As a result of these settlers moving into the area the forests underwent many changes. Today the predominant forest type in southern parts of the region are maple-beech (the oak-hickory and elm-ash-cottonwood forests being considerably reduced). In northern areas pine-oak-aspen and aspen-birch forests have replaced many of the old northern hardwood and boreal forests. Still, despite extensive logging, the area remains heavily forested. The settlers also encountered small prairies in west central Ohio, southwestern Michigan, northwestern Indiana, and northern Illinois that were remnants of the more widespread grasslands of previous postglacial warm spells.

By the early 1800s an estimated 300,000 people lived in the Great Lakes region; today the region is home to over 37 million people (29.5 million Americans and 7.5 million Canadians). Nine percent of the U.S. population and 29 percent of the Canadian population live in the Great Lakes region.

This figure does not include the 5 million people who live west of Chicago, outside the actual Great Lakes basin, but depend on the Great Lakes for their water supply. On average there are 393.6 people per square mile. Of course the population is not evenly distributed throughout the region, and upwards of 73 percent of the population live in three major urban centers (Chicago, Detroit, and Toronto) and about two dozen other cities. The vast northern reaches of the region are sparsely populated in most areas.

Agricultural use of the land is on the decline in Michigan, shrinking from 42.5 percent in 1949 to 30 percent in 1982. As of 1977, 56 percent of Indiana's acreage was in cultivation. The forest, although different than the original presettlement forest, now covers more than 50 percent of the region. Between 1952 and 1987 there was a net increase in forested acres of 2,242,000 in Michigan alone. One hundred and sixty years ago Indiana was about 87 percent forested and the rest was tallgrass prairie. Today most of the land is in agricultural cultivation, and the once vast prairies have been reduced to a limited number of acres in fewer than six preserves. The Grand Marsh of the Kankakee, once 50,000 acres, was reduced by drainage to 28,500 acres by 1834 and to 10,000 acres by 1917, and it has since disappeared (Mumford and Whitacker 1982). Development of land for housing, shopping centers, resorts, highways, and water diversion projects has dramatically changed the face of the landscape throughout the region.

Humankind has also brought many new plants and animals into the region: birds, fish, and insects, to name a few. Many of these introduced plants and animals have become serious pests and have replaced less competitive native species, changing the bioscape dramatically.

Chapter 2

Entomological History of the Great Lakes Region

The record of studies on the postglacial insect fauna begins less than 250 years ago. Insect specimens were undoubtedly collected from the frontier areas of the Great Lakes region between 1750 and 1850, although there are very few written records of their collection. In the tradition of the period, nearly all specimens collected on the North American continent were sent back to the Old World for study and description by the leading European entomologists of the time, including C. Linnaeus, J. C. Fabricius, P. A. Latreille, C. DeGeer, W. Kirby, P. F. M. A. Dejean, G. A. Olivier, and others. Because these specimens from the New World typically had vague locality data, little information on insects that might have come from the Great Lakes region is available from this early period. One of the few papers with descriptions and distribution information was William Kirby's "Insects of British America" in Richardson's *Fauna Boreali-Americana*, published in 1837. Kirby provided sketchy information on 447 species known from eastern Canada at that time.

Two of America's most eminent entomologists put an end to this European exportation of entomological specimens for study. In 1806 Frederick V. Melsheimer, a Lutheran clergyman from Lancaster County, published *A Catalog of Insects of Pennsylvania*. Twenty-five years later, in the period between 1824 and 1828, Thomas Say published his *American Entomology or Descriptions of Insects from North America*, a work based on his many travels throughout the eastern United States. From this point on, American entomologists took a more prominent role in the description and identification of their own insect fauna.

Some of the earliest known collections of insects from the northern Great Lakes region were made in the mid-1800s. In 1844 Dr. John L. LeConte visited the Lake Superior region, traveling along the entire southern shore of the lake and heading west to the headwaters of the Mississippi River. The results of this expedition were published in 1850. In the period between 1842 and 1882 he assisted the Geological Survey of Canada with the identification of beetles collected by its survey teams in many remote parts of Canada, including the Lake Superior region.

In 1867 Professor Albert J. Cook began teaching a course in entomology—one of the very first to be offered anywhere in the United States—at the Michigan State Agricultural College. With the help of his students he began an insect collection at the college. Many of his first specimens came from within walking distance of the campus and are still housed in the Entomology Museum at Michigan State University. Henry G. Hubbard and Eugene A. Schwarz collected in the area in and around Detroit during 1874. Two years later, in 1876, they traveled all around Lake Superior by schooner and collected beetles. They repeated the trip the following year. The results of their efforts, "The Coleoptera of Michigan," were published in the *Proceedings of the American Philosophical Society* in 1878. This was later followed by Charles H. T. Townsend's "Contributions to a List of Coleoptera of the Lower Peninsula of Michigan" (1889) and "Hemiptera Collected in Southern Michigan" (1890).

During the last 40 years of the nineteenth century the insects of Illinois were studied, thanks primarily to the efforts of State Entomologists Benjamin Dann Walsh, William LeBaron, and Cyrus Thomas. Stephan A. Forbes, another leading entomologist from Illinois, worked as a curator for the Illinois State Natural History Museum (1872–76) and published many fine papers.

As a student at Indiana University, Willis S. Blatchley wrote papers on the butterflies of Indiana; in 1891 he produced a thesis on that subject. In 1894 he was elected the Indiana state geologist (on the Republican ticket), and he held the office for 16 years (until he was replaced by a Democratic candidate). Each year, he produced an extensive annual report on the botany, zoology (especially entomology), or geology of the state. From 1906 to 1910 he worked on his *Coleoptera of Indiana* (Blatchley 1910). He published many significant works between 1912 and 1926; in all he produced 80 papers on insect topics, describing 14 new genera and subgenera and 470 new species and subspecies, primarily in Coleoptera, Orthoptera, and Heteroptera (Favinger 1984).

In the later part of the nineteenth century a large number of entomological papers on insects of the Great Lakes region were published by the prominent entomologists of the period. Shortly after the turn of the century the rate of publication increased dramatically and many new and interesting papers on insects of the Great Lakes region were published. Complete citations to the articles of early entomologists who worked in the Great Lakes region can be found in Dunn (forthcoming). Readers wishing to know more about early entomologists of the Great Lakes region can find a great deal of interesting information in *American Entomologists* (Mallis 1971).

In recent times the studies and investigations of the insect fauna of the Great Lakes region has been concentrated in only a handful of different areas, including the Douglas Lake area (Emmet and Cheboygan Cos.), Beaver Island (Charlevoix Co.), Isle Royale (Keweenaw Co.), the Edwin S. George Reserve (Livingston Co.), the Huron Mountains (Marquette Co.), and Whitefish Point

(Chippewa Co.) in Michigan; Buffalo and vicinity in western New York; Byron Bog and Point Pelee in southwestern Ontario; the Pine-Popple River in northeastern Wisconsin; Cedar Point (Sandusky Co.) in Ohio; and Presque Isle (Erie Co.) in Pennsylvania.

CHANGES IN THE INSECT FAUNA OF THE GREAT LAKES REGION

Introduced Insects

Humans have had a significant impact, both positive and negative, on the insect fauna of the Great Lakes region during the past 100 years. However, not all the changes affecting the insect fauna can be neatly categorized as either distinctly positive or negative. Some changes have benefited some species of insects at the expense of others.

Deforestation, for example, has negatively affected leaf-feeding insects but has benefited certain wood-boring insects. The replacement of natural vegetation with "plantations" of both trees and crops, which create large monocultural acreages, has benefited selected crop-feeding insects at the expense of native woodland and grassland insects. Draining swamps has an adverse effect on aquatic insects to the benefit of terrestrial insects. Damming rivers creates new habitat for some aquatic and semiaquatic insects at the expense of terrestrial insects. Almost any change to the landscape, including urbanization, has adverse effects on one group or species of insects and beneficial effects for other insects.

Humans have also directly and indirectly introduced new insects into the region. Many of these introduced insects have become serious economical and ecological problems (e.g., the gypsy moth). Still other predatory and parasitic insects have been introduced to combat the pest insects. Many insects are excellent hitchhikers and have been introduced into the region as passengers on automobiles, trucks, and ships or have been dispersed here after establishing themselves elsewhere in the United States or Canada. Back in the middle of this century, when the European earwig (*Forficula auricularia* L.) was still in its early stages of dispersing throughout the Great Lakes region, distribution maps of its range showed an interesting correlation with major shipping ports of the Great Lakes. Since this insect commonly hides in bundles of commercial goods, it is easy to see how it would have been literally shipped around the region. Some introduced insects, freed from their native predators and parasites, became pests of significant magnitude and economic importance. Some of the notorious introduced pests include the

European earwig (*Forficula auricularia* L.), German yellowjacket (*Vespula germanica*), gypsy moth (*Lymantria dispar*), Japanese beetle (*Popillia japonica*), cabbage butterfly (*Pieris rapae*), European corn borer (*Ostrinia nubilalis*, and European pine sawfly (*Neodiprion sertifer*), to name a few. Other insects, such as the European praying mantid (*Mantis religiosa* L.), Chinese mantid (*Tenodera aridifolia sinensis* Saussure), caterpillar hunter (*Calosoma sycophanta*), European ladybird beetle (*Scymnus suturalis* Thunberg), parasitic wasps (families Braconidae and Ichneumonidae), and parasitic flies (Tachinidae), have been deliberately introduced because of their beneficial roles as predators and parasites.

Many other introduced insects have gone virtually unnoticed over the years and have become naturalized additions to our native fauna. Some of these insects include the little earwig, *Labia minor* (L.); European katydid, *Metrioptera roeselii* (Hagenbach); the spittlebug, *Aphrophora alni* (Fallen); European robber fly, *Dioctria baumhaueri* Meigen; black ground beetle, *Pterostichus melanarius* Illiger; European flea beetle, *Psylliodes affinis* (Paykull); European scarab beetle, *Valgus hemipterus* (L.); alfalfa leafcutter bee, *Megachile centuncularis* (L.); and pale leafcutter bee, *Megachile concinna* Smith. One word of caution is in order: not all European and Asian insects are nonnative. There are many insects that have a circumpolar distribution and occur naturally in the northern regions of both the Eurasian and the North American continents. The scientific literature is full of names that were proposed for the North American populations of such insects.

Endangered Insects

Changes to the environments of the Great Lakes region have brought some insects closer to the brink of extinction. These insects are faced with the wholesale loss or pollution of their native habitat and/or food plants, and their populations have been reduced to dangerously low levels. The widespread application of pesticides against agricultural and forest insect pests is having a deleterious effect on nontarget insects. The habitats of many insects have been lost at an alarming rate. The region now has a growing list of insects that are endangered, threatened, or special concern species.

An *endangered species* is one whose survival is in immediate jeopardy because of a clearly identifiable threat. A *threatened species* is one whose survival is not in immediate danger, but for whom an identifiable threat exists. Species of *special concern* are very rare: they may appear to be undergoing an unexplained decline (with no identifiable threat), they may be seldom seen or recorded, or they may be infrequent residents. The bulk of the endangered, threatened, and special concern species currently listed are from insect groups that are well known or well studied in the Great Lakes region (such as butterflies and moths, beetles, and dragonflies). The status of many other lesser-known insect groups has yet to be investigated. When such groups are studied, the list will undoubtedly increase dramatically. A list of endangered, threatened, and special concern insects of the Great Lakes region can be found in appendix F.

STATE INSECTS

Many states, including some of the states in the Great Lakes region, have adopted official state insects. They have done so to remind their citizens of the vital role that insects play in our daily lives. As of 1995, 35 states have adopted such symbols. Among the Great Lakes states, Illinois has chosen the monarch butterfly, Wisconsin has selected the honey bee, and Pennsylvania has chosen the lightningbug. The ladybug/ladybird beetle was adopted by both New York and Ohio.

Chapter 3

Distributional Patterns of Insects in the Great Lakes Region

HABITAT ZONES

To better understand how various insects come to inhabit certain areas, we must first look at the bigger picture—the habitat zones that occur in the Great Lakes region. These zones determine what types of habitats (macro and micro) can occur in any given area. After understanding these zones we can shift our attention to specific habitats.

The combination of climate, topography, and biological organisms has produced distinct habitat zones within North America, each with its own insect fauna. Four of these North American habitat zones are found in the Great Lakes region. The zones are arranged more or less latitudinally because of the significant influence of climate, which gets progressively cooler from south to north. In general there is a trend from species richness in the southern habitat zones to species poorness in the northern habitat zones. This trend was demonstrated for beetles in "The Coleoptera of Isle Royale, Lake Superior, and Their Relation to the North American Centers of Dispersal" (Adams 1909b).

Habitat zone 1—the Carolinian zone. The Carolinian zone includes portions of southern Wisconsin, Illinois, Indiana, southern Michigan, southern Ontario (the area north of Lake Erie), Ohio, northwestern Pennsylvania, and western New York (the areas immediately south of Lakes Erie and Ontario). The topography of this zone is more or less flat, is often low-lying, and has little relief (mostly lacustrine and till plains). The Carolinian zone is dominated by the great hardwood forests of the eastern United States and its associated fauna. In the Great Lakes region the climax beech-maple, oak-hickory, and maple-basswood forests are predominant in this zone. The undergrowth is well developed and rich in understory species.

The Carolinian zone has an abundant insect fauna and is especially abundant in those species associated with hardwood trees. This zone is an ancient center of insect origin, preservation, and dispersal. Many of the insect species in the Great Lakes basin have close relatives in temperate Europe, which was separated from North America by "continental drift" in the distant past. Most of the species found in the Carolinian zone have extensive ranges throughout the eastern United States and southern Canada. There is generally a large number of species at any given site, but species from the east and the north usually occur in relatively low numbers. This zone favors southern species, and many species have spread here from the southern parts of the habitat zone. Some of these species reach their maximum northern range here, such as the spicebush swallowtail (*Papilio troilus*). The extensive cultivation in this zone has created many field habitats, which favor open-land species. The aquatic insect fauna is also well developed, with the major rivers and lakes inhabited by large populations of caddisflies (Order Trichoptera), mayflies (Ephemeroptera), aquatic bugs (Hemiptera), midges (Diptera: especially the family Chironomidae), and aquatic beetles (Coleoptera). The occurrence of endemic bog-inhabiting species is infrequent because the number of suitable bogs is fewer than in the zones to the north (the Alleghenian and Canadian zones).

Habitat zone 2—the Alleghenian zone. The Alleghenian zone includes more northerly areas of central Wisconsin, central and northern lower Michigan, east central Ontario, and northern New York. These areas are higher and drier than many areas in the Carolinian zone. This zone is a transitional zone between the Carolinian zone to the south and the Canadian zone to the north. It is dominated by the hemlock-white pine-northern hardwood forest, and the understory is reasonably well developed. There are fewer open areas of cultivated land because of generally poorer soils, but there are often large numbers of forest clearings created by logging. The aquatic insect fauna is extremely well developed, with the major rivers and lakes inhabited by large populations of caddisflies (Trichoptera), mayflies (Ephemeroptera), aquatic bugs (Hemiptera), midges (Diptera: e.g., Chironomidae), biting flies (Diptera: e.g., Simuliidae), and aquatic beetles (Coleoptera)—more than in the zones to the north or south. There is a significant number of bog habitats in this zone, most with a well-developed endemic insect fauna.

Since this area is a transition zone between the southern Carolinian zone and the northern Canadian zone, many of the plants and animals found in this zone have affinities with and overlap with either the Carolinian or the Canadian zones. A large number of southern insects reach their maximum northern range in this zone, while a smaller number of northern species reach their maximum southern range here. Examples of southern insects reaching their northern limit of distribution include the black swallowtail (*Papilio polyxenes asterius*), West Virginia white (*Pieris virginiensis*), luna moth (*Actias luna*), dobsonfly (*Corydalus cornutus*), and argus caddisfly (*Astenophylex argus*). Northern insects reaching their southern limit include the modest hawk moth (*Pachysphinx modesta*) and concatenate caddisfly (*Banksiola concatenata*).

Habitat zone 3—the Canadian zone. The Canadian zone includes the far northern areas of western upper Michigan, northern Wisconsin, northern Minnesota, and Ontario (north of Lake Superior). The whole area is characterized by moderate moisture and a cool, short growing season. The topography of this zone features some highlands of exposed Precambrian and Cambrian bedrock with considerable relief, but much of the area is low-lying with poor drainage. The dominant trees are spruce and balsam fir, at times mixed with or replaced by aspen and birch. The undergrowth is typically poorly developed.

Generally the plants and animals of this region have a strong affinity with the northern, boreal regions of North America, and many species have extensive transcontinental or circumpolar distributions (often including the montane forests of the Rocky Mountains). The diversity of species is often low, but the populations are often quite high in any given locality.

Many of the insects that inhabit this region are associated with conifers, aspens, or birch trees, including the hemlock loopers (*Lambdina* species), budworms (*Choristoneura* species), spruce sawflies (*Neodiprion* and *Pikonema* species), pine weevils (*Pissodes* species), bark beetles (family Scolytidae), long-horned beetles (Cerambycidae), forest tent caterpillars (*Malacosoma disstria*), leaf roller caterpillars (*Archips* species), leaf skeletonizer caterpillars (*Buccalatrix* species), birch sawfly (*Arge pectoralis*), and American aspen beetle (*Gonioctena americana*). There is an exceptionally large number of lakes and streams in this northern zone, so there are well-developed aquatic insect communities. The black flies (Diptera: Simuliidae) and *Aedes* mosquitoes (Diptera: Culicidae) are especially well represented, but many species of midges (Diptera: especially Chironomidae), caddisflies (Trichoptera), mayflies (Ephemeroptera), stoneflies (Plecoptera), aquatic bugs (Hemiptera), and aquatic beetles (Coleoptera) are abundant as well. There are a large number of specialized habitats, such as bogs, that serve as refugia for arctic and subarctic species.

Habitat zone 4—the Prairie zone. The prairie zone occurs within the Great Lakes states but is not well developed in the Great Lakes basin. It occurs chiefly in the central United States and southcentral Canada. However, there is an eastern extension of prairie (prairie peninsula) that extends from central Illinois into southern Manitoba and western Ontario (Lake of the Woods District) and includes parts of western Wisconsin and western Minnesota. Small prairie remnants can still be found in parts of northeastern Illinois (e.g., Goose Lake Natural Area), northwestern Indiana (e.g., Hoosier Prairie), southwestern Michigan (e.g., Newaygo Prairie), northwestern and central Ohio (e.g., Bigelow Cemetery, Smith Cemetery, and Irwin Prairies), and southeastern Wisconsin (e.g., Chiwaukee Prairie); these small areas are all that remain of the expansive prairie areas that occurred in the Great Lakes region before settlement.

The availability of local habitats within these broad habitat zones is most directly influenced by physiography and climate. Thus these large habitat zones are actually a mosaic of distinctly different macrohabitats and microhabitats, each created by slight variations in elevation, aspect, slope, soil type, soil pH, soil moisture, relative humidity, or vegetation. These factors often work in concert, but sometimes one factor may exert a greater influence than the others. Areas of strong topographic relief afford greater variety of habitats than areas of low topographic relief because of localized differences in temperature, moisture, relative humidity, and soil. It is therefore safe to say, generally speaking, that areas with significant differences in relief will have a greater diversity of plant and animal life than areas with little relief.

Within the four Great Lakes habitat zones many unique habitats that are highly attractive to insects have developed. In low-lying areas, where the water table is high, wet meadows, marshes, swamps, and bogs have formed. The Great Lakes region is blessed with an abundance of riparian (stream) and lacustrine (lake) habitats—there are tens of thousands of miles of streams and approximately 80,000 inland lakes in the region. Then, of course, there are the Great Lakes themselves and the thousands of miles of shoreline (beaches and dunes). All the islands in the Great Lakes, even those a considerable distance from the present-day mainland, are inhabited by insects. In areas where the soil is very course and porous, pine-oak forests have formed. In areas where the soil holds more moisture, large forests of deciduous trees (in the south) and conifers (in the north) have developed. Many insects have adapted to habitats created by humans, such as roadsides, agricultural lands, suburban yards and gardens, urban environments, and buildings.

Populations of some northern species of plants and animals remain in little pockets of suitable habitat located in the southern part of the region. In effect, these populations were left behind when the species advanced northward or westward as the glaciers retreated. These plants and animals are known as *relics*. Relic plants and animals can be found in locally suitable habitats (north-facing cliffs, high elevations, canyons, and bogs for northern species; prairie remnants for dry, grassland species). Examples of sites where relic populations can be found include Spruce Lake Bog, Osceola County, Wisconsin; Cedar Bog, Champaign County, Ohio; Volo Bog, Lake County, Illinois; Blue Creek Fen, St. Joseph County, Michigan; and the remnant prairies in Illinois, Indiana, Michigan, Ohio, and Wisconsin mentioned previously in this section.

INSECT DISTRIBUTIONS IN THE
GREAT LAKES REGION

When looking at insect distributions in relation to habitat zones, we often find that insects seem to inhabit large geographical areas in one or more habitat zones. The Carolinian zone of the eastern United States and southern Canada is a very large area and many insect species occur throughout the entire zone. However, many species also extend into the Alleghenian zone and can be found throughout the whole of eastern North America south of the Canadian zone (boreal forest). Other species are limited to the Alleghenian zone (with no populations to the south), whereas some inhabit the entire combined Alleghenian/Canadian zones. In the transcontinental boreal forests of the Canadian zone many insects are well distributed from eastern Canada westward to Alaska, and in some cases even southward in the Rocky Mountains. In this regard some of the insect fauna of Wyoming, Colorado, and even New Mexico is very similar to that of the northern Great Lakes region. Many insects inhabit the central parts of North America—the Great Plains of Saskatchewan, Manitoba, the Dakotas, Nebraska, Kansas, Oklahoma, and Texas— and some of them range into areas along the western edge of the Great Lakes region in Minnesota, Wisconsin, and Illinois. There are even some insect species, few in number, that inhabit the combined Carolinian, Alleghenian, Canadian, and prairie zones (this distribution is frequently described as "North America east of the Rockies").

It is because of these distributional trends in North America that the Great Lakes region has such a diverse and abundant insect fauna. The Great Lakes region is in a unique geographical position; it is at the crossroads of several overlapping distributional patterns, which is made possible by the presence of four major habitat zones in the region. Therefore, most of the species that occur in the Great Lakes region are typically widely distributed and also occur in the appropriate habitats of far away places, such as boreal Alaska, the Great Plains, the Rocky Mountains, and eastern/southeastern forests. There is only a small handful of insect species that are endemic to the Great Lakes region.

One word of caution is warranted: when describing the distribution of a widely distributed species of insect it is an oversimplification to state that the species is found throughout the state or region. It is more accurate to say that the species is found at suitable sites in widely scattered areas of the correct habitat. We must remember that no species is uniformly distributed over its entire range; its occurrence depends on the availability of the suitable habitat. We must also keep in mind that habitats are not static. They are very dynamic and always changing.

The degree to which an insect species occupies an area of suitable habitat, in the terms of its regularity and population size, is also variable. Insects may be permanent inhabitants (in their primary and most favored habitat, where they can be found at all times of the season in one stage or another), occasional inhabitants (in a habitat in which conditions are favorable enough to permit multiseasonal occupation but may change to conditions that eliminate the species from the area), or impermanent inhabitants (in a generally unfavorable habitat that allows only temporary occupancy for part of a season). Species that are permanent inhabitants of a particular habitat are said to be characteristic of that habitat, while those that are impermanent inhabitants of a particular habitat are called "erratics" (migrants or wanderers) and have come usually from nearby (though sometimes from distant) areas of permanent habitat.

INSECTS AND MICROHABITATS

Insects are very responsive to variations in living conditions and the resources available in different microhabitats. Some insects are so dependent on specific conditions that they cannot survive outside the appropriate microhabitat. The distribution of many insects is often determined by the distribution of certain host/food plants. Other insects utilize very specialized microhabitats, such as subterranean cavities, beaver lodges and other animal nests, treeholes, and roadkills.

OPPORTUNITIES FOR INSECT STUDY
IN THE GREAT LAKES REGION

There are still many opportunities for observing, photographing, collecting, and studying insects in the Great Lakes region. When it comes to investigating and documenting the insect fauna of the Great Lakes region, we have only begun to scratch the surface. There is ample opportunity for insect enthusiasts of all ages and expertise to help gather important information on the insects of this vast region. So, next time you have an urge to spend some time outdoors, why not use it to investigate insects that live in the Great Lakes region?

You might think that there is nothing left to investigate. Not so! We have abundant information on the insect fauna of only a few restricted areas, rather than an evenly distributed knowledge of the region as a whole. In the entire Great Lakes region there are only a handful of areas that have been seriously investigated over the years.

Most areas in the region have received very little study and vast sections are grossly understudied, while many other areas still remain totally unexplored. In one case, a search of the literature records revealed that there were about 350 species of ground beetles (Coleoptera: Carabidae) known from Michigan. However, collecting efforts in the field have discovered at least another 100 species not previously recorded from the state (Dunn, unpublished data). Furthermore, it is not necessary to collect in virgin forests, in wilderness areas, or on remote islands to make new discoveries. Many collectors must think so, because among entomologists there seems to be a tendency to travel as far away from their own home state as possible before beginning to collect and study insects. However, the odds are very good that there are interesting discoveries to be made right in your own backyard— both literally and figuratively.

The need for continued entomological investigations by both professional and amateur entomologists in the Great Lakes region is clear. There is so much we do not yet know about the insects of the Great Lakes region, their distributions, habitat associations, life histories, food plants, plant interactions, host animals, and behaviors. One of the greatest needs is for habitat and resource requirements information on endangered, threatened, and special concern insects. Discovery of the limiting factor(s) for rare and endangered insects would be a major step forward in preservation of invertebrate populations.

LOCATING INSECTS FOR STUDY

When to Look

The spring, summer, and fall months are generally the most suitable time for observing insects in the Great Lakes region, primarily because of the favorable conditions. Most insects reproduce during the spring and summer months, so active adult and/or immature insects are generally easy to find at this time.

However, many interesting insects can also be found during the winter months, if you are willing to spend time looking for them. Some species are hibernating under the snow in leaf litter, rotten wood, and soil. Dozens of species of ground beetles (Carabidae) can be collected in Michigan from soil, leaf litter, and rotten wood (Dunn 1983). Similarly, H. T. Fay (1862) reported collecting 129 species of beetles in the Columbus, Ohio, area. Winter stoneflies, springtails, snow scorpionflies, and wingless crane flies are examples of insects with regular periods of adult activity during the winter months. Other insects may be active during warm spells and can be found on the ground/snow or in the air as conditions permit. Some collectors have had luck catching owlet moths (Noctuidae), oecophorid moths, tortricid moths, butterflies, water striders, and lesser dung beetles (*Aphodius finementarius* (L.)) during warm spells in the middle of winter.

Insects also vary in their daily activity periods, and searching for insects at various times during day and night (even in the exact same area or habitat) will produce different results. Some insects, like butterflies, bees, and biting flies are most active during the daylight hours. Other insects (like mosquitoes) are only active for a short period of time at dawn or dusk. Still other insects are active during the nighttime hours, when their activity is masked by darkness. It is at night that moths, many beetles, and other nocturnal insects make their appearance.

Where to Look

The search for insects often begins in the library. Checking appropriate publications and articles is a good place to start when seeking information on certain species or groups of insects for study. After all, it is easiest to start looking in places where the insects have been collected at some point in the past. The written information of others will provide you with useful information on periods of adult (or immature) activity, host organisms (plants and animals), habitat preferences, and distribution. The bibliography located in this book will help you locate information specifically on the insects of the Great Lakes region. You should be able to find many of these books and articles at your local public library, but some may require visiting the library of a major scientific institution, such as a university or natural history museum. The entomologists at your state universities will know where you might be able to find some of the older or hard-to-find periodicals. Also, many of the more recent issues of publications are available for purchase as back issues from the publishing organization (see app. B).

The specimens housed in the insect collections of various public and private institutions (see app. C) are an also excellent source of distribution information. It is always best to contact the collection curator or manager in advance to make arrangements for your visit, rather than just dropping by.

However, keep in mind that habitats change over time, and former habitats may be long gone by the time you get there. Once you attain a "feel" for the appropriate habitat (by actual fieldwork or by consulting the literature), you are ready to search for new areas of habitat. You can locate prospective new collecting sites with the help of maps—road maps, topographic maps, county maps, and atlases. Maps will help you identify access roads, prominent natural features, waterways and wetlands, public lands, and other useful information. Topographic maps have the most detailed information but are costly and difficult to obtain, while road maps, road atlases, and county maps are not detailed enough to be useful. The solution to this problem can be found in a collection of topographic-like maps for each state published in book form (*Atlas and Gazetteer*, DeLorme Mapping Co., Freeport, ME). Look for these at your local bookstore, or you can purchase them by mail from the Young Entomologists' Society (Y.E.S. Buggy Bookstore, 1915 Peggy Place, Lansing, MI 48910–2553).

Most private property is open to insect observation and collection, provided you seek permission in advance. Public lands are open to everyone, but there may be restrictions on the collection (e.g., removal) of insects and other arthropods. See appendix E for regulations concerning the collection of insects and arthropods from public lands in the Great Lakes region. Please note that there are no restrictions on the observation or photography of insects, except perhaps in fragile habitats that are off-limits to visitors. Once again, seeking permission from the proper authority may overcome even this obstacle to your studies.

Where to specifically search for insects can be summed up in one word: everywhere! Insects utilize every macrohabitat imaginable. Look for them in woodlands and forests, clearings, roadside ditches, swamps, marshes, bogs, rivers and streams, lakes and ponds, fields and grasslands, sand dunes, caves, cultivated lands, roadsides, right-of-ways, and vacant lots. Once you have found an appropriate area of suitable macrohabitat, check the microhabitats. Look for insects around flowers, fruits, berries and seeds, foliage, tree sap, galls, tree stumps, rotten wood, loose bark, logs, leaf litter, humus, mosses and lichens, fungi, dung, roadkills and carrion, debris, puddles, tree holes, stream banks and lakeshores, aquatic plants, rocks and stones, and animal nests and burrows, and on other animals.

Typical Insects of Aquatic Environments

Rainwater collects in woodland depressions, roadside ditches, and other low areas during the spring and fall months. These temporary waters are home to such insects as mosquitoes, midges, water beetles, water bugs, and springtails. Discarded containers, especially old automobile tires, provide excellent habitats for many species of mosquitoes.

Tree holes are another unique aquatic habitat, and several species of mosquitoes—*Anopheles barberi* Coquillett, *Orthopodomyia signifera*

(Coquillett), *Orthopodomyia alba* Baker, *Toxorhynchites rutilis septentrion-alis* (Dyar and Knab), and *Aedes triseriatus* (Say)—specialize in utilizing this microhabitat.

Ponds and lakes, ranging in size from under an acre all the way up to the depth and width of the Great Lakes themselves, are home to many species of aquatic insects, such as stonefly nymphs, mayfly nymphs, dragonfly and damselfly nymphs, midge larvae, caddisfly larvae, aquatic beetles and their larvae (e.g., whirligig beetles, predaceous diving beetles, and water scavenger beetles), aquatic bugs and their nymphs (e.g., giant water bugs, water boat-men, backswimmers, water scorpions, and creeping water bugs), and even aquatic moth caterpillars.

Swift-flowing streams are home to stonefly nymphs, mayfly nymphs, caddisfly larvae, black fly larvae, water pennies, and ripple bugs. The slower moving streams are home to many species of mayfly nymphs, caddisfly lar-vae, midge larvae, mosquito larvae, dragonfly and damselfly nymphs, alderfly larvae, dobsonfly larvae, and water striders. The adults of many species, such as the black-winged damselfly (*Calopteryx maculata*), clubtail dragonflies, stoneflies, caddisflies, alderflies, and dobsonflies, linger nearby. In one small stream in northeastern Ohio, 73 species of aquatic insects (in 60 genera and 7 orders) were found (Robertson 1984), and these results are probably typical for the region's Carolinian zone. However, in all likelihood, additional species would be found if the area were investigated further.

The thick vegetation of marshes, swamps, and bogs provides cover and food for many types of insects. Even in small marshes, such as the Dundas Marsh near Hamilton, Ontario, there can be great diversity. During 1946 and 1947, 51 species (18 families and 9 orders) were collected at that marsh (Judd 1949a), and during 1947 and 1948, 32 species of insects (from 19 families and 5 orders) were collected there (Judd 1949b). Typical insects of the marshes include dragonflies and damselflies, mosquitoes, water bugs (e.g., marsh treaders, creeping water bugs, and velvet water bugs), caddisflies, butterflies and skippers (the least skipper, *Ancyloxypha melinus* (Huebner), for instance), water beetles, and long-horned leaf beetles (*Donacia* species). The sedges and rushes provide food and shelter for the spine-tailed earwig, *Doru aculeatum* (Scudder); the ornate grouse locust, *Acrydium ornatum* Say; the meadow locust, *Chorthippus longicornis* (Latreille); the fork-tailed bush katydid, *Scudderia furcata* Brunner; the short-winged meadow grasshopper, *Conocephalus brevipennis* (Scudder); and such plant bugs as *Teratocoris dis-color* Uhler and *Mimiceps insignis* Uhler.

The tamarack (larch) bogs in the southern portion of the region hold small relic populations of boreal insects, and many interesting (and rare) insects can be found there, such as Mitchell's satyr, *Neonympha mitchellii* French; the bog copper, *Lycaena epixantha* (Boisduval and LeConte); the

Columbia silk-moth, *Hyalophora columbia* (S. I. Sm.); the sphagnum cricket, *Nemobius cubensis palustris* Blatchley; the pine tree cricket, *Oecanthus pini* Beutenmuller; the larch tree cricket, *Oecanthus laricis* Walker; the larch plant bugs (*Deraeocoris laricicola* Knight, *Pilophorus uhleri* Knight, and *Plagiognathus laricicola* Knight); the shore bug *Saldoida turbaria* Schuh; the huckleberry planthopper, *Oliarus cinnamonensis* Provancher; the heath spittlebug, *Clastoptera saint-cyri* (Provancher); the parallel spittlebug, *Philaenus parallelus* Stearns; and the meandering ground beetle, *Carabus maeander* Fischer.

The pitcher plant, *Sarracenia purpurea* L., is present in many bogs throughout the region. This plant is famous for its ability to capture and "eat" insects, but there are actually several insects that reside inside this insectivorous plant. The pitcher plant midge, *Metriocnemis knabi* Coquillett, the pitcher plant mosquito, *Wyeomyia smithii* (Coquillett), and the flesh fly *Fletcherimyia flettcheri* (Aldrich) all reside within the water-filled leaves of the pitcher plant. They feed on the mass of dead insects that accumulates in the bottom of the "pitcher." Insects that cannot survive in the pitcher plant and that serve as food for the plant include blow flies, horse flies, ants, lady beetles, weevils, leaf beetles, ichneumon wasps, click beetles, small moths, small katydids, plant bugs, froghoppers, leafhoppers, aphids, and millipedes (Judd 1959).

Typical Insects of Semiaquatic Environments

The Great Lakes region has more than 12,000 miles of shoreline and lake beaches, plus countless small beaches along streams and major rivers. These microhabitats are some of the most productive in the entire region and provide an excellent place to look for insects. Insects inhabiting the shorelines of streams, rivers, ponds, and lakes include many species of ground beetles (*Elaphrus*, *Brachinus*, *Agonum*, and *Bembidion*), rove beetles, variegated mud-loving beetles, shore bugs, toad bugs, shore flies, pygmy locusts, and ants.

The Great Lakes have often been referred to as "freshwater seas" because of their immense size. They are similar in many ways to the saltwater oceans and have powerful currents, waves, and even tides. These conditions combine to create one of the most unique insect phenomenon seen anywhere in the world: insect drift. The abundance of insect drift in the Great Lakes region was noted as early as 1900 by prominent entomologists, such as James G. Needham. Needham was one of the first to report on the tremendous numbers of insects that wash ashore on the beaches of Lake Michigan, and he wrote about the subject on several occasions (Needham 1900, 1904, and

1917). L. N. Snow (1902) and R. F. Hussey (1922) also reported on insects of the Lake Michigan drift line, and A. W. Andrews (1923) mentioned his experiences on the beaches of Whitefish Point on Lake Superior.

The number and variety of insects can, at times, be unbelievable. Many rare insects have turned up in beach drift, making places like Warren Dunes, Sturgeon Bay, and Whitefish Point in Michigan and Point Pelee in Ontario into meccas for insect collectors. The types of insects reported from beach drift include beetles (weevils, whirligig beetles, lady beetles, ground beetles, diving beetles, stag beetles, may beetles, leaf beetles, click beetles, long-horned beetles, hister beetles, soldier beetles, fireflies), true bugs (shore bugs, water striders, giant water bugs, waterscorpions, assassin bugs, damsel bugs, stink bugs, plant bugs, leaf bugs), flies (midges, crane flies, robber flies, flesh flies, flower flies, house flies, anthomyid flies), butterflies (swallowtails, monarchs, painted lady butterflies, question marks, cabbage butterflies), moths (owlets, underwings, tiger moths, sphinx moths), homopterans (cicadas, aphids, planthoppers, leafhoppers, froghoppers, treehoppers), neuropterans (green lacewings, brown lacewings, dobsonflies), grasshoppers and crickets (pygmy grasshoppers, locusts, katydids, crickets), ants, bees (leaf-cutter bees, bumble bees, honey bees), wasps (ichneumonids, spider wasps, paper wasps, yellowjackets), dragonflies, damselflies, caddisflies, and mayflies (Dunn, unpublished data).

Most observers concur that the amount of insect drift is greatest following the passage of strong weather fronts (winds and waves) through the area at a time when insects are actively flying about. The insects fly out over the water (or are carried out by the wind) and get dunked in the water (or fall exhausted on the surface of the water). Only the hard-bodied species survive their immersion in the water and the pounding of the surf. If a storm front passes at a time when a certain insect is in the process of swarming or migrating, the resultant drift is often composed of tremendous numbers of that one species, with only a few other insects mixed in. Needham (1900) counted 2,520 crickets per meter in the windrow at Lake Beach, Illinois; he estimated that the drift line could have easily stretched for at least 50, if not as much as 100, miles. (Incidentally, that figures out to be 403,200 crickets per mile.) At other times there can be a wide variety of insects, for example 114 species in 31-plus families were reported by Snow (1902). Certain insects show up with regularity at prescribed times of the year—for example, May beetles in early summer; dragonflies, butterflies, and crickets in midsummer; and grouse locusts in the fall.

Not all the insects found in the drift line got there by washing up on the beach. Some scavenger species are attracted there by the abundance of food. It is likely that certain insects, such as ground beetles, rove beetles, dermestid beetles, hister beetles, carrion beetles, flies, ants, and spiders, feed on the dead and dying insects. Some birds and other animals may partake of the "feast" also. Dead insects that are not eaten dry up and blow inland, accumulating in piles at the base of dunes, trees, and other objects.

Not all the insects that wash ashore die. Many can be seen crawling up the beach, seeking shelter and a place to dry out. Of course, aquatic forms generally fair better than terrestrial forms, but survival in the water is in great part determined by the length of exposure and the severity of the wave action. Needham (1900) made several interesting observations about the insect drift. Perhaps one of the most interesting was his observation that the species that turned up most often in the drift were the most active, hardy, dominant species of their respective groups. It makes sense that the weaker, less capable species spend more time under cover and therefore are less likely to be caught out in the weather. The stronger and more venturesome species are the ones that occasionally fall as casualties to the ever changing weather conditions.

Typical Insects of Terrestrial Environments

The entomological exploration of subterranean habitats is frequently overlooked, probably because the insects that reside there are often small and dull colored. However, many insects make their homes in soil, humus, caves, and rodent tunnels and beneath boards, stones, and debris. Soil inhabitants include springtails, diplurans, proturans, root aphids, burrowing bugs, and mole crickets. Beneath stones, logs, and boards and in rodent burrows, you will find wood cockroaches, beetles, camel crickets, earwigs, termites, and fleas. Cave inhabitants include springtails, diplurans, camel crickets, and blind ground beetles (some of which are endangered species).

Sand dunes and other expanses of open sand are common in the Great Lakes region. This habitat imposes harsh demands on insect inhabitants. Very little cover (shade) is available, and the temperature of the soil's surface can easily reach 130°F on a sunny, summer day. Furthermore, very little water is available most of the time. It is often windy and the sandy substrate is far from stable, constantly blowing and shifting. Still, many unique insects make their homes here. Some, like the Lake Huron locust, the seaside locust, stink bugs, and plant bugs feed on the sparse vegetation. Others, like the velvet ants, tiger beetles, and sand wasps are predators that roam the sand in search of prey. The predatory larvae of tiger beetles and antlions conceal themselves in the sand and wait for their prey to wander by. Other insects, like ants, hister beetles, and antlike flower beetles, are scavengers that feed on the unfortunate creatures that were unable to survive. Some species of springtails can be abundant in sand dunes, suddenly emerging when the sand is moistened. It is still unknown what they utilize for their food.

Fields, meadows, grasslands, and prairies, unlike the sand dunes, have a rich assortment of grasses and herbaceous plants.

Often abundant there are plant-feeding insects, such as grasshoppers—especially the spring yellow-winged locust, *Arphia sulphurea* (Fabricius), and the Carolina locust, *Dissosteira carolina* (L.)— field crickets, plant bugs, spittlebugs—especially the two-lined spittlebug, *Philaenus bilineatus* (Say), and the meadow spittlebug, *Philaenus spumarius* L.—leafhoppers, planthoppers, aphids, ground beetles (*Amara* and *Harpalus* species), caterpillars, adult butterflies and skippers, thrips, bee flies, robber flies, sawflies, ants, honey bees, and bumble bees. Predatory insects, such as praying mantids, ground beetles, rove beetles, soldier bugs, lacewings, dragonflies, and wasps, also inhabit these areas. Some insect inhabitants of tallgrass prairies still survive in localized areas within the Great Lakes region. Prairie insects that still inhabit remnant prairies of the Great Lakes region include the plant bug *Miris dolobratus* (L.), the gibbose spittlebug (*Philaneus gibbosa* Ball), the powesheik skipperling (*Oarisma powesheiki* Parker), and the ottoe skipper (*Hesperia ottoe* W. H. Edwards).

One of the most productive places to observe and collect insects is at flowers. When we think of insects and flowers we instinctively think of bees and butterflies. It is no surprise then that 201 species of bees are known to visit Indiana flowers (Montgomery 1956). But many other insects are attracted to and visit flowers, either for the nectar and pollen offered there or for the opportunity to ambush unsuspecting prey. For example, in one study the flowers of marsh marigold (*Caltha palustris*) in southern Ontario were visited by 39 species of insects (in 20 families and 6 orders) (Judd 1964), while in another study flowers of dandelion (*Taraxacum officinale*) were visited by 25 species of insects (15 families in 7 orders) (Judd 1970). Among the beetles, at least 33 species of the Michigan long-horned beetles (Cerambycidae) visit flowers (Gosling 1984), including the black-and-yellow locust borer, which is abundant on goldenrod in the late summer and early fall.

Many studies show that despite frequent disruption (tilling, cultivating, and harvesting), agricultural lands support large populations of insects (not all of them pest species either). For example, 26 species of ground beetles (Carabidae) were found in Michigan small-grain fields (Dunn 1982), 28 species of ground beetles were found in a small asparagus field in Michigan (Dunn 1984), and at least 10 common species of ground beetles were found in apple orchards using integrated pest management in southern Ontario (Rivard 1964 and 1965). In a study of the hymenopterous insects of sweet corn fields in Ohio, 64 species of insects (in 21 families) were found (Everly 1939).

Researchers studying sweet corn, field corn, and soybean fields in Michigan have been amazed by the number and diversity of insects throughout the course of the growing season, including dragonflies, field crickets, tree crickets, short-horned grasshoppers, stink bugs, plant bugs, leafhoppers, spittlebugs, aphids, thrips, tiger beetles, ground beetles, sap beetles, ladybird beetles, rove beetles, fireflies, soldier beetles, click beetles, June bugs, false Japanese beetles, milkweed beetles, dogbane beetles, green lacewings, antlions, monarch butterflies, cabbage butterflies, hopmerchants, painted lady butterflies, mourning cloaks, Milbert's tortoise shells, tiger swallowtails, giant swallowtails, black swallowtails, skippers, owlet moths, crane flies, mosquitoes, deer flies, flower flies, robber flies, blow flies, muscid flies, ichneumon wasps, honey bees, field ants, Allegheny mound ants, carpenter ants, yellowjackets, and spider wasps (Dunn, unpublished data collected in 1990–92).

A small number of interesting insects inhabit animal nests, either as ectoparasites of the hosts or as scavengers on debris in the nest. Some of the scavengers include psocids, hister beetles, and the ground beetles *Platypatrobus lacustris* Darlington and *Pterostichus castor* Goulet (both in beaver lodges). Some of the ectoparasites include fleas, chewing lice, sucking lice, and cimicid bugs (bat bugs and swallow bugs). The beaver parasite beetle, *Platypysllus castoris* Ritsema (Leptinidae), is an obligate parasite of the beaver and probably occurs throughout the range of the beaver in the Great Lakes region. The mouse-nest beetle, *Leptinus testaceus* Müller, and the beaver-nest beetles, *Leptinus validus* Horn, inhabit the nests of their respective hosts.

The best techniques for locating, observing, photographing, collecting, and studying insects are well documented and the reader is referred to the following publications: *A Field Guide to the Insects of America North of Mexico* (Borror and White 1970), "Collecting, Preparing, and Preserving Insects, Mites, and Spiders" (Martin 1977), *A Field Guide to the Moths of Eastern North America* (Covell 1984), *A Field Guide to the Beetles of North America* (White 1983), *A Field Guide to Eastern Butterflies* (Opler and Malikul 1992), *The Practical Entomologist* (Imes 1992), and *A Beginner's Guide to Observing and Collecting Insects* (Dunn 1994).

Chapter 4

Insect Classification and Identification

The first step in learning how to identify insects is to become familiar with the basics of animal classification. Next, you need to have a working knowledge of the names and locations of primary insect body parts. Then you can concentrate on understanding and recognizing the significant features for each insect order. After becoming familiar with the insect orders, you might want to turn your attention to the significant features of the families within the larger orders, especially in the more frequently encountered orders, such as Odonata, Orthoptera, Hemiptera, Homoptera, Coleoptera, Neuroptera, Lepidoptera, Diptera, and Hymenoptera.

Since insects are the largest group in the animal kingdom (there are at least 10 million species, making up about 85 percent of all animals), you may expect to have difficulty identifying some insects to the family level and some even to the order level. Some of the larger orders have hundreds of families, and some of these families have several thousand species. Insects have been an incredibly successful group because of their ability to adapt to many habitats, both as immatures and adults, and to utilize special microhabitats. As a result, insects have evolved into a seemingly endless variety of sizes, shapes, and colors. This incredible variation is a main reason for the difficulties frequently encountered in insect identification. It is no wonder that insect identification can be frustrating at times. Occasionally, the characteristics used to differentiate between some orders or families are difficult to see or interpret.

PRINCIPLES OF INSECT CLASSIFICATION

All insects of the Great Lakes region are members of the Phylum Arthropoda (commonly referred to as the arthropods). The arthropods comprise a very large group of related animals, about 90 percent of all known animal species. The insects belong to the class Insecta (sometimes called Hexapoda). The class Insecta is distinguished from other arthropod groups (and most other animals) by the following characteristics:

1. an exoskeleton divided into three parts (head, thorax, and abdomen)
2. three pairs of jointed legs (as adults)
3. one pair of antennae
4. wings on most adults (never on immatures)
5. one pair of compound eyes (in adults and some immatures)

The class Insecta is organized into 28 different orders. Each order is further divided into families. Some of the smaller orders contain only a handful of families, while some of the larger orders contain hundreds of families. All but a few of the orders are represented in the Great Lakes region, and there are over 16,000 species recorded from the region (see table 1 at the beginning of chapter 5), although the actual number occurring there is probably closer to 20,000.

To many people the term *bug* is used to describe many small, crawly creatures (and not necessarily limited to insects). Those of us with a scientific mind are quick to tell people that "bugs are members of the order Hemiptera, and characterized by piercing/sucking mouthparts, half leathery and half membranous wings (hemelytra), and gradual metamorphosis." Entomologists refer to these insects as the true bugs in an attempt to minimize the confusion.

So, when you're talking to an entomologist and you hear the terms *waterbug, ladybug, lightningbug, billbug,* or *greenbug,* they must be talking about true bugs, right? Wrong! The waterbug is actually a cockroach, the ladybug, lightningbug, and billbug are actually beetles, and the greenbug is an aphid! Even the professionals seem to be all mixed up.

Names like lightningbug and ladybug are firmly rooted in our language, and, despite their apparent incorrectness, are here to stay. Fear not, for there is a simple way to distinguish between true bugs and other "bugs" when communicating with other insect enthusiasts. If the word *bug* is part of the insect name, such as in ladybug, then you know that the insect is not a true bug. When referring to a true bug the word bug stands alone, as in giant water bug.

Similar confusion occurs with the word *fly,* which may be either a part of an insect name or a specific group of insects—the true flies. As with the word *bug,* the same basic rule applies here. For example, we would know that a butterfly, dragonfly, or caddisfly are not true flies, while a house fly or soldier fly are true flies.

Each insect described in this book is referred to by a common name followed by a scientific name. All insect groups (orders and families) have standardized common names, but because of the influence of local customs or language, the individual species often have more than one common name in usage. As you might imagine, the use of two or more different common names to refer to one insect can easily lead to some confusion.

Fortunately, all insects known to science have official, internationally recognized scientific names that are unique to each insect, which allows scientists of all nations and of all time periods to communicate with one another. A scientific name is binomial—composed of both a genus and a species name. The genus name is capitalized, while the species name is not. Because they are Latin terms, both names are either underlined or, as in this book, italicized. While some of these scientific names are tongue twisters, most are quite easy to pronounce and roll off the tongue with only a little practice.

BASIC INSECT ANATOMY

A working knowledge of the names and locations of insect body parts is an important foundation for learning how to identify insects. This section is intended to give you the bare-bones version of basic insect anatomy. When you encounter unfamiliar terms in the identification keys or insect descriptions featured in this book, consult the glossary. For more complete information, refer to *A Beginner's Guide to Observing and Collecting Insects* (Dunn 1994) or *The Practical Entomologist* (Imes 1992).

External Anatomy

Insects are covered with an armorlike shell called an exoskeleton that is divided into three distinct body sections. The three body sections are (1) the head, which bears mouthparts, compound eyes, and antennae; (2) the thorax, a midsection that bears wings (if present) and three pairs of legs; and (3) the abdomen, a capsulelike container for the major internal organs.

Insects vary considerably in body size. In this book the following terms are used to describe the relative size of insects:

minute: 0.5–2 mm   *large:* 21–35 mm
small: 3–10 mm   *very large:* greater than 36 mm
medium: 11–20 mm

Insect Mouthparts

Insects use a wide variety of plant and animal materials, living and dead, as their food. As a result insects have developed specialized mouthparts for eating their food. There are two basic types of mouthparts: chewing and sucking.

Chewing mouthparts are characterized by a pair of prominent mandibles with associated accessories: an upper lip (labrum), fingerlike palps, and a lower lip (labium). The accessories are used to manipulate solid foods, and the mandibles grind food into small pieces.

Sucking mouthparts come in a variety of configurations, all used to ingest liquid foods. A beaklike piercing/sucking mouthpart is used to extract sap from plants or blood from animals; a sponging mouthpart is used to soak up semiliquid foods; and a long, coiled, butterfly proboscis is used to draw nectar out of flowers.

Insect Antennae

The principle sensory organs of the insect, the antennae, are located on the head. Insect antennae come in a variety of distinctive shapes and sizes and can be a useful identification clue. Insect antennae may be threadlike (filiform), beadlike (moniliform), swollen at the tip (clubbed), sawlike (serrate), comblike (pectinate), featherlike (plumose), or elbowed (geniculate).

Insect Legs

Insect legs are composed of five distinct segments; they are (in order from the body outward): coxa, trochanter, femur, tibia, and tarsus. Insect tarsi are composed of two to five segments, and the exact number of tarsal segments is often an important feature used in identification. Since insects have three pairs of legs, the tarsal count may be summarized like this: 1–2–3. The first digit is the number of tarsal segments on the front pair of legs, the second digit is the number of tarsal segments on the middle pair of legs, and the third digit is the number of tarsal segments on the hind pair of legs.

Insect legs have evolved into many distinctive types, such as the running legs of cockroaches and beetles; the swimming legs of aquatic insects; the jumping legs of grasshoppers, flea beetles, and fleas; the grasping legs of the praying mantid; the clinging legs of lice; and the digging legs of the mole cricket.

Insect Wings

Insects are the only winged invertebrates. Wings are only present during the adult part of the life cycle and are never found on any of the immature stages. Most adult insects have two pairs of wings, although some have only one pair and others have none. A few adult insects lack wings, either because they are primitive types of insects, or because they are advanced forms that have evolved into a wingless form of their winged ancestors. Examples of wingless adult insects include the proturans (Protura), diplurans (Diplura), springtails (Collembola), bristletails (Thysanura), chewing lice (Mallophaga), sucking lice (Anoplura), termites (Isoptera), booklice (Psocoptera), thrips (Thysanoptera), sheep keds (Diptera), fleas (Siphonaptera), ants and some wasps (Hymenoptera), and some scale insects, mealybugs, and aphids (Homoptera). Since wing characteristics are constant at the order level of classification, wings are a primary characteristic used to make insect identifications.

Many insects have membranous wings composed of a thin, usually transparent, membrane and supporting veins. Most of the veins run lengthwise (longitudinally), but there are also cross veins. The areas surrounded by these wing veins are called cells. This type of wing is found in the mayflies (Ephemeroptera), dragonflies and damselflies (Odonata), stoneflies (Plecoptera), lacewings (Neuroptera), scorpionflies (Mecoptera), true flies (Diptera), and sawflies, ants, wasps, and bees (Hymenoptera).

Some insects have wings that are thickened and leathery in the basal two-thirds and membranous at the tip. This type of wing, known as a *hemelytron*, is found only in the true bugs (Hemiptera).

Some insects have wings that are uniformly thickened and leatherlike. Remnants of the wing venation pattern is usually visible and the color of the wing usually matches that of the insect body. Examples of insects with this type of wing, called a *tegmina*, are grasshoppers and crickets (Orthoptera), cockroaches (Blattaria), praying mantids (Mantodea), walkingsticks (Phasmida), and some leafhoppers, froghoppers, and planthoppers (Homoptera).

Two groups of insects have a highly modified, armorlike forewing called an *elytron*. This type of wing is thicker than a tegmina and the venation pattern is usually not visible. Earwigs (Dermaptera) and beetles and weevils (Coleoptera) have this type of forewing.

A few other insects have unique wings. For example, butterflies and moths (Lepidoptera) have scaly wings, caddisflies (Trichoptera) have hairy wings, and thrips (Thysanoptera) have featherlike wings.

Other Features

In addition to distinctive wings, mouthparts, and antennae, some insects
have unique body appendages, such as *cerci* (abdominal filaments or pincers),
cornicles (tubular abdominal structures), *furculae* (springlike mechanisms),
halteres (knobbed balancing organs), or distinctive shapes (compressed from
side to side—for example, fleas—or flattened from top to bottom—for example, lice and bedbugs).

TIPS FOR INSECT IDENTIFICATION

The trick to insect identification is learning which characteristic(s) to look
for and which ones to ignore. In insect identification, you must be able to recognize *significant* characteristics, such as the wings, mouthparts, antennae,
and other unique body appendages. One problem is that insect body parts are
frequently adapted to specific habits and lifestyles, so the mouthparts, wings,
and legs are often highly modified. Fortunately, some modifications are constant at the order and family level and serve as significant characteristics for
distinguishing among the various groups of insects, enabling us to classify
and identify them.

Some insect orders have a single definitive characteristic —for example,
the single pair of wings in true flies (Diptera); the coiled proboscis in butterflies, skippers, and moths (Lepidoptera); and the fringed wings in thrips
(Thysanoptera). There are even some families, including those with thousands of different species, that have a single definitive characteristic that
immediately separates them from all other families, such as the snout of the
weevils (Curculionidae) or the spurious wing vein of the flower flies (Syrphidae). Sometimes just a general appearance —a combination of size, shape, coloration, antennal structure, tarsal structure, and other features—gives a com-
posite picture of a family and helps identify a family that has many species,
such as the ants (Formicidae) or ground beetles (Carabidae).

The ability to make accurate identifications takes time and experience.
Eventually, you will develop a "feel" for some of the more difficult orders and
families. The rapid and accurate identification of insects depends on recognition and recall of significant characteristics that are associated with each particular order and/or family. Therefore, in the key to orders, quick guides to
identification, and descriptions of insects in this book, major characteristics
are always given before the minor ones. You must learn to distinguish
between "trivial" characteristics, those that belong to a single genus or

species, and general characteristics that apply to most or all of the species in a family.

The "Keys" to Identification

Entomologists use a reference called a "key" to make orderly decisions about the identification of insect specimen unknown to them. Most identification keys are dichotomous, or double- branched, by design. Dichotomous keys offer a choice of two alternatives (these paired choices are called couplets) at each step along the way. By choosing the half of the couplet that best describes the specimen at hand, you are directed to other couplets and finally to an answer (the identification).

A dichotomous key to the adults of the insect orders found in the Great Lakes region follows. This key should assist you in making accurate identifications of adult insects. For keys to insect families, genera, and species, refer to some of the many keys published in the articles cited as identification resources in this book. After you have mastered the identification key and begun to recognize the major insect groups, the quick guides to identification featured in this book will help you identify insects more quickly by reminding you of the key diagnostic character for commonly encountered insect groups.

KEY TO ORDERS OF ADULT INSECTS OF THE GREAT LAKES REGION

See the Glossary for definitions of unfamiliar terms.

1a.	Without wings 2
1b.	With wings (including elytra) 19

2a.	Waxy or shell-like cover over body; legs and antennae not present; sucking mouthparts concealed under body . . . . . scale insect, **Homoptera**
2b.	Body without shell-like cover; legs and antennae usually present . . . 3

3a.	With six or fewer abdominal segments, often indistinct; suckerlike collophore on first abdominal segment and forked furcula on fourth or fifth segment . . . . springtails, **Collembola**
3b.	With ten or more distinct abdominal segments; no collophore or furcula present 4

4a. Abdomen with three long filaments at tip 5
4b. Abdomen with two (or fewer) filaments at tip 6

5a. Body covered with tiny scales; mouthparts exposed and protruding, at least at tip . . . . bristletails, **Thysanura**
5b. Body not covered with scales; mouthparts withdrawn into head with only palps exposed . . . . diplurans, **Diplura**

6a. Antennae not easily seen (shorter than head or of fewer than five segments) or lacking 7
6b. Antennae prominent and longer than head or more than five segments 11

7a. Antennae lacking; front legs directed forward in their place . . . .
. . . . proturans, **Protura**
7b. Antennae present 8

8a. Body flattened from side to side; legs long and muscular; body hard . . .
. . . . fleas, **Siphonaptera**
8b. Body flattened from top to bottom; legs short; body soft 9

9a. Abdomen saclike, without distinct segments; tarsi with five segments
. . . . sheep ked, **Diptera**
9b. Abdomen with distinct segments; tarsi with one or two segments 10

10a. Sucking mouthparts (hard to see); head usually pointed, much narrower than thorax; claws large . . . . sucking lice, **Anoplura**
10b. Chewing mouthparts; head usually blunt, rounded, as wide as or wider than thorax; claws small . . . . chewing lice, **Mallophaga**

11a. Body strongly constricted between thorax and abdomen . . . .
. . . . ants and some wasps, **Hymenoptera**
11b. Thorax and abdomen broadly joined 12

12a. Tarsi with five segments 13
12b. Tarsi with fewer than five segments 14

13a. Body elongate and rounded, sticklike . . . . walkingsticks, **Phasmida**
13b. Body flattened from top to bottom, not sticklike . . . .
. . . . cockroaches, **Blattaria**

14a. Sucking mouthparts, with jointed "beak" beneath head 15
14b. Chewing mouthparts, with movable mandibles 16

15a. Small, globular, soft-bodied insects, usually with a pair of tubelike projections ncar tip of abdomen . . . . aphids, **Homoptera**
15b. Usually larger insects with more or less hardened body; without tubelike projections . . . . true bugs, **Hemiptera**

16a. With large compound eyes; hind legs large and used for jumping; generally large, hard-bodied insects . . . grasshoppers and crickets, **Orthoptera**
16b. With small compound eyes; legs not modified for jumping; generally small, soft-bodied insects 17

17a. Tarsi with four segments; antennae short, beadlike . . . . termites, **Isoptera**
17b. Tarsi with less than 4 segments 18

18a. Tarsi with one or two segments; antennae short; body elongate . . . . thrips, **Thysanoptera**
18b. Tarsi with two or three segments; antennae long and threadlike; body more or less oval . . . . booklice, **Psocoptera**

19a. With only a single pair of membranous wings; hind wings reduced to a small pair of knobbed structures . . . . true flies, **Diptera**
19b. With two pairs of wings 20

20a. Wings partially or entirely covered with scales; coiled proboscis . . . . moths, skippers, and butterflies, **Lepidoptera**
20b. Wings not covered with scales 21

21a. Wings long and narrow, fringed with hairs; minute insects . . . . thrips, **Thysanoptera**
21b. Wings not fringed with hairs 22

22a. Front wings small and knoblike, hind wings broad and fanlike . . . . twisted-wing parasites, **Strepsiptera**
22b. Front wings not reduced in size 23

23a. With sucking mouthparts (beak on lower front part of head) 24
23b. With chewing mandibles 25

24a. · Wings held rooflike over body or outstretched; both pairs of wings uniformly textured throughout, often with greatly reduced venation; head lengthened beneath body so beak appears to come out from between front legs .... froghoppers, leafhoppers, treehoppers, planthoppers, cicadas, whiteflies, and aphids, **Homoptera**

24b. Wings overlap over back of body; front half of wings leathery in texture; beak begins at front part of head .... true bugs, **Hemiptera**

25a. Front wings hard and of same texture as body, meeting in a straight line down back and hiding a back pair of wings 26

25b. Front wings of a different texture than body, not meeting in a straight line down middle of back 27

26a. Abdomen ending in a pair of prominent pincerlike appendages; front wings short, not covering entire body .... earwigs, **Dermaptera**

26b. Abdomen not ending in a pair of prominent pincerlike appendages; front wings almost always covering entire body .... beetles and weevils, **Coleoptera**

27a. Front wings more or less leathery in texture, different from hind wings 28

27b. Both pairs of wings membranous, not thickened 31

28a. Body elongate and sticklike .... walkingsticks, **Phasmida**

28b. Body not elongate and sticklike 29

29a. Forelegs modified for grasping prey .... mantids, **Mantodea**

29b. Forelegs not modified for grasping prey 30

30a. Body flattened from top to bottom .... cockroaches, **Blattaria**

30b. Body rounded, not flattened from top to bottom .... grasshoppers and crickets, **Orthoptera**

31a. Antennae very short, bristlelike, and hardly visible 32

31b. Antennae easily seen and of various shapes, but never bristlelike 33

32a. Hind pairs of wings much larger than front pair; abdomen ending in two or three long filaments .... mayflies, **Ephemeroptera**

32b. Wings of more or less equal size; no long filaments at tip of abdomen .... damselflies and dragonflies, **Odonata**

33a. Hind legs enlarged and fitted for jumping .... some crickets, **Orthoptera**
33b. Hind legs either not enlarged or only slightly larger than middle legs 34

34a. Wings lie flat over the body when at rest 35
34b. Wings held outstretched or folded rooflike over back at rest 36

35a. Hind wings about same size or smaller than front wings; veins of wings indistinct; antennae beadlike .... termites, **Isoptera**
35b. Hind wings larger than front wings; wing veins distinct; antennae of various shapes, but never beadlike .... stoneflies, **Plecoptera**

36a. Face long and beaklike; abdomen of male scorpionlike ....
.... scorpionflies, **Mecoptera**
36b. Face not long and beaklike; abdomen of male not scorpionlike 37

37a. Body and wings heavily clothed with hairs; wings folded rooflike over back .... caddisflies, **Trichoptera**
37b. Body and wings not clothed with hairs 38

38a. Antennae long and slender (filamentlike); small, soft-bodied and less than 5mm in length; face somewhat bulging .... booklice, **Psocoptera**
38b. Antennae not long and slender (may be knobbed, serrate or elbowed); generally larger than 10 mm 39

39a. Hind wings distinctly smaller than front wings ....
.... sawflies, ants, bees, and wasps, **Hymenoptera**
39b. Hind wings about same size as front wings ....
.... antlions, snakeflies, dobsonflies and lacewings, **Neuroptera**

Chapter 5

Insect Fauna of the Great Lakes Region

The insect fauna of the Great Lakes region is both numerous and diverse, with at least 16,000 specics inhabiting the area (see table 1). A discussion of all the insects occurring in the region is well beyond the scope of this book. Therefore, it is only possible to include representative species of the Great Lakes insect fauna, with special emphasis on those that are commonly seen, those with economic significance (pest or beneficial insects), those with unusual or interesting habits, or those that live exclusively in the region. For readers wanting more in-depth information on select insect species or insect groups of the Great Lakes region, bibliographic references are included after the discussion of each insect order.

SIX-LEGGED ARTHROPODS

(Class Insecta)

Proturans
(Order Protura)

Description. Proturans are minute (0.5–2.5 mm), elongate, and pale whitish or yellowish. They lack wings, compound eyes, and antennae. The front pair of legs, which are covered with many hairs, bristles, and setae, protrude forward and serve as "antennae." Proturans have a simple type of piercing/sucking mouthpart that is used to feed on fungal threads (mycelia). The first three segments of the abdomen have appendages on the underside and there is often a tail (telson) at the tip of the abdomen. Members of two families (Protentomidae and Acerentomidae) lack breathing "pores" on the outside of the body (spiracles) and internal breathing tubes (trachea) for respiration. This order of insects was not officially recognized until 1907.

Quick Guide to Identification

Six-Legged Arthropods
(Class Insecta)

Wingless (Mouthparts Withdrawn into Head)

Diagnostic Characteristics	Common Name(s)	Scientific Order
Springlike furcula; collophore	springtails	Collembola
No antennae or compound eyes	proturans	Protura
Two "tails"; no compound eyes	diplurans	Diplura

Wingless (Mouthparts Exposed)

Diagnostic Characteristics	Common Name(s)	Scientific Order
Three "tails"; body with scales	silverfish/firebrat	Thysanura

Advanced Wingless (Winged Ancestors or Relatives)

Diagnostic Characteristics	Common Name(s)	Scientific Order
Long antennae; large; humpbacked	camel crickets	Orthoptera
Long antennae; small; wings held rooflike	barklice	Psocoptera
Flattened, broad head; large claws	chewing lice	Mallophaga
Flattened, narrow head; large claws	sucking lice	Anoplura
Flattened, narrow head; no large claws	bed bugs/bat bugs	Hemiptera
Abdomen with cornicles; sucking beak	homopterans (aphids)	Homoptera
Flattened, scalelike body	homopterans (scales)	Homoptera
Compressed body; jumping legs; biting	fleas	Siphonaptera

Primitively Winged (No Wing Folding Mechanism)

Diagnostic Characteristics	Common Name(s)	Scientific Order
Two "tails"; hind wings small	mayflies	Ephemeroptera
Long body; big eyes; small antennae	dragonflies/damselflies	Odonata

Advanced Wings (Folding/Hooking Mechanism)

Diagnostic Characteristics	Common Name(s)	Scientific Order
Two "tails"; hind wings large	stoneflies	Plecoptera
Sticklike body	walkingsticks	Phasmida
Leathery forewings; chewing mouthparts	grasshoppers/crickets	Orthoptera
Abdominal "pincers"	earwigs	Dermaptera
Leathery wings; raptorial forelegs	mantids	Mantodea
Leathery wings; flat body	roaches	Blattaria
Beaded antennae; front and hindwings similar	termites	Isoptera
Small; long antennae; wings held rooflike over body	barklice	Psocoptera
Hemelytra; sucking beak	true bugs	Hemiptera
Leathery wings, held rooflike over body; sucking beak	homopterans (hoppers)	Homoptera
Membranous wings; sucking beak	homopterans (others)	Homoptera
Fringed wings; conical face	thrips	Thysanoptera
Membranous wings; chewing mouthparts	lacewings/antlions	Neuroptera
Elytra; chewing mouthparts	beetles and weevils	Coleoptera
Membranous wings; narrow waist	ants/bees/wasps	Hymenoptera
Hairy wings, held rooflike over body	caddisflies	Trichoptera
Scaly wings; coiled mouthpart	butterflies/moths	Lepidoptera
Scorpionlike abdomen; long face	scorpionflies	Mecoptera
Two membranous wings; halteres	true flies	Diptera

Life Cycle and Habits. Very little is known about the life cycles, behavior, and food habits of proturans.

Habitats. Proturans are relatively uncommon, but this may be, at least in part, because they are easily overlooked. They can be found under bark and in moist soil, humus, rotting wood, moss, and lichens.

Distribution and Faunistics. Three families of proturans (Eosentomidae, Acerentomidae, and Protentomidae) occur in the Great Lakes region. Bernard 1975a and 1975b lists 14 species (in seven genera) in Michigan. Arnett 1985 reports two additional species from the region, for a total of 16 species.

Table 1   Summary of the Insect Fauna of the Great Lakes Region (GLR)

Insect Order Name		Number of Species	
Scientific	Common	U.S./Canada	GLR
Protura	proturans	20	16
Collembola	springtails	677	246
Diplura	diplurans	64	6
Thysanura	silverfish/firebrats	18	2
Ephemeroptera	mayflies	611	222
Odonata	dragonflies/damselflies	407	220
Plecoptera	stoneflies	465	126
Phasmida	walkingsticks	29	4
Orthoptera	grasshoppers/katydids/crickets	1,080	205
Dermaptera	earwigs	20	6
Blattaria	cockroaches	49	18
Mantodea	mantids	20	3
Isoptera	termites	44	7
Psocoptera	booklice/barklice	245	56
Mallophaga	chewing lice	943	118
Anoplura	sucking lice	56	30
Hemiptera	true bugs	3,587	900
Homoptera	cicadas/hoppers/aphids/scales/whiteflies	6,359	1,515
Thysanoptera	thrips	694	223
Neuroptera	dobsonflies/lacewings/antlions	349	65
Coleoptera	beetles/weevils	23,701	3,750
Strepsiptera	twisted-wing parasites	109	10
Mecoptera	scorpionflies/hangingflies	68	24
Trichoptera	caddisflies	1,261	250
Lepidoptera	butterflies/moths/skippers	11,286	2,500
Diptera	true flies	16,914	2,930
Siphonaptera	fleas	325	46
Hymenoptera	sawflies/ants/wasps/bees	17,777	2,705
Total number of species		87,178	16,203

Acerentomid Proturans
(Family Acerentomidae)

Members of this family (and Protentomidae) lack spiracles and
trachea.

The common acerentomid proturan, *Amerentulus americanus* (Ewing), is minute with an elongate body. It is apparently the most common species of this family found in the Great Lakes region and ranges throughout most of the eastern United States and southern Canada.

Eosentomid Proturans
(Family Eosentomidae)

Members of this family have functional breathing pores (spiracles) and trachea.

The pale eosentomid proturan, *Eosentomon pallidum* Ewing, is known from Illinois, Indiana, and Pennsylvania and appears to be the most common representative of this family. Tomlin 1978 reports the existence of undetermined species of Eosentomidae collected in southern Ontario.

Identification Resources. For identification keys to families of proturans, see Bland 1978 and Arnett 1985.

Springtails
(Order Collembola)

Description. Springtails are minute to small (less than 10 mm), wingless, and generally pale (though occasionally spotted, striped, or colored in hues of gray, green, yellow, orange, red, lavender, blue, or purple). They have simple mouthparts (mostly chewing, sometimes piercing/sucking), four-segmented antennae, and an eye patch with eight simple eyes (ommatidia) on the head.

Though they lack wings, most springtails are extremely mobile, thanks to a special set of appendages. On the underside of the fourth abdominal segment there is a forked appendage known as a furcula. The furcula bends forward, under muscular tension, and is held in place by the tenaculum on the underside of the third abdominal segment. When the furcula is suddenly released the springtail is catapulted into the air for distances of up to 100 times the length of the body. There is only one problem with this method of locomotion—there is no directional control. The springtail is likely to land anywhere (and not always in a better spot than before).

Springtails also have another unique appendage, the collophore, a suckerlike organ located on the underside of the first abdominal segment. This multipurpose organ can provide adhesion (it works like a wet suction cup), absorbs oxygen, obtains drinking water, and secretes special chemical substances.

Life Cycle. Springtails develop through a very simple type of metamorphosis; they molt with very little outward change. They can reach maturation in as little as three weeks. Some species are parthenogenetic (their eggs develop without fertilization), but most reproduce by sexual means.

Habits and Habitat. Springtails inhabit cool, moist habitats— soil, humus, leaf litter, fungi, bacteria, algae, moss, pollen, sap, decaying wood, feces (even their own), and debris. A few live in ant nests, bumble bee nests, termite nests, and mammal nests. Others live on water and snow, some live in caves, and many even live high off the ground in vegetation (including trees). Some species are known to inhabit sand dunes, but there is no documentation of species utilizing this habitat in the Great Lakes region. Generally speaking those species living in exposed locations are highly mobile, are colorful, and have well-developed eye patches and antennae, while the species living in soil or other protected situations are less mobile (even springless), are pale colored, and have poorly developed eye patches and antennae.

Ecological and Economic Status. Because of their unique abilities to obtain food, water, and oxygen, springtails are able to occupy ecological niches that would otherwise be void of life. Springtails are very resistant to cold temperatures and can reproduce at temperatures as low as 40°F. In fact, springtails are the dominant organism in the soil ecosystem of the tundra zone. Populations of springtails can reach enormous numbers, and populations in excess of several billion per acre is not uncommon. Usually these tiny insects go unnoticed, but the emergence of the "snow flea," *Hypogastrura nivicola* (Fitch), a bluish springtail seen on top of snow in the late winter and early spring, often attracts considerable attention.

Springtails serve important roles as decomposers and recyclers of organic materials. They are food for many species of ants, beetles, and other small animals. A few springtail species that feed on germinating seeds and seedlings are considered pests in mushroom cultures, greenhouses, and earthworm beds.

Distribution and Faunistics. Springtails are abundant and common throughout the Great Lakes region. The Michigan species are especially well known, with 171 species reported from the state. Hart 1972 lists a total of 114 species for Indiana. Altogether, a total of 246 taxa (species and subspecies) are known to occur in the Great Lakes region.

Fig. 1. The granulate blind springtail, *Tullbergia granulata* Mills. (From Christensen 1992.)

Blind Springtails
(Family Onychiuridae)

The blind springtails are generally minute, whitish, and slender. The prothorax is well developed, but the furcula is short and does not extend past the tip of the abdomen. Eyes are absent. Most species are soil inhabitants. The granulate blind springtail, *Tullbergia granulata* Mills, is minute (1 mm) and has an elongate, whitish body with short antennae. They are common in soil, especially loose, moist soils of fields and woodlots, but they can also be quite abundant in agricultural soils, as they do not seem to be adversely affected by cultivation.

Elongate-Bodied Springtails
(Family Hypogastruridae)

The elongate-bodied springtails are similar to blind springtails, but eyes are usually present.

Fig. 2. The snow flea, *Hypogastrura nivicola* (Fitch), a type of elongate-bodied springtail. (From Christensen 1992.)

The "snow flea," *Hypogastrura nivicola* (Fitch), is so named because of its frequent appearance on the surface of snow in late winter and early spring. It is minute (2 mm) and uniformly dark or slate blue, with an elongate body. The armed springtail, *H. armatus* (Nicolet), is also minute but is uniformly slate gray. It has been collected from under piles of leaves, in rotting wood, and in other moist vegetation. It is occasionally a pest of cultivated plants.

Slender Springtails
(Family Entomobryidae)

Fig. 3. The armed springtail, *Hypogastrura armatus* (Nicolet). (From Christensen 1992.)

This is the largest family of springtails. Slender springtails are usually elongate, with short appendages. The pronotum is poorly developed and the antennae are four-segmented. Scales and clubbed-shaped setae (bristlelike hairs) are lacking.

The many-spotted slender springtail, *Entomobrya multifasciata* (Tullgren), is small (2.5 mm) and yellowish with purplish bands on each body segment that are very nearly interrupted at the center. This cosmopolitan species has been found on window ledges, in leaf litter, in bird nests, under bark, in grass, and in loose soil and drifting sand.

Smooth Springtails
(Family Isotomidae)

These springtails are closely related to the slender springtails, but the body is frequently covered with scales or clubbed setae. The pronotum is also poorly developed, but the third and fourth abdominal segments are nearly equal in size. There are four to six antennal segments.

The marsh springtail, *Isotomurus palustris* (Müller), has

Fig. 4. The marsh springtail, *Isotomurus palustris* (Müller). (From Folsom 1937.)

various color patterns (but is never banded or unicolorous) and an elongate body (2 mm). Associated with aquatic situations, it can be found in roadside ditches, swamps, lakeshores, wet meadows, and soggy soil. The smooth springtail *Folsomia fimetaria* (L.) is minute and whitish, with a distinctly segmented, elongate body. This species is common in soil and leaf litter. It has even been found in mouse nests (Snider 1967).

Water Springtails
(Family Poduridae)

Water springtails are usually whitish and slender and have short appendages, including four-segmented antennae. The pronotum is well developed and the body is granulate (covered with small granules).

The aquatic springtail, *Podura aquatica* L., is the only species in this springtail family. It is active in the spring and can be found on the surface of ponds and slow-moving streams. It is minute (1.5 mm) and bluish, black, or reddish brown, with a furcula that is long and flattened.

Globular Springtails
(Family Sminthuridae)

The globular springtails are suitably named: the abdomen is large and glob shaped, with the segments fused and indistinct. The eyes are often black.

The three-lined globular springtail, *Sminthuris trilineatus* Banks, is small (3 mm) and globular and has three distinct lon-

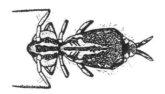

Fig. 5. The three-lined globular springtail, *Sminthuris trilineatus* Banks. (From Pedigo 1966.)

gitudinal stripes. Once thought to be uncommon, it is readily taken in sweep nets and pitfall traps. The garden or seedling springtail, *Bourletiela hortensis* (Fitch), is minute (1.5 mm) and mottled bluish or purplish brown, with the top of the abdomen covered with scattered pale spots (the underside is completely pale). This springtail can occasionally be a pest of vegetable crops and greenhouse plants. It is a sun-loving species and is found in soil, in leaf litter, and on plants.

Identification Resources. For an identification key to springtail families in North America, see Scott 1961; for a key to Michigan species, see Snider 1967; for a key to aquatic species, see Lehmkuhl 1979.

Diplurans
(Order Diplura)

Description. Diplurans are small (8–10 mm), pale insects with elongate, flattened bodies that are tapered at both ends. The head is conical and bears chewing mouthparts but lacks compound eyes. There are two threadlike processes (caudal filaments) at the tip of the abdomen, each with ten segments. (In members of one family, Japygidae, the adults have the caudal filaments replaced by forcepslike appendages.) The tarsi are simple and one-segmented.

Life Cycle and Habits. Very little is known about the life cycle and habits of diplurans.

Habitats. Diplurans are generally found in damp habitats—under stones, bark, soil, rotting wood, humus, and debris.

Distribution and Faunistics. Since this group is poorly known there is very little detailed information available on their distribution. In the Great Lakes region members of the genus *Campodea* (family Campodeidae) are the most likely to be encountered. As best as can be determined, a total of six species are found in the Great Lakes region.

Slender Diplurans
(Family Campodeidae)

Members of this family have cerci that are as long as or longer than the antennae. They lack eyes and occur in damp locations.

Folsom's slender dipluran, *Campodea folsomi* Silvestri, is thought to be one of the most widely distributed species in North America, and it can no doubt be found in the region. Other species known to occur in the Great Lakes region are the fragile slender dipluran, *C. fragilis* Meinert, and *C. plusiochaeta*

Silvestri. Tomlin 1978 reports the existence of undetermined species of Campodeidae collected in southern Ontario. A fourth species, *Plusiocampa cookei* (Packard), has been taken in some of the caves of southern Indiana (Chandler 1956).

Earwiglike Diplurans
(Family Japygidae)

Members of this dipluran family have short, pincerlike cerci.

Two species, the subterranean earwiglike dipluran, *Metajapyx subterraneus* (Packard), and the small earwiglike proturan, *Parajapyx minimus* (Swenk), are known from Indiana (Chandler 1956). The small earwiglike dipluran was collected from ant nests and soil in West Lafayette, Indiana.

Identification Resources. For identification keys to the dipluran families, see Bland 1978 and Arnett 1985.

Silverfish and Firebrats
(Order Thysanura)

Description. Most members of this order are covered with flattened hairs or scales and have three "tails" (caudal filaments) at the tip of the abdomen. They are usually white, grey, silvery, or brown, and they range in size from 2 to 20 millimeters. The head bears long, slender antennae, compound eyes (occasionally absent), and chewing mouthparts.

Life Cycle. Thysanurans have a simple metamorphosis; they grow by molting, but the body generally does not change shape. The young thysanurans grow slowly and may require up to two years to complete their development.

Habits. They feed on many types of starchy materials—such as paper and glue—and also on other vegetable materials (cloth, leaves, fungi, etc.).

Habitats. Most members of the Thysanura live outdoors and can be found in leaf litter, under bark, in rotting wood, or in caves. Some members of this order have adapted well to human structures. Two species are commonly encountered in buildings.

Ecological and Economic Status. Several Thysanurans are occasional household pests, damaging paper, book bindings, wallpaper, starched cloth, photographs, and stored foods. Outdoor species are important decomposers and recyclers.

Distribution and Faunistics. Two species of Thysanura are widely distributed throughout the Great Lakes region because of their close association with humans.

Silverfish and Firebrats
(Family Lepismatidae)

The silverfish, *Lepisma saccharina* L., is small to medium sized (8–13 mm) and silvery gray. They are usually found in cool, damp, dark locations and are especially common in basements, attics, and bathrooms. The firebrat, *Thermobia domesticus* (Packard), is similar in size and shape but is tan or yellowish brown in color. Firebrats favor hot, humid locations near furnaces, stoves, and steampipes.

Identification Resources. For an identification key to the families and genera of the order Thysanura, see Jackson 1906.

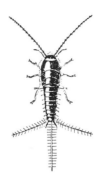

Fig. 6. The silverfish, *Lepisma saccharina* L. (From Marlatt 1915.)

Mayflies
(Order Ephemeroptera)

Introduction. Adult mayflies are common and abundant in areas near water and the immature (nymphal) stage is associated with relatively clean water—lakes, ponds, streams, and rivers (both warm and cold). They are considered to be one of the most primitive surviving winged insects.

Description. The adult mayfly has an elongate body with two long filaments at the tip of the abdomen. The midsection (thorax) bears two pairs of membranous wings; the hind pair are significantly smaller than the front pair (less than one-third the size). The front pair of legs are exceptionally long compared to the middle and hind legs. The head has nonfunctional (vestigial) chewing mouthparts, well-developed compound eyes, and short, bristlelike antennae.

The immature mayfly (naiad, or aquatic nymph) is more or less flattened and elongate. The ten-segmented abdomen bears leaflike tracheal gills attached to the outer, upper edge. There are three (rarely two) caudal filaments.

Life Cycle. Mayfly eggs are laid in the water and hatch in about two weeks. The aquatic nymphs develop underwater, where they feed on algae, diatoms, aquatic plants, detritus (tiny bits of organic material), and other small invertebrate animals. The aquatic nymphs may be swimmers (with elongate bodies), crawlers (with well-developed legs), clingers (with flattened bodies and short legs), or burrowers (with molelike legs and jaw tusks). As the mayfly nymphs grow they shed their exoskeleton frequently, and 30 to 50 molts is not uncommon for some species. After completing their nymphal development a winged form (the subimago, or preadult) emerges and seeks a sheltered spot to rest. Within a short period, no more than a day or two, the subimago molts into the adult stage (imago). It is believed by

Fig. 7. An adult mayfly (Ephemeroptera), showing large front wings, small hind wings, and two caudal filaments. (From Kansas State University Extension 1962.)

some that this transition stage is necessary so that the aquatic respiratory system of the immature may be converted into the terrestrial system of the adult. This final molt makes mayflies unique in the insect world—they are the only insects that are capable of molting after the formation of the wings. After reaching adulthood mayflies are completely preoccupied with the mating process, and they die soon after mating.

The scientific name for this order was derived from the word *ephemeral*, in reference to the supposedly short life of the mayfly. But only the adult stage is ephemeral, "lasting but a day." The immature nymphal stage is quite long-lived and the mayfly life cycle requires one or more years for completion.

Habits and Habitats. Mayfly nymphs utilize a variety of aquatic habitats, both flowing and still waters, and they are not dependent on cold water. Most species live among gravel riffles and rocks of streams or in the mud and bottom sediments of lakes. Some genera of mayflies are adapted to specific aquatic habitats, while others are more ubiquitous.

To ensure successful mating, the adults of many mayflies emerge in synchrony and occasionally may become extremely abundant. The mating swarms of some mayfly species leave life-long impressions for human observers and have to be experienced to be believed. Some areas adjacent to breeding habitats may be inundated with adult mayflies. Cars inadvertently driven through swarms are quickly covered with smashed bodies and sticky egg masses. Reports of dead mayflies up to two feet deep have been reported in some places, especially in the vicinity of Lake St. Clair and Lake Erie. Large windrows of dead bodies may wash up on beaches or accumulate on streets and bridges. The stench from the dead mayfly bodies is often overpowering.

Ecological and Economic Status. Mayflies are an important and substantial part of the diet of many fish species (and other aquatic organisms). In many areas, the bottom-dwelling burrowing mayflies are harvested for sale as fish bait. Since the adults are also consumed by fish, fly fishermen have learned to create artificial mayflies out of bits of feathers and hairs. Mayflies, like most aquatic insects, are somewhat sensitive to water pollutants, and their numbers and distribution often aid in assessing the severity of pollution. Indeed, many mayfly populations have been greatly reduced, even eliminated, in some bodies of water. The nuisance that they may become during mating is the only negative aspect to these important aquatic insects.

Distribution. Few comprehensive works have been published on the mayfly fauna of the Great Lakes region, but a review of the literature reveals that at least 222 taxa (species and subspecies) in 12 families have been reported from the Great Lakes region to date.

Primitive Minnow Mayflies
(Family Siphlonuridae)

These mayflies have relatively large hind wings and the upper part of the compound eye has larger facets (ommatidia) than the lower part. Many species have white legs. The nymphs are elongate and minnowlike and are found mostly in streams.

Sand Minnow Mayflies
(Family Ametropidae)

These mayflies are very similar to the flat-headed mayflies, but the hind tarsi are four-segmented. The nymphs inhabit stream bottoms. There are only three species in this small family of mayflies, although they are widely distributed.

Small Minnow Mayflies
(Family Baetidae)

The hind wings are reduced in size with only one to three longitudinal veins, or sometimes they are entirely absent. The hind tarsi are four-segmented. The males are characterized by turban-shaped heads (the top half of the compound eyes are greatly enlarged, creating a turbanlike shape). The forward part of the abdomen is light colored and the rear part is generally darker in males, but often similar in females. Nymphs are elongate and minnowlike but differ from members of the primitive minnow mayflies (Siphlonuridae) by having much longer antennae. They inhabit a wide variety of warm waters.

Flat-Headed Mayflies
(Family Heptageniidae)

Flat-headed mayflies are medium sized (up to 15 mm) with five-segmented hind tarsi (most other families have four, or even three, hind tarsal segments). The wing venation is distinctive, with the base of the second medial vein (one of a set of longitudinal veins located near the middle of the forewing) nearly straight. The nymphs are dark and distinctly flattened, with two or three tails at the tip of the abdomen. They inhabit streams or lakes.

The Canadian flat-headed mayfly, *Stenonema canadense* (Walker), is medium sized (10–12 mm) and variable in color, but the head is generally pale amber with a brown line from the rear of the head to the simple eyes (ocelli). The thorax is generally pale amber with dark brown stripes on the side margins and along the rear margin and with a dark brown stripe from the wing base to the top of the middle leg. The abdomen is medium brown above with pale bean-shaped spots on each side of the midline and triangular spots on the sides.

Prong-Gill Mayflies
(Family Leptophlebiidae)

The prong-gill mayflies are similar to the flat-headed mayflies, but the second cubital vein (one of a set of longitudinal veins located between the middle of the wing and the hind angle) is strongly arched toward the rear margin of the wing. There are three long "tails" (caudal filaments) and the hind tarsi are four-segmented. The compound eyes of the male have larger facets (ommatidia) on the top than on the bottom. The nymphs are mostly stream inhabitants.

The adoptive prong-gill mayfly, *Paraleptophlebia adoptiva* (McDunnough), is small (7–8 mm) and has a dark brown head and shiny reddish brown thorax with pale sides. The adbomen is dark reddish brown and has a pale basal band and a pale stripe down the center, with triangular spots below this midline. This species is commonly attracted to lights.

Hacklegill Mayflies
(Family Potamanthidae)

The adults of this family are predominantly pale white to creamy white in color and the wings are transparent. This small family includes only eight species, of which at least the vertical hacklegill mayfly, *Potamanthus verticalis* (Say), occurs in the Great Lakes region. The nymphs inhabit fast-flowing streams.

Pale Burrowing Mayflies
(Family Polymitarcidae)

The adults of these pale, whitish mayflies can be recognized by the stunted (atrophied) middle and hind legs. The pale burrowing mayfly *Ephoron leucon* is an eastern species that most certainly occurs in the Great Lakes region. Nymphs are found in swiftly flowing streams or rivers, where they make tubular burrows or burrow into clay banks.

Common Burrowing Mayflies
(Family Ephemeridae)

These are the largest mayflies to occur in the Great Lakes region, and some may reach lengths of up to 1.3 inches (33 mm).

The second medial vein (one of a set of longitudinal veins located in the middle of the forewing) has an abrupt bend at its base. The hind tarsi are four-segmented. The nymphs inhabit sand and silt of large streams or lakes.

The brown drake, *Ephemera simulans* Walker, is medium-sized (10–14 mm) and dark brown to dark reddish brown with a yellowish to reddish brown abdomen. The wings have many dark spots and dark cross veins. The Michigan caddis or golden mayfly, *Hexagenia limbata* (Serville), is medium to large sized (17–33 mm) and varies from yellowish (marked with purple) to dark purplish (marked with olive) with the leading edge of the forewing purplish. There are two "tails" (caudal filaments).

Spiney Crawlers
(Family Ephemerellidae)

These mayflies are small to medium sized and usually brownish in color. The males have large compound eyes that meet along the top of the head (the eyes are smaller in the female). The hind tarsi are four-segmented and there are three "tails" (caudal filaments). Nymphs are common inhabitants of trout streams.

One of the most common species is *Ephemerella invaria* (Walker). It is small (7–9 mm) and has a reddish-brown thorax and abdomen. The sides of the thorax and abdomen are yellowish brown. The eyes of the male are deep orange. Another common species is the midboreal mayfly, *E. subvaria* McDunnough. It is slightly larger and darker than *E. invaria*, with the middle of the thorax, near the base of the front wings, tinged with purple. The eyes are shiny reddish brown.

Armored Mayflies
(Family Baetiscidae)

Adult armored mayflies have irregularly spaced cross veins in the forewing. All known species (only 12 in North America) belong to the genus *Baetisca*.

The nymphs of this family are unique. The one-piece thorax is irregularly shaped and covers the entire front two-thirds of the top surface. The nymphs inhabit stream edges and lakeshores.

Identification Resources. The identification of mayfly adults is accomplished through examination of the wing venation (primarily) along with a few other characteristics. For identification keys to mayfly families (adults and nymphs), see Borror, DeLong, and Triplehorn 1976, Arnett 1985, Lehmkuhl 1979, and Leonard and Leonard 1962; for keys to some of the species of the Great Lakes region, see Leonard and Leonard 1962, Bergman and Hilsenhoff 1978, Burks 1953, and Flowers and Hilsenhoff 1975.

Dragonflies and Damselflies
(Order Odonata)

Introduction. These familiar insects are a conspicuous part of the insect fauna of the Great Lakes region. They comprise an ancient group (with fossils dating back to 300 million B.P.), and many specialized features have contributed to their continuing success. The Odonata were probably at their peak during the Permian period (250 million B.P.), and fossils indicate that some species had wingspans of 29 inches. Many folktales are associated with this particular group of insects, and such local names as snake doctors, mosquito hawks, and sewing needles are commonly used for members of this group.

Description. The Order Odonata is divided into two suborders, the dragonflies (Anisoptera) and the damselflies (Zygoptera). Both have elongate, slender bodies (ranging in size from 20 to 190 mm) and two pair of narrow, membranous wings. Many species have brightly colored bodies and the wings may be tinted or patterned as well. The antennae are inconspicuous and bristlelike. The compound eyes, however, are far from inconspicuous. In many species (especially in the dragonflies) the eyes virtually occupy the entire head capsule except for the lower front of the face. The legs are generally long, spiny, and directed forward, forming baskets for catching their aerial prey and for clinging to plants. Damselflies, which tend to be more slender and delicate, have the base of both the front wings and the hind wings narrowed, and at rest they hold their wings above the body. Dragonflies, which can often be large and robust, have the base of the hind wings broad, and at rest the wings are held outstretched.

The aquatic nymphs of the two odonate groups are also dissimilar. The damselfly nymphs are generally elongate and have a set of three leaflike gills at the tip of the abdomen. The

Fig. 8. An aquatic damselfly nymph, showing three caudal gills. (From Howe 1899.)

Fig. 9. An aquatic dragonfly nymph (genus *Macromia*); note the hinged labium for prey capture and the absence of caudal gills. (From Howe 1899.)

dragonfly nymphs are generally more robust and there are no signs of any tracheal gills. They have rectal gills, which are located within the tip of the abdomen and therefore are not visible. Odonate nymphs have a unique way of catching their prey. Their lower lip (labium), which is long and hinged, is armed with a set of pincers. In the blink of an eye this weapon can be thrust forward, almost always capturing the unsuspecting prey.

Life Cycle. The eggs of dragonflies and damselflies are laid in the water. The aquatic nymphs are voracious predators and will eat anything they can catch and hang on to, including insects, snails, tadpoles, or small fish. They grow through a series of molts and then the nymphs crawl out of the water and rest on any nearby object (a plant stem, stick, bridge abutment, etc.). Soon the adult emerges and the old nymphal skin is left behind as evidence of their transformation into the aerial adult stage. Adults generally live for a period of two to three weeks.

Habits and Habitats. The nymphs of damselflies and dragonflies are restricted to aquatic habitats—temporary and permanent pools, marshes, swamps, ponds, lakes, streams, and rivers. The adult dragonflies are accomplished aerialists; the damselflies are feeble fliers, at best. The airborne adult dragonflies may patrol large areas and roam great distances from their nymphal home in search of prey. Male dragonflies are frequently territorial and will chase away other males of the same species, but any females that enter the territory are usually courted. They are particularly effective in consuming mosquitoes, midges, and other small flying insects.

Ecological and Economic Status. The odonate insects are extremely valuable, both as food for aquatic organisms and as predators of mosquitoes and other pests.

Distribution. The dragonflies and damselflies of the Great Lakes region are relatively well known, at least in the eastern two-thirds of the region. For example, 117 species are known to occur in New York state (Needham and Betten 1901); 86 in Pennsylvania (Beatty and Beatty 1968); 155 in Ohio (Alrutz 1992); 152 in southern Ontario (Corbet 1967); 150 in Michigan (Kormondy 1958); 144 in Indiana (Lawrence 1967); 39 in Illinois (Needham and Hart 1901); 117 in Wisconsin (Ries 1967); and 62 in southern Minnesota (Whedon 1914). Altogether, approximately 220 taxa (species and subspecies) are reported to occur within the region.

Graybacks
(Family Petaluridae)

The eastern grayback, *Tachopteryx thoreyi* (Hagen), occurs in the Great Lakes region (Indiana, Michigan, Ohio, and New York), although it is rare and local. It is most likely to be found near small streams in wooded valleys. The adults are large (about 3 in) and grayish brown. The compound eyes do not meet on the top of the head. The grayback frequently perches on tree trunks and stones, and its gray coloration is an effective camouflage.

Darners
(Family Aeschnidae)

Fig. 10. The big green darner, *Anax junius* (Drury). (From USDA 1952.)

Darners are swift and agile fliers and can frustrate even the most experienced entomological "dragon slayer." The big green darner, *Anax junius* (Drury), is common around ponds and sometimes swamps. The eyes meet on the top of the head and there is a distinctive targetlike marking on the upper part of the face. The thorax is light green and the abdomen is bluish. This big dragonfly (68–80 mm) has been observed migrating in large swarms.

The blue darners (13 species in the genus *Aeschna*) are also common in the Great Lakes region. *Aeschna constricta* Say (a blue or brownish species with greenish face and pale yellowish markings on the sides of the thorax) and *A. umbrosa* Walker (with yellow or brown markings on the back of the head and pale spots on the underside of the middle abdominal segments) are very large (68–72 mm) and occur throughout the region.

Clubtails
(Family Gomphidae)

The common name for this family comes from those species that have the tip of the abdomen enlarged and clublike. Most species are large and darkly colored, with green or yellow stripes. The compound eyes do not meet on the top of the head. Clubtails are most often encountered near streams and along lakeshores, where they frequently rest on exposed rocks.

Commonly encountered species include *Gomphus exilis* Selys (a large, greenish yellow species with pale brown stripes and the top of the eighth abdominal segment blackish), which is commonly found around ponds and streams; and *Gomphus* (*Gomphurus*) *externus* Hagen (a slightly larger, yellowish green species with brown stripes on the thorax and a wide yellow band at the ninth abdominal segment), which is found near large streams and rivers. The black clubtail, *Hagenius brevistylus* Selys, is primarily black in color, although the sides of the thorax are marked with yellow. It frequents large streams and rivers. *Ophiogomphus rupinsulensis* (Walsh) also occurs throughout the region. It is greenish with brown stripes at the "shoulders" of the thorax. It frequents rapid streams. The white-lined clubtail, *Gomphus* (*Gomphurus*) *lineatifrons* Calvert, is listed as a species of special concern.

Fig. 11. A clubtail dragonfly (family Gomphidae). (Courtesy of the Young Entomologists' Society.)

Biddies
(Family Cordulegasteridae)

The biddies are large brownish or blackish dragonflies with yellow markings. They are found around small woodland streams and are generally not common. Four species in the genus *Cordulegaster* are known from the Great Lakes region. The spotted biddie, *Cordulegaster maculata* Selys, has the widest distribution in the region and is known from Wisconsin, Indiana, Michigan, Ontario, Ohio, and New York. Several species of clubtail dragonflies are currently listed as species of special concern, including the false spiketail (*C. erronea* Hagen).

Belted and River Skimmers
(Family Macromiidae)

The belted skimmer, *Didymops transversa* Say, is a brownish dragonfly and is known from Illinois, Indiana, Michigan, Ohio, Ontario, and New York. It frequents the shorelines of boggy ponds.

The river skimmers, *Macromia* species, are generally black with yellow markings and bright green eyes (when alive). Seven species are known from the region and they frequent large streams and lakes.

Green-Eyed Skimmers
(Family Corduliidae)

The green-eyed skimmers have brilliant green eyes (when alive); their bodies are generally blackish or metallic with conspicuous markings. The hind margin of the compound eyes is slightly lobed. They frequent swamps and ponds (and occasionally streams).

One commonly encountered species, at least in the northern portions of the region, is *Cordulia shurtleffi* Scudder, which is an attractive dragonfly with an orange-brown face, bronze-green thorax, and shiny black abdomen. The water prince, *Epicordulia princeps* (Hagen), is yellowish brown or olive colored (although the top of the abdomen is black) and occurs throughout the region. *Somatochlora tenebrosa* (Say) is dark brown with metallic green or bronze and has a distinctive orange labrum. One species of green-eyed skimmers, the warpaint emerald, *S. incurvata* (Walker), is currently listed as a species of special concern.

Common Skimmers
(Family Libellulidae)

Libellulidae is one of the most common families of dragonflies, and its species can be found around just about any pond or swamp. They are small to medium sized with wings that are longer than the length of the body. The wings are often banded or spotted. The bodies are brightly colored (but not metallic). Females lack a well-developed ovipositor. Males are aggressive and may drive most other dragonflies from their territory.

Both the widow, *Libellula luctuosa* Burmeister (a blackish species with broad, dark basal wing bands), and the ten-spot dragonfly, *L. pulchella* Drury (a striking species with three large brown spots on each wing, the basal spots united across the thorax), occur throughout the Great Lakes region. Skimmers of the genus *Sympetrum* (10 species in the region) are mostly small and reddish. They are particularly abundant in the later part of the summer, and since they are slow fliers they are easily caught. The white-tailed dragonfly, *Plathemis lydia* (Drury), is easily recognized by the frosted, whitish abdomen and banded wings (a broad band near the wing tip and a small band at the base of the wing). It is commonly found near ponds and other

Fig. 12. The white-tailed dragonfly, *Plathemis lydia* (Drury), a commonly seen member of the common skimmer family (Libellulidae). (From USDA 1952.)

bodies of still water. The amberwing, *Perithemis domita* (Drury), is small and yellowish brown and has two irregular brownish spots on each amber-colored wing. Shy and reclusive, it is frequently found on low aquatic plants and away from water. It is known from both Minnesota and New York, so it probably occurs elsewhere in the region in areas between these two states.

Broad-Winged Damselflies
(Family Calopterygidae)

Males of the common black-winged damselfly, *Calopteryx maculata* Beauvois, are metallic green with bluish reflections (black underneath) and their wings are velvety black throughout. Females are similar, but with a brassy brown abdomen that has a white band near the tip; their wings are dark gray with a white spot (stigma). This species is encountered commonly near small streams (and occasionally near larger waters) where overhanging vegetation covers the banks. The American rubyspot, *Hetaerina americana* Fabricius, is greenish bronze, coppery bronze, reddish, or dark brown. The males have spots of ruby red covering parts of the wing bases; females have amber yellow wing spots. This species is commonly found around both still and moving water. The nymphs are often found clinging to plants in the rapids of streams or to the bulrushes of wave-churned lakes.

Narrow-Winged Damselflies
(Family Coenagrionidae)

Members of the narrow-winged damselflies are commonly encountered around ponds and streams. They are easily identified in the field by their wing posture. The wings are held together above the abdomen. Most species are brightly colored in blue, black, or violet.

The violet dancer, *Argia violacea* Hagen, is 29–34 millimeters long with a violet and black body. It is found in marshy zones of ponds and lakes, cedar-spruce bogs, and sluggish streams. *A. apicalis* Say is slightly longer (33–38 mm) but has a light blue thorax with some black markings and a bluish or brownish abdomen striped with black. The wings are clear except for a brown spot at the base. It is found only in the southern part of the region (in Canada it is restricted to extreme southern Ontario), and it is generally on the wing from June to September. The smokey-winged dancer, *A. fumipennis* (Burmeister), at 50+ millimeters in length is the largest damselfly species found in the Great Lakes region. It is primarily a southern species, so it is rare in the Great Lakes region, but it

Fig. 13. *Argia apicalis* Say, a common narrow-winged damselfly. (Courtesy of the Young Entomologists' Society.)

has been reported from Ohio (Wright 1939). It is brown and black with uniformly tinted wings of smoky brown.

Hagen's bluet, *Enallagma hageni* Walsh, is 26–33 millimeters long and blue with some black on the head and thorax (male) or dull green with black on the abdomen (female). The blue civit damselfly, *E. civile* Hagen, is long and slender (31–39 mm) and similar to *hageni*, but with more black (especially on top of the sixth abdominal segment). The female is dull bluish or yellowish green. It is common around the margins of streams, lakes, and pools. Adults fly low, near the surface of the water, frequently stopping to rest on nearby vegetation. The common forktail damselfly, *Ischnura verticalis* (Say), is 23–33 millimeters long and is often extremely common. The male has a yellowish green and black body with the eighth and ninth abdominal segments bright blue. The female has a greenish black or reddish yellow body. *Nehalennia irene* (Hagen) is 25–28 millimeters long with a greenish black head, metallic green thorax, and metallic green abdomen with bluish on the sides and at the base. It is found near marshes and bogs.

Spread-Winged Damselflies
(Family Lestidae)

These large damselflies are common and can easily be identified in the field by their wing posture. The wings are held apart over the body when at rest. They are commonly encountered around swamps and ponds.

Some of the more common species include the following. *Lestes disjuncta* Selys is 33–41 millimeters long with a blackish brown head and thorax and metallic green abdomen with a black tip. Parts of the head, thorax, legs, and abdomen are yellow. There are two subspecies: *disjuncta australis* Walker, which occurs only in the southern part of the region (southeastern Michigan, extreme southern Ontario, and south); and *disjuncta disjuncta*, which occurs in the northern part of the region. *L. unguiculatus* Hagen is 34–38 millimeters long with a blackish brown forebody (head and thorax) and metallic green or brown abdomen. Both the thorax and abdomen have yellow markings. The male's eyes are blue. This species inhabits marshy beach pools of the Great Lakes and other marshy ponds. *L. rectangularis* Say is long bodied (37–47 mm) and short winged. It is blackish brown with yellow markings on the head and thorax; the abdomen is green-gray to pale brown and pale yellow at the base and blackish at the tip.

Identification Resources. For keys to Odonata of North America, see Lehmkuhl 1979, Needham and Westfall 1955, Walker 1953 and 1958, and Walker and Corbet 1975. Supplemental keys to adults (and occasionally nymphs as well) are available in some of the older annotated lists (but some of these keys are often poorly constructed or out of date): for Illinois, see Needham and Hart 1901 on dragonflies and Garman 1917 on damselflies; for Indiana, see Williamson 1900; for Ohio, see Kellicot 1899. For a key to the genera of dragonfly nymphs of the United States and Canada, see Wright and Peterson 1944.

Stoneflies
(Order Plecoptera)

Description. Stoneflies are medium- to large-sized, soft-bodied, more or less flattened insects. They are generally greenish, yellowish, brownish, or grayish in color. They have two prominent "tails" (cerci) at the tip of the abdomen, although in some species these "tails" are short and concealed by the wing tips when the wings are folded. The membranous forewings fold over the back, appearing either flattened or "rolled" (curved); the hind wings are usually much larger than the front wings. In a few cases the wings of some males may be reduced or lacking. The head bears chewing (or nonfunctional) mouthparts and long, slender antennae (with 25 to 100 segments).

The aquatic nymphs are flattened, with broadly spaced legs (an adaptation for clinging to the smooth surface of stones and

Fig. 15. An adult stonefly with wings in typical resting position. The two caudal filaments are exposed beyond the tip of the folded wings. (Courtesy of Michigan State University Extension.)

Fig. 14. An adult stonefly with wings spread; note the long, narrow front wings and the large, rectangular hind wings. (From Newcomber 1918.)

logs). The featherlike, external breathing organs (tracheal gills) may be located on the thorax or abdomen. There are two stout cerci at the tip of the abdomen. Stonefly nymphs require well-aerated water because their gills are small or in some cases nonexistent. Those species without gills must resort to breathing through their cuticle. Even at the maximum degree of development, stonefly gills are nothing more than small tufts of respiratory filaments. With less surface area to absorb oxygen, they are forced to occupy only those habitats rich in dissolved oxygen. In contrast, mayfly, dragonfly, and damselfly nymphs have flat, platelike gills that offer greater surface area for contact with the water, allowing them greater utilization of the full range of aquatic habitats.

Life Cycle. The eggs, up to 6,000 of them, are deposited directly in the water by the female stonefly. The aquatic nymphs hatch from the eggs and begin feeding. Some feed on vegetable matter like algae, diatoms, moss, or detritus. Others are carnivorous and feed on small insects, especially immature midges and mayflies. It generally takes 12 to 36 molts and one to three years of development to reach the final nymphal stage. Mature nymphs crawl from the water and adults emerge from within the nymphal skin. Many male stoneflies "drum" to attract a potential mate, and after completing the mating process, the life cycle starts anew.

Habits. Stoneflies inhabit rapid streams and wave-swept shores of ponds and lakes, where the water is clean and well aerated. The flattened nymphs cling to the undersides of stones and wood and move about by slowly crawling (they are not particularly good swimmers). The adults are slow, feeble flyers that generally stay close to breeding areas. They rest on vegetation, buildings, bridge abutments, trees, or any more or less vertical objects. They feed on algae, tree buds, and pollen. Many species of stoneflies are active during the late fall, winter, and early spring months. Since this is the off-season for most insect investigators, the fall and winter stoneflies frequently get overlooked and as a result are not particularly well known.

Ecological and Economic Status. Due to their high requirements for clean, well-aerated water, stonefly nymphs are an excellent biological indicator of the environmental quality of aquatic habitats. Stoneflies are sensitive to even low levels of pollution, and their absence from aquatic habitats is usually the first indication of deterioration. The aquatic nymphs are also an extremely important source of food for fish.

The adults can occasionally be a nuisance when they gather in large numbers on the sides of buildings (especially under the lights) or drum on walls. Adult stoneflies are also known to feed on tree buds and have on occasion caused minor damage.

Distribution. The stonefly fauna of the Great Lakes region has received some attention and investigation. Studies have revealed 30 species from northeastern Minnesota (Lager et al. 1979), 75 species from Indiana (Bednarik and McCafferty 1977), 40 species from Illinois (Frison 1935), and 56 species from Ohio (Gaufin 1956). In total, at least 126 species (in nine families) have been reported from the Great Lakes region.

Spring Stoneflies
(Family Nemouridae)

Spring stoneflies are medium to large insects, up to 15 millimeters in length. The forewings are held flat over the back at rest. The second tarsal segment is short, less than half the length of either the first or third segments. They inhabit small streams with sandy bottoms. One common species is *Nemoura venosa* Banks, which is 7–7.5 millimeters long and black or dark brown.

Winter Stoneflies
(Family Taeniopterygidae)

Winter stoneflies are mostly dark brown or black in color, are less than 15 millimeters in length, and have short cerci.

The second tarsal segment is nearly as long as either the first or third segment. Common species include *Taeniopteryx nivalis* Fitch (dark brown or blackish, 11–17 mm in length), which is active from February to April, and *Strophoteryx fasciata* (Burmeister) (dark brown or blackish, 10–15 mm in length), which is active from March to April.

Small Winter Stoneflies
(Family Capniidae)

Small winter stoneflies are small (less than 12 mm) and blackish, with long antennae. The "tails" (cerci) are long and four-segmented. The second tarsal segment is short. The adults are active from December to April. *Allocapnia pygmaea* (Burmeister), a common species, is dark brown or blackish and 3–8 millimeters in length.

Rolled-Wing Stoneflies
(Family Leuctridae)

These small stoneflies (less than 10 mm) appear very slender because the edges of the wings are rolled around the sides of the abdomen. They are generally dark brown or blackish. The second tarsal segment is short. Adults are on the wing during the spring and summer months.

Giant Stoneflies
(Family Pteronarcyidae)

The largest of our stoneflies are members of this family. Most are greater than one inch (25 mm) in length, and some species reach two inches (50 mm). Most species are brownish or grayish. They inhabit small rivers, and the adults are active from May to July. The adults are nocturnal and frequently fly to lights. *Pteronarcys dorsata* Say, which occurs in the Great Lakes region, is the largest North American stonefly species.

Roachlike Stoneflies
(Family Peltoperlidae)

These small stoneflies (10–12 mm) are brownish in color and active in the early summer months. The "tails" (cerci) are short and there are only two ocelli on the head (most stoneflies have three ocelli). The aquatic nymphs are flattened and roachlike and also only have a pair of ocelli.

Fig. 16. An aquatic stonefly nymph; note the "feathery" gills at the base of the legs and the two "tails," or caudal filaments. (From Frison 1938.)

Predatory Stoneflies
(Family Perlodidae)

These medium to large (10–25 mm) stoneflies inhabit moderate to large rivers. Most adults are brownish or blackish, although members of the genus *Isoperla* are greenish or yellowish. At least 21 species of green-winged stoneflies (*Isoperla* species) are found in the Great Lakes region. These resemble the green stoneflies, but the corners of the pronotum are nearly squared. Adult predatory stoneflies are active in the spring and summer.

Green Stoneflies
(Family Chloroperlidae)

These small- to large-sized (6–24 mm) greenish or yellowish stoneflies are delicate in appearance. The front and rear corners of the pronotum are broadly rounded (rather than nearly squared as in the green-winged stoneflies of the predatory stonefly family). They inhabit small streams and the adults are active in the spring and summer. Two common species are *Alloperla atlantica* Bauman, which is 10–12 millimeters long and light green, and *Hastaperla brevis* (Banks), which is 6–9 millimeters long and yellowish green to bright green, with a small hind wing (it lacks the enlarged anal area that is typical of most stoneflies).

Common Stoneflies
(Family Perlidae)

This is the largest and most common family of stoneflies. Adults are yellowish or brownish, with two or three ocelli, and with remnants of the nymphal gills on the underside of the thorax near the leg bases. The nymphs inhabit streams, ponds, and lakes; adults are active during the spring and summer months. Common species include *Phasganophora capitata* (Pictet) (16–24 mm, brownish body with the leading edge of the wings yellow); *Neoperla clymere* Newman (9–14 mm, yellow or yellowish brown body with darkened wings and two ocelli surrounded by a black spot on the head); and *Perlesta placida* Hagen (10–18 mm, brown or blackish body with a yellow stripe down the middle of the thorax and the leading edge of the wings yellowish).

Identification Resources. For identification keys to families of stoneflies, see Arnett 1985, Bland 1978, and Lehmkuhl 1979; for keys to some genera and species of stoneflies, see Frison 1935 for the Illinois fauna, Hilsenhoff and Billmyer 1973 for the predatory stoneflies (Perlodidae) of Wisconsin, and Surdick 1985 for the green stoneflies (Chloroperlidae) of North America.

Walkingsticks
(Order Phasmida)

Introduction. The stick insects, also known as walkingsticks, are closely related to the mantids, cockroaches, grasshoppers, and crickets. Their unusual body shape affords them effective protection against many types of predators.

Description. Stick insects have an elongate, cylindrical (sticklike) body, 50 to 100 millimeters in length, with long, slender legs and antennae; there are generally no wings. The color is typically greenish or brownish and the mouthparts are chewing.

Life Cycle. Stick insects develop through gradual metamorphosis, with egg, nymph, and adult stages. The populations of most species are composed mainly of females (males being rare or even nonexistent) and viable eggs are produced without fertilization. Parthenogenetic reproduction is therefore quite common among the stick insects. The eggs are deposited in the late summer and early fall. The females simply drop their eggs on the forest floor while moving about in the tree tops. The eggs hatch the following spring and early summer (sometimes the second season after being laid). The young nymphs feed on the leaves of shrubs and small trees, and as they mature they move higher up into the forest canopy. The life cycle requires one or two years for completion.

Habits. Stick insects feed at night and rest during the day. Because of their highly effective camouflage they are rarely detected by other animals. When frightened, stick insects straighten out to closely resemble dead twigs. The nymphs are capable of regenerating lost limbs (at the time they molt), an attribute not held by any other species of immature or adult insect.

Habitat. Walkingsticks are forest insects. They are foliage feeders, preferring the leaves of hardwood trees.

Ecological and Economic Status. Many years ago there were occasional reports of widespread forest defoliation caused by stick insects; however in recent years this has not occurred. The walkingstick is little more than a natural curiosity these days.

Distribution and Faunistics. Except for the common walkingstick, which is found throughout the region, the other species are uncommon and often local. Only four species are known from the region.

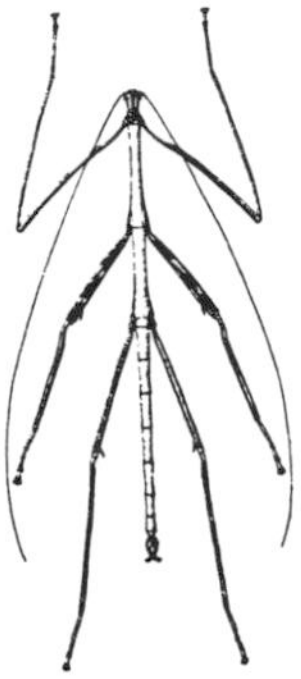

Fig. 17. A male adult of the common walkingstick, *Diapheromera femorata* (Say). (From Hebard 1934.)

Common Walkingsticks
(Family Heteronemiidae)

This family is the predominate one in the Great Lakes region. The common walkingstick, *Diapheromera femorata* (Say), occurs from New York to southern Minnesota (including Michigan and southern Ontario) and southward. It feeds on the leaves of oak, black locust, and cherry. Adults are present in late summer and early fall (August to October), but because of their camouflage and preference for feeding high up in trees, they are seldom seen. Two other walkingsticks found in the Great Lakes region are Blatchley's walkingstick, *D. blatchleyi blatchleyi* (Caudell), an arboreal species that occurs in northern Illinois and southeastern Wisconsin; and the prairie walkingstick, *D. velii* Walsh, an inhabitant of prairie grasses and herbaceous vegetation that occurs in southern and western Minnesota (and possibly in parts of Illinois).

Striped Walkingsticks
(Family Pseudophasmatidae)

Members of this walkingstick family are not as slender as the members of the common walkingstick family. Their bodies are thicker, a little more soft bodied, and slightly tapered at each end. The brown walkingstick, *Anisomorpha ferruginea* (Beauvois), is very large (30–36 mm) and brownish with inconspicuous longitudinal stripes. It is primarily southern in distribution and reaches its northern limit in southern Illinois (and possibly Indiana).

Identification Resources. For a key to subfamilies and tribes of walkingsticks, see Bradley and Galil 1977. For an excellent picture key to the walkingsticks, see Helfer 1987; for a key to the species found in Illinois, see Hebard 1934.

Grasshoppers, Katydids, and Crickets
(Order Orthoptera)

Description. Order Orthoptera includes small to large (6–90 mm) insects, usually brownish or greenish (blending in with the background), with a more or less elongate, sometimes slightly compressed body. Orthopterans have two pairs of wings: narrow front wings that are leathery in texture and have many veins (tegmina), and broader and more membranous hind wings that fold fanlike beneath the front wings. Some species have shortened wings. The head bears chewing mouthparts, well-developed compound eyes, and long, slender antennae. Most species have enlarged, muscular hind legs for jumping. Those species

---

Quick Guide to Identification

Grasshoppers, Katydids, and Crickets
(Order Orthoptera)

Diagnostic Characteristics	Common Name(s)	Scientific Family Name
Small size; elongated pronotum	pygmy grasshopper	Tetrigidae
Larger; antennae shorter than body	locusts	Acrididae
Antennae long; elongate-oval; flattened	bush katydid	Tettigoniidae
Antennae long; oval and convex	true katydid	Tettigoniidae
Antennae long; more slender	meadow katydid	Tettigoniidae
Antennae long; cone-shaped head	coneheads	Tettigoniidae
Antennae long; shieldlike pronotum	shield-bearers	Tettigoniidae
Antennae long; humpbacked; wingless	camel cricket	Gryllacrididae
Black, heavy bodied; with ocelli	field crickets	Gryllidae
Brownish; thinner bodied; without ocelli	ground crickets	Gryllidae
Pale green; elongate; more or less flattened	tree crickets	Gryllidae
Small; front legs not spadelike	pygmy mole cricket	Tridactylidae
Large; front legs spadelike	mole cricket	Gryllotalpidae

that "sing" (stridulate) have eardrums (auditory organs) located usually on the thorax or abdomen, but sometimes on the front legs. Female orthopterans have a conspicuous egg-laying organ (ovipositor) on the end of the abdomen.

Life Cycle. Grasshoppers, katydids, and crickets reach adulthood through a gradual (simple) metamorphosis, with egg, nymph, and adult stages. The eggs are laid either in the soil (by grasshoppers and ground crickets) or in plant tissue (by katydids and tree crickets) and often serve as the overwintering stage. The eggs hatch the following spring and the nymphs molt five to seven times before reaching the adult stage.

Habits. Members of this order may be either burrowing (subterranean), ground dwelling (terrestrial), or tree dwelling (arboreal). They are omnivorous, feeding primarily on vegetation, but they may become carnivores, cannibals, or scavengers when vegetation is unavailable because of drought or overpopulation. Most species depend on protective coloration (mimicry of foliage and camouflage) to escape danger, but some species are capable of producing irritating fluids (digestive juices, the so-called tobacco juice produced by certain grasshoppers) or foul-smelling odors to repel enemies.

Orthoptera, at least the males, have a special ability to "sing." The "singing" is more properly referred to as stridulation—rubbing one part of the body against another to produce vibrations. These vibrations are frequently amplified with special resonating surfaces or chambers. Some species stridulate in chorus, and their songs fill the night air during the summer months. Each species has its own unique song, and it is possible to identify most species by sound alone (after some practice, of course). The females generally do not sing (they either can not or will not), although a few have soft, quiet songs.

Habitat. Orthoptera dwell in various habitats, usually those associated with certain plants or plant communities.

Ecological and Economic Status. In southern France a cave drawing of the cave cricket (*Trogophilus*) is the oldest recognizable picture of an insect drawn by a human. It dates back to the Magdalenian culture (17,900 B.P. or earlier) (Vickery and Kevan 1985).

Some Orthoptera are plant pests, and periodic outbreaks can result in significant damage to pastures, crops, and ornamental plants. Grasshoppers also make great fishing bait, and orthopterans are readily consumed by fish when they have the misfortune of falling into the water.

Distribution and Faunistics. The Orthoptera of most Great Lakes states are well known. For example, 124 species of Orthoptera are known from Minnesota (Hebard 1932); 124 species from Michigan (Cantrall 1968); 83 species from Indiana (McCafferty and Stein 1976); 154 taxa (species and subspecies) from Illinois (Hebard 1934); and 104 species from Ontario (Vickery and Kevan 1985). Unfortunately, very little has been published on the Orthoptera of Ohio, Wisconsin, Pennsylvania, and New York. In all, at least 205 taxa (species and subspecies) are known to inhabit the Great Lakes region.

Pygmy Grasshoppers, or Grouse Locusts
(Family Tetrigidae)

The pygmy grasshoppers, or grouse locusts, are the smallest members of the order. They range from 6 to 13 millimeters in length. The pronotum is elongated and directed rearward over the thorax and abdomen, reaching to the tip of the abdomen. The front wings (tegmina) are greatly reduced, but the hind wings are well developed and functional. The antennae are short.

Fig. 18. A grouse locust, or pygmy grasshopper, *Acridium* species; note the enlongate pronotum that extends beyond the tip of the abdomen almost to the wing tips. (From Hebard 1934.)

Some species live near the water and are capable of swimming. They feed on algae, moss, fungi, and other plant materials. They hibernate as adults.

The sedge grouse locust, *Tettigidea lateralis* (Say), is frequently encountered along damp, flat shorelines of pools (gravel pits, marshes, etc.) and shallow lakes where the shoreline vegetation is composed of sedges (*Carex*), grasses, and mosses (Bland 1989). The color is variable, from light brown to blackish.

Short-Horned Grasshoppers and Locusts
(Family Acrididae)

Short-horned grasshoppers are medium to large insects that are generally brownish or greenish in color. The antennae are short, less than half the length of the body. The hind leg is well developed for jumping. The large hind wing (which is folded up beneath the front wing when not in use) is often brightly banded with red, pink, or yellow.

Most species are active during the daytime hours. They are

capable of stridulation and rub a keellike (occasionally peglike) ridge on the legs against a thickened vein on the wings. Some species are migratory, including the classical "locusts" known for their plaguelike invasions that cause considerable economic damage.

Fig. 19. The clear-winged locust, *Camnula pellucida* Scudder. (From USDA 1954.)

Many species are common. The meadow grasshopper, *Chorthippus curtipennis* (Harris), is common throughout the region. It is a large grasshopper (13–24 mm) and is light brown and greenish in color with a blackish bar on each side of the pronotum. It inhabits lowland meadows and wet areas, including swamps and bogs. The red-winged locust, *Arphia pseudonietana* (Thomas), inhabits open upland fields and grassy places. It is large (22–33 mm), blackish or brownish, and speckled with yellow. The hind wings are orange-red (or yellowish in the western portions of the region). This grasshopper, like a few others, makes a loud crackling noise when it flies. The clear-winged locust, *Camnula pellucida* Scudder, is often abundant enough to be destructive. It tends to be more common in the north than in the south. The body is brownish and the tegmina have a pale stripe near the "shoulder" (or humerus) and large, rounded dark spots. The antennae are yellow at the base and darkened toward the tips.

One of the most common grasshoppers in the region favors dry, barren soils and is frequently encountered along dusty paths, trails, dirt roads, and open areas. It is the Carolina locust, *Dissostiera carolina* (L.), and is easily recognized by the hind wings, which are black with a pale border. The cracker locust, *Trimerotropis verruculata* Kirby, is found in northern areas where there are rocky uplands and other open areas of barren soil. The body is dark brown or blackish and the wings are grayish and blotched with blackish. The hind wing is more or less clear with a dark cross band. This species also makes a loud crackling sound in flight.

The Lake Huron locust, *Trimerotropis huroniana* E. M. Walker, is a rare grasshopper that inhabits small areas of open sand dunes in the northern Great Lakes. It is known from a few areas along northern Lake Huron (in Michigan and Ontario), the northern Lake Michigan shoreline and islands (in Michigan and

Fig. 20. The Carolina locust, *Dissostiera carolina* (L.). The right wings are spread to show the black hind wing with pale border. (From Lugger 1897.)

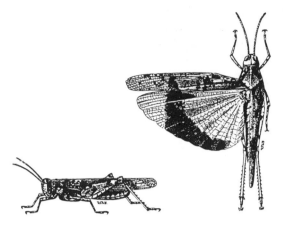

Fig. 21. The seaside locust, *Trimerotropis maritima interior* E. M. Walker. *Left,* male, side view; *right,* female with left wings spread. (From Lugger 1897.)

Wisconsin), and southern Lake Superior (in Michigan). The restricted distribution and limited dune habitat have made this grasshopper eligible for listing as a species of special concern. The closely related seaside locust, *T. maritima interior* E. M. Walker, occurs along the shorelines of Lake Ontario (New York and Ontario), western Lake Erie (Ontario), southern Lake Huron (Michigan and Ontario), and southern Lake Michigan (Michigan, Indiana, Illinois, and Wisconsin). It has also been recorded from the beaches of northern Indiana lakes and from the banks of the Mississippi River in western Minnesota.

The beautiful, short-winged secretive locust, *Appalachia arcana* Hubbell and Cantrall, is the only orthopteran insect known to be endemic to Michigan; it inhabits the sand plains of the northern lower peninsula and can be found in open areas with shrubby vegetation and full sun throughout most of the day. Because of its limited distribution it is currently listed as a species of special concern in Michigan..

The northern locust, *Melanoplus borealis* (Fieber), is widely distributed, although it is most abundant in the northern sections of the region, where it can be found in wet areas and bogs. In southern areas it is usually restricted to isolated bogs and cool, shady marshes. The tegmina are shorter than the abdomen and the legs are uniformly yellowish to brownish.

Another common, frequently destructive grasshopper is the differential grasshopper, *M. differentialis* (Thomas). It is dark yellowish green and the hind legs bear a series of black, chevron-shaped markings. The range of this species is expanding northward, and it is at home in all types of weedy fields and roadsides. The red-legged grasshopper, *M. femur-rubrum* (De Geer), is another widespread and common species—one of the most

Fig. 22. The differential grasshopper, *Melanoplus differentialis* (Thomas); note the black, distinctly chevron-shaped markings on the hind leg. (From Lugger 1897.)

Fig. 23. The red-legged grasshopper, *Melanoplus femur-rubrum* (De Geer). (From USDA 1954.)

abundant in the region. It is capable of utilizing a variety of habitats, from moist lowlands to dry uplands. The body is brownish and the hind legs are distinctively red.

For identification keys and extensive descriptions of short-horned grasshoppers of the United States, see Otte 1984a and 1984b.

Long-Horned Grasshoppers
(Family Tettigoniidae)

Many of these grasshoppers are also known as katydids because of the characteristic song of one of the large green North American species. This song is a familiar sound from dusk until dawn during the summer months. For a descriptive key to the songs of Michigan katydids, see Alexander et al. 1972. The laterally compressed bodies, the very long, thin antennae (longer than the length of the body), and the well-developed, sword-shaped ovipositor of the female readily distinguish this group. Ears are located on the front legs.

Many species are arboreal, and despite their large size (up to 90 mm) they are rarely seen because of their cryptic green coloration. They feed on leaves and other vegetation. Males stridulate at night, rubbing a rough file on one wing against a ridgelike scraper on the other wing. Eggs are laid either in the soil or in plant tissue.

Bush Katydids. Members of the genus *Scudderia* are collected from shrubs and broadleaf herbaceous plants in upland and lowland grassy fields (Bland 1989). The fork-tailed bush katydid, *Scudderia furcata* Brunner, is a medium-sized katydid (14–22 mm) that is dark green with a yellow head, pronotum, and

Fig. 24. The fork-tailed bush katydid, *Scudderia furcata* Brunner. *Left,* adult male, side view; *right,* tip of male abdomen showing notched abdominal tip. (From Lugger 1897.)

Fig. 25. A male broad-winged bush katydid, *Microcentrum rhombifolium* (Saussure). (From Hebard 1934.)

undersides. The narrow tegmina are shorter than the hind wings and can be seen protruding from beneath. The broad-winged bush katydid, *Microcentrum rhombifolium* (Saussure), is a large (25–30 mm) katydid with a dark green body and yellowish head, pronotum, and underside. The tegmina are broad, with the upper (rear) edge obtusely angled. The hind wings are longer and protrude from beneath the tegmina. This southern species, which has been expanding its range northward, inhabits shrubs, bushes, and the upper foliage of deciduous trees.

True Katydids. The only true katydid—the actual namesake for the entire group—is the northern katydid, *Pterophylla camellifolia* (Fabricius). It is large (25–34 mm), robust, and uniformly green. It is more often heard than seen, because it spends most of its time in the upper foliage of deciduous trees. The song, which is heard on summer nights, is a loud and harsh "katy-did, katy-she-did."

Meadow Katydids. At least four species of meadow katydids are common and widespread in the Great Lakes region. The delicate meadow katydid, *Orchelimum delicatum* Bruner, is greenish and occurs in grassy swales near sand dunes and beaches. The gladiator meadow katydid, *O. gladiator* (Bruner), is also greenish but inhabits moist meadows and wetlands. The black-legged meadow katydid, *O. nigripes* Scudder, is greenish brown or reddish brown with black legs and is found in tall shrubs near lakes, ponds, and streams. The common meadow katydid, *O. vulgare* Harris, is pale green or reddish brown with a green face and inhabits drier marshes and meadows, including weedy upland habitats. The short-winged meadow katydid, *Conocephalus brevipennis* (Scudder), is medium sized (11–23 mm)

Fig. 26. The common meadow katydid, *Orchelimum vulgare* Harris. *Left,* male; *right,* female (note ovipositor). (From Lugger 1897.)

Fig. 27. The sword-bearing cone-headed katydid, *Neoconocephalus ensiger* (Harris). *Left,* male; *right,* female (note ovipositor). (From Lugger 1897.)

and pale greenish brown or reddish brown, with one dark brown and two yellow longitudinal stripes on the pronotum. The tegmina are short. They are found in thick vegetation (grasses, sedges, or herbaceous plants) in wet areas.

Cone-Headed Katydids. Cone-headed katydids have the top of the head more or less pointed and distinctively cone shaped. One of the more commonly encountered species in the Great Lakes region is the sword-bearing cone-headed katydid, *Neoconocephalus ensiger* (Harris). It is large (24–30 mm) and green (occasionally brownish). The female has a very long ovipositor, as long as or longer than the length of the body.

Shield-Bearing Katydids. The shield-bearing katydids, members of the genus *Atlanticus*, have an enlarged shieldlike pronotum that covers the thorax and part of the abdomen. Two species are common in the region: Davis's shield-bearer, *Atlanticus monticola* Davis, which inhabits pine barrens and dry woodlands; and the short-legged pale shield-bearer, *A. testaceous* (Scudder), which inhabits the herbaceous growth and shrubs of forest margins and fallow fields.

Camel and Cave Crickets
(Family Gryllacrididae)

These crickets are distinctive and easy to recognize. They are robust, humpbacked, and wingless. The antennae are long and slender, much longer than the body. They are generally brownish or grayish, and they are frequently mottled or blotched with darker shades. The ovipositor of the female is distinct and cylindrical.

They are carnivorous and live in both moist woodlands and caves. They are active at night and can be found under leaf piles, stones, boards, or debris, or in animal burrows, hollow logs, or stumps during the day. They do not stridulate.

Fig. 28. The spotted camel cricket, *Ceuthophilus maculatus* Harris. (From Hebard 1934.)

There are many common species in the Great Lakes region. The spotted camel cricket, *Ceuthophilus maculatus* Harris, inhabits dry woodlands. It is brown with mottled spots of yellow and a more or less distinctive middorsal stripe. The short-legged camel cricket, *C. brevipes* Scudder, is common in or near moist or wet forests. It is dull brown, mottled, and without a middorsal stripe. The woodland camel cricket, *C. sylvestris* Bruner, is small with pale legs. It can be quite numerous but is seldom seen because it spends most of its time in burrows created by moles, shrews, and mice.

Field, Ground, and Tree Crickets
(Family Gryllidae)

These crickets are small- to medium-sized insects with bodies that are usually compressed or flattened from top to bottom. The wings lay flat on the back (often curling around the sides of the abdomen). The antennae are long and slender, much longer than the body. Ears are located on the forelegs. The female has a well-developed, cylindrical ovipositor.

Most species are nocturnal and hide under many kinds of objects during the day. Males stridulate at night (or on cloudy days). They rub a file on one wing against a scraper on the other wing. The wings are elevated when stridulating to create a large resonating chamber to amplify the sound. The crickets may create ventriloquial effect by changing the angle of the tegmina, thus projecting the sound away from the source to another location. The intervals between notes (chirps) in the song of most species is dependent on the temperature, and it is possible to ascertain the temperature by counting the number of stridulations over a specific interval, generally 15 seconds. For example, the snowy tree cricket (*Oecanthus fultoni* T. J. Walker) chirps at a rate such that the number of chirps in 14 seconds plus a factor

Fig. 29. A female fall field cricket, *Gryllus pennsylvanicus* Burmeister. (From Lugger 1897.)

Fig. 30. The striped ground cricket, *Nemobius fasciatus* (De Geer). (From Hebard 1934.)

of 40 gives the approximate air temperature in degrees Fahrenheit.

Two field crickets are commonly found in a wide variety of open upland habitats in the Great Lakes region: the fall field cricket, *Gryllus pennsylvanicus* Burmeister; and the spring field cricket, *G. veletis* (Alexander and Bigelow). Both are medium to large (greater than 14 mm), robust, and black in color; and both have three ocelli on the head. They are nearly identical in appearance and are best told apart by differences in their life cycle. Adults of the spring field cricket are active from May to July, while adults of the fall field cricket are active from July to October.

The ground crickets are medium sized (less than 12 mm) and lack ocelli. The striped ground cricket, *Nemobius fasciatus* (De Geer), is dark reddish brown and the head is swollen and marked with dark and light longitudinal stripes on the top.

The tree crickets are medium sized (12–15 mm), are thinner with an elongate head, and lack ocelli. The snowy tree cricket, *Oecanthus fultoni* T. J. Walker, is whitish or pale green with the top of the head yellowish. The wings are broad (widest toward the tip) and lay flat over the back. They are arboreal (favoring deciduous woodlands) but are commonly encountered in raspberry and blackberry thickets. The black-horned tree cricket, *O. nigricornis* F. Walker, has a greenish yellow body with a completely or partially blackish head, pronotum, and legs. They are most common in old weedy fields and bramble thickets.

Two other species deserve a brief mention. The pine tree cricket, *O. pini* Beutenmuller, is green with a brownish head, pronotum, and legs. They can be found on pine and fir trees. The tamarack tree cricket, *O. laricis* T. J. Walker, is dusky green with a brown head. They can be found on tamarack (larch) or hemlock trees. Both species have restricted distributions and host plant requirements and are listed as species of special concern.

Pygmy Mole Crickets
(Family Tridactylidae)

Members of this family are small (less than 10 mm) and molelike. The front tarsi are not as noticeably enlarged and spadelike as in the mole crickets, and the femur of the hind leg is greatly enlarged and flattened. The hind tarsi are very small and seldom noticed because of the presence of enlarged, flattened tibial spurs.

Two species, *Ellipes minutus* (Scudder) and *Neotridactylus apicalis* (Say), are known from Minnesota, Illinois, Michigan, and Ontario. They can be found burrowing in sparsely vegetated, damp, sandy soils near streams, roadside ditches, ponds, or lakes.

Fig. 32. The pygmy mole cricket, *Ellipes minutus* (Scudder). *Left,* top view; *right,* side view. (From Lugger 1897 and Hebard 1934.)

Mole Crickets
(Family Gryllotalpidae)

These are some of the most unusual members of the Orthoptera that are found in the Great Lakes region. Their brownish bodies are modified for a subterranean existence—with a robust, velvety body, spadelike front legs, and a large head for burrowing through the soil. The antennae are short and there are no jumping hind legs. The tegmina are short, exposing the abdomen, but the hind wings are well developed and mole crickets are accomplished aerialists.

Fig. 31. A female black-horned tree cricket, *Oecanthus nigricornis* F. Walker. (From Hebard 1934.)

Because mole crickets are nocturnal and have burrowing habits, they are seldom seen. They frequent moist soils near streams, ponds, and marshes. Their song, a series of one to three loud, low chirps each second, is commonly heard near such habitats at night. Mole crickets tend their offspring, at least for a short while, and use special recognition calls to communicate with their brood.

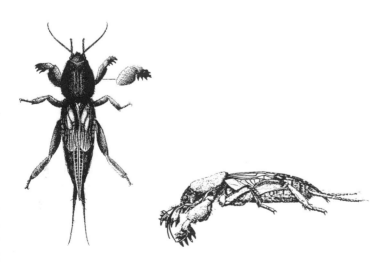

Fig. 33. The northern mole cricket, *Neocurtilla hexadactyla* (Perty). *Left,* top view; *right,* side view. (Courtesy of Michigan State University Extension and from Hebard 1934.)

The northern mole cricket, *Neocurtilla hexadactyla* (Perty), is the only species found in the region.

Identification Resources. For identification keys to the Orthoptera of Michigan, see Pettit and McDaniel 1918; for keys to Orthoptera of Minnesota, see Hebard 1932; for keys to Orthoptera of Illinois, see Hebard 1934; for keys to species of Canada and the adjacent United States, see Vickery and Kevan 1985; for keys to species of the United States, see Helfer 1987.

Earwigs
(Order Dermaptera)

Introduction. Earwigs are primarily a tropical and subtropical group of insects, and they are only sparingly represented in the Great Lakes region. There are only 28 species of earwigs in the United States, and at least several of these are introduced from other continents. The name earwig is derived from the ancient belief that these insects would crawl into the ears of sleeping people and bore into the inner ear and brain, causing insanity. There is, of course, no truth to this legend.

Description. Earwigs are medium sized (10–13 mm), elongate, flattened, and somewhat roachlike in appearance. They can be immediately identified by the pincerlike appendages at the tip of the abdomen. These cerci are used for defense, prey capture, mating, and wing unfolding. The forewings of most earwigs are short and hard in texture (and are often referred to as elytra). The membranous hind wings, which are much larger, are completely hidden beneath the front wings. A few species are wingless.

Life Cycle. Earwigs pass through three stages of development (gradual metamorphosis). The rate of development, which ranges from 20 to 70 days, is greatly influenced by the temperature. Male and female earwigs spend the winter as pairs or small groups in protected quarters, often in underground burrows. In the spring the female evicts the male and egg laying begins. The female watches over her 20 to 50 small, white eggs, chasing away all intruders, including the male. She also protects them from desiccation and mold by licking them repeatedly. Even after the nymphs hatch from the eggs the female will continue to care for her brood. As the nymphs approach maturity they gain independence from their mother and by late summer and early fall they are fully grown. In the temperate regions there are generally two generations each year.

Habits. Earwigs are predominantly nocturnal scavengers, feeding mostly on soft vegetative matter—such as decaying foliage, algae, fungi, sprouts and seedlings, flower petals, pollen, and corn silks—but also on other insects and small invertebrates (dead or alive). When hunting live prey an earwig uses its cerci to grasp the prey and pass it over its back to its mouth. When the sun comes up earwigs gather in groups beneath stones, logs, boards, and other objects on the ground.

Despite the large hind wings and the ability to fly, earwigs prefer to escape danger by running (usually with the pincerlike cerci arched over their back).

Ecological and Economic Status. Some earwigs can occasionally cause damage to flowers and vegetables (especially corn), ornamental trees and shrubs, and even the honey in beehives. Also, when their numbers reach high levels near houses they can become a considerable nuisance problem inside the home. They can be found in and around homes in household articles, foodstuffs, crawl spaces, debris, cut flowers, produce, birdhouses, and shrubbery. Their grotesque appearance (abdominal pincers), secretive habits, and occasional foul odor are the primary reasons for homeowner disdain.

Nonnative earwig species have spread rapidly throughout many parts of eastern and western North America, despite the fact that they rarely fly (even though they have fully functional flight wings) and are unable to crawl long distances. This rapid spread has been possible because they are accomplished hitchhikers. They wander about at night and crawl into any available hiding place at dawn. Thus they may get carried long distances in bundled newspapers, luggage, cut flowers and produce, crated merchandise, lumber and building supplies, pallets, automobiles, and even mail.

Distribution and Faunistics. The earwig fauna of the Great Lakes region is small, with only six species known to occur (and most of them are introduced or adventive species from the south).

Ring-Legged Earwigs
(Family Carcinophoridae)

The ring-legged earwig, *Euborellia annulipes* (Lucas), is a wingless adventive from the southern United States that may be established in parts of the Great Lakes region (Ontario in particular). It is 9–13 millimeters and the yellowish legs are striped with darker rings.

Long-Horned Earwigs
(Family Labiduridae)

The seashore earwig, *Anisolabis maritima* (Bonelli), is most common along the Atlantic coast but is also known from two localities in Ontario. During the day it hides under beach debris and at night it forages for food.

Little Earwigs
(Family Labiidae)

The little earwig, *Labia minor* (L.), was introduced into the Great Lakes region from Europe as early as 1870. This small earwig (4–5 mm) is dull black with pale legs and pubescent body (and can sometimes be mistaken for a rove beetle). Because of its small size and nocturnal habits it is seldom seen, although it flies well and is occasionally attracted to lights.

Fig. 34. An adult female little earwig, *Labia minor* (L.). (From Hebard 1934.)

Common Earwigs
(Family Forficulidae)

The spine-tailed earwig, *Doru aculeatum* (Scudder), is a native species that occurs in southern Ontario, southern Michigan, Indiana, and northern Illinois (where it is at the northern extreme of its range). During the day it can be found hiding head-down in the leaf axils of sedges (*Carex*) in marshes bordered by low, moist woodlands. At night they are active, climbing around on the sedges and nearby shrubs looking for food. The linear earwig, *D. lineare* (Eschscholtz), is a southern species that has been occasionally reported as an adventive in Illinois and Ontario.

The most common earwig in the Great Lakes region is the European earwig, *Forficula auricularia* L. This earwig is generally about one-half inch long and is reddish brown, dark brown, or black in color. It has slender antennae that are only half as long as the body. The immatures (nymphs) resemble the adults in appearance, except for their smaller size, grayish brown color, and lack of wings. European earwigs were not known to occur in the United States prior to 1900. They are a native of Europe, western Asia, and north Africa and were first reported from the Pacific Northwest and southern New England shortly after the turn of the century. They have occurred in the eastern sections of the Great Lakes region (Ontario) since 1937 and in the western sections since 1948. By 1964 they were well established in Michigan, and one of the first notable outbreaks was reported from Saginaw in 1967 (Cantrall 1968 and 1972).

Fig. 35. An adult male European earwig, *Forficula auricularia* L. (Courtesy of Michigan State University Extension.)

Identification Resources. For identification keys to the families of earwigs, see Arnett 1985 and Bland 1978; for keys to the species of Illinois, see Hebard 1934; for keys to the species of the United States, see Helfer 1987; for keys to the species of Canada and the adjacent United States, see Urquhart 1942 and Vickery and Kevan 1985.

Cockroaches
(Order Blattaria)

Introduction. Most people are acquainted with the cosmopolitan species of cockroaches that frequently inhabit homes, factories, restaurants, hospitals, and hotels.

Description. Cockroaches are broadly elongate insects with flattened bodies. Color varies and may be predominantly yellowish, dark brown, or even blackish. The head bears chewing mouthparts and long, slender antennae. The wings are thickened and leathery (tegmina) or, in a few species, greatly reduced. The legs are generally slender and designed for rapid mobility. They are medium to large sized (ranging from 15 to 60 mm).

Life Cycle. Like their close relatives, the cockroaches develop by gradual metamorphosis, with egg, nymph, and adult stages. The eggs are sealed inside bean-shaped capsules that are casually deposited by the female (except in the case of the German cockroach, where the female retains the capsule at the tip of the abdomen until the eggs are ready to hatch). The nymphs grow rapidly, and some species are capable of completing several generations in a season (at least in a heated building). Outdoor cockroaches overwinter as immature nymphs, resume their activity in the spring, and mature by midsummer. By early winter another generation is ready to hibernate.

Habits. Cockroaches are nocturnal, and during the day they seek refuge in cracks or crevices (in buildings) or in leaf piles, rotten wood, or debris (outdoors). Male wood cockroaches are agile flyers and frequently fly to lights during the summer months. As scavengers cockroaches feed on all types of organic materials, from foodstuffs to the corns on the feet of sleeping humans.

Habitat. Some cockroaches are "domesticated" and can only survive in heated buildings. Other native species are adapted to an outdoor existence and can be found in all types of woodland settings.

Ecological and Economic Status. Domestic species of cockroaches are a big pest problem. They destroy stored products, contaminate food, spread disease, and create nuisance problems. We spend millions of dollars each year trying to eliminate them from our homes and businesses. Outdoors, however, cockroaches and wood cockroaches have a valuable role as recyclers and decomposers of organic materials.

Fig. 36. A female German
cockroach, *Blattela ger-
manica* (L.). (From
Michelbacher and
Furman 1951.)

Fig. 37. A male
Pennsylvania wood cock-
roach, *Parcoblatta penn-
sylvanica* (De Geer).
(From Lugger 1897.)

Fig. 38. The Oriental
cockroach, *Blatta orien-
talis* L. *Left,* adult female;
right, adult male. (From
Lugger 1897.)

Distribution and Faunistics. Fourteen cockroach species are cosmopolitan and are widely distributed in urban, suburban, and even rural areas as domestic pests or turn up as adventives hitchhiking in produce and goods shipped from tropical locations. An additional 8 species are native wood cockroaches. Therefore, 22 species of cockroaches and wood cockroaches are either permanent or temporary inhabitants of the region.

Wood Cockroaches
(Family Blatellidae)

The German cockroach, *Blatella germanica* L., is one of the most aggravating insects in the world. These small, yellowish tan roaches (with two longitudinal black markings on the pronotum behind the head) are common in homes and restaurants, where they favor the moist conditions of kitchens and bathrooms. The brown-banded cockroach, *Supella longipalpa* (Fabricius), is another small species (yellowish brown with two dark brown markings across the base of the wings). It favors other areas of buildings away from moisture and is particularly fond of the warm dark conditions associated with electronic equipment (televisions, radios, computers, etc.).

The Pennsylvania wood cockroach, *Parcoblatta pennsylvanica* (De Geer), is the most widely distributed native species and is found in woodland settings from Ontario to Minnesota and southward. The males are slender, straw colored, and fully winged. Females are more robust, are darker in color, and have shortened, nonfunctional wings. During the day nymphs and adults can be found hiding beneath logs, stones, leaf piles, and other materials.

Cockroaches
(Family Blattidae)

The Oriental cockroach, *Blatta orientalis* L., the American cockroach, *Periplaneta americana* (L.), and the Australian cockroach, *P. australasiae* (Fabricius) are all large-sized cosmopolitan species closely associated with humans. The Oriental cockroach is blackish and has short front wings and hind wings. The American cockroach is more or less uniformly reddish brown (usually darker on the pronotum), while the Australian cockroach is brownish with yellow markings near the base of the wings, at the "shoulders."

Identification Resources. For an identification key to genera of cockroaches (including adventives), see Rehn 1950. Additional

Fig. 40. The Australian cockroach, *Periplaneta australasiae* (Fabricius). *Left,* adult male with wings spread; *right,* adult female. (From Lugger 1897.)

keys and information for species occurring in Illinois can be found in Hebard 1934. Consult Vickery and Kevan 1985 for information on the Canadian species and Helfer 1987 for keys to species occurring in the United States (and adjacent Canada). Also, cooperative extension services and other public agencies have fact sheets on domestic cockroaches (many of which include color illustrations).

Fig. 39. The American cockroach, *Periplaneta americana* (Fabricius). (From Lugger 1897.)

Mantids
(Order Mantodea)

Introduction. The praying mantids (also known as soothsayers or rear-horses) are familiar to most people. The mantids are primarily a tropical group, but well-meaning individuals introduced two nonnative species into the region many years ago. These mantids have faired well and have increased their range even without the direct assistance of humans.

Description. Mantids are large (50 to 100 mm), slender insects with characteristic forelegs. These raptorial forelegs are specifically designed for grasping unsuspecting prey. Color is variable—green, green-brown, or brown. It was found that green individuals outnumbered brown individuals by two to one for both males and females in southern Ontario (Judd 1950a).

Life Cycle. Mantids develop through gradual metamorphosis, with egg, nymph, and adult stages. They overwinter in the egg stage. The mantids mature by late summer and will survive until they succumb to the frost and cold weather in the late fall and early winter. Egg masses are deposited by early fall on any convenient surface.

Habits. In the late spring the young mantids emerge from their egg mass and begin their search for food. The nymphs are highly carnivorous, even cannibalistic under crowded conditions, and will eat any other type of insect they can get their tarsi on. They usually wait motionless for their prey to get within grabbing distance, but they are also known to stalk their prey when exceptionally hungry.

Mantids appear to be fearless. When approached they rear up on their back pairs of legs and partially raise their wings in a threatening manner (no doubt this behavior led to the common name "rear-horses"). Some species (like the European mantid) have black spots that resemble eyes on the inner face of the forelegs near the base, and these spots can be used to startle enemies. Mantids may even stridulate and produce a rasping noise.

The mating behavior is bizarre, at least from a human standpoint. Unless the male takes the female by surprise the female is more likely to eat him than mate with him. In most cases the female severs the head of the male before the mating act is completed. Even if the male is successful in keeping his head during mating he still has to escape without being caught after leaving the female.

Habitat. Mantids favor areas with abundant vegetation, especially grasses, herbaceous plants, and shrubs. They are commonly found along roadsides, pastures, and upland fields.

Ecological and Economic Status. Mantids are highly beneficial as predators of noxious insects, but their actual value in the battle against pests is often overrated. In view of their nonselective feeding habits they are as likely to eat other insects as they are to eat pest insects.

Distribution and Faunistics. One native and two introduced species of mantids may be found in the Great Lakes region.

Praying Mantids
(Family Mantidae)

The European praying mantid, *Mantis religiosa* L., was introduced into northwestern New York shortly after the turn of the century. Conditions were favorable for the survival of this hardy insect, and by 1915 the first mantids were detected on the other side of the St. Lawrence River in Ontario. The many published reports on the spread of the European mantis in Ontario make this one of the best documented accounts on the dispersal of an introduced insect (for example, Walker 1915, Urquhart and Corfe 1940, Judd 1947 and 1950a, James 1949, and Smith 1958). The Chinese mantid, *Tenodera aridifolia sinensis* Saussure, was also deliberately introduced into the United States near Philadelphia and spread to the Great Lakes region. This mantid is larger than the European mantis and measures 70 to 104 millimeters in length (vs. 47–56 mm for the European species). The only native species of mantid in the Great Lakes region is the Carolina mantid, *Stagmomantis carolina* (Johannson). This southern species is known to occur in the southern portion of Ohio, Indiana, and Illinois.

Fig. 41. An ault female Carolina mantid, *Stagmomantis carolina* (Johannson). (From Hebard 1934.)

Identification Resources. For an identification key to mantids of Illinois, see Hebard 1934; for a key to the species found in southern Canada, see Vickery and Kevan 1985; for a key to the species found in the United States, see Helfer 1987.

Termites
(Order Isoptera)

Introduction. Most termites live in tropical and subtropical environments. They build nests in soil, earthen mounds, or wood. Of the 41 species of termites that live in the United States, most of them occur in the southwest. The spread of termites into the temperate region has been assisted by human desire to construct wood buildings and other structures that provide termites with food, moisture, and shelter.

Description. At first glance many people think that termites are "white ants." A closer inspection reveals several prominent differences. Termites have a thorax that is broadly joined to the abdomen (ants have a constricted, hourglass waist); termite antennae are straight and composed of beadlike segments (ants have slender antennae that are "elbowed"); and termite reproductives have two pairs of wings that are of equal size and lack cross veins (ants have hind wings that are significantly smaller than the forewings and there are some cross veins). Termites are small- to medium-sized (7–15 mm), soft-bodied, pale (dirty white) insects. Soldier termites are heavily armored, with darkly pigmented heads and jaws. The reproductives, in contrast, are usually dark brown or black. Worker and soldier termites also generally lack eyes.

Life Cycle. Termites develop through gradual metamorphosis (egg, nymph, and adult stages), with five to six molts required for maturation. This process usually stops in the last nymphal stage and the individuals become workers. Those that complete the passage into adulthood but do not become sexually mature become either workers or soldiers. The life cycle of individual termites requires about one year for completion. It is not uncommon for queens to survive for several years, and a colony may exist in the same spot for many years.

After a brief mating flight in the spring the fertile males (kings) and females (queens) shed their wings and establish a new colony in the soil. The eggs and nymphs are cared for and protected by special workers and soldiers. Though these sterile worker and soldier termites are of both sexes, their reproductive development is suppressed by hormones secreted by the queen. If the original king and queen die the consequent lack of these hormones will permit more reproductives to develop very quickly. Also, colonies may produce supplemental reproductives at other times to boost a colony's reproduction, if conditions warrant. Female termites are known to lay millions of eggs each year and may live for up to 20 years (probably less than 10 years for the subterranean termites). Nonreproductive termites usually live for only a couple of years. Individual termite colonies have been known to exist for up to 100 years.

The termites are closely related to the cockroaches, but like the ants, termites are highly social (eusocial) insects. However, the evolutionary origin of termite social behavior is entirely different from that of ants, bees, and wasps. Termites live in large colonies composed of cooperating individuals from three different castes. A caste is a group of insects that performs certain highly specialized tasks. The three termite castes are reproductives, workers, and soldiers. The termite castes, unlike those of the ants, are composed of both sexes (not just females). Termite reproductives are the elite and are provided with every necessity of life by the soldiers and workers. The soldiers are sterile adults that are specialized for defending the colony. They have large mandibles and heavily armored heads. The workers are the most populous caste, and they are responsible for building the nest and galleries, cleaning and maintaining the nest, collecting food, and feeding members of the soldier and reproductive castes.

The termites' greatest hazard is a harsh environment; they are very susceptible to desiccation, so they remain secluded in their damp, dark, climate-controlled nests, avoiding exposure to the air. They often build special additions onto their nests (in the form of mud tubes) to reach new sources of food. Ants are the major animal predators of termites, and sometimes entire termite colonies are wiped out by tiny parasitic nematodes (roundworms).

Ecological and Economic Status. Termites are an important source of food for many animals, including humans. In many tropical areas, termites (especially the queens) are a tasty delicacy that provide valuable dietary protein.

Termites are important recyclers of wood in nature. By tunneling through wood, termites create entrances for moisture and decay fungi to speed the decomposition of wood into humus. Termites derive their nutrition from the cellulose that is contained in the wood. The digestion of the wood is accomplished by microbial bacteria and protozoans that reside in the intestinal tract of the termites. Young termites are "inoculated" with cultures of this bacteria and protozoans through feeding by the worker termites. If these microorganisms are artificially removed from the termite gut, the termite will starve to death, even if it continues to feed on wood.

Termite colonies cause extensive damage to buildings and other structures made of wood, and homeowners in the Great Lakes region spend millions of dollars annually to control termite infestations and repair termite-damaged wood.

Distribution and Faunistics. Since termites are primarily tropical and subtropical insects, their distribution in the Great Lakes region is quite limited. A total of five species (three native and two adventive) have been documented as occurring in the Great Lakes region.

Subterranean Termites
(Family Rhinotermitidae)

Fig. 42. The eastern subterranean termite, *Reticulitermes flavipes* Kollar. *Left,* adult reproductive (note wings); *right,* adult worker. (From USDA 1952.)

The eastern subterranean termite, *Reticulitermes flavipes* Kollar, is the most common species of termite found in the Great Lakes region. They normally live in nests located in the soil, traveling out to nearby sources of cellulose (trees, logs, stumps, woodpiles, fence posts, lumber, paper, and cardboard) to feed. The termites remain hidden within the moist, protective confines of the soil and wood. If they must travel across stone, masonry, or concrete surfaces, they construct special mud tubes. In recent years subterranean termite colonies have been found in sites where there was no soil-wood contact. These nests apparently get their moisture from nearby leaks, and humidity levels are maintained by the moist termite bodies that make up the colony.

Three other species, the sand dune termite (*Reticulitermes arenicola* Goellnert), the Virginia termite (*R. virginicus* Banks), and *R. tibialis* Banks have been reported from the region. The sand dune termite was described from specimens taken prior to 1931 in southern Michigan and northern Indiana, near the shore of Lake Michigan. The Virginia termite occurs primarily in the southeastern states but is also known from southern Ohio and Illinois and possibly from Indiana as well. *R. tibialis* is primarily a southwestern species, but its range extends northeastward into Illinois and southwestern Michigan.

Drywood Termites
(Family Kalotermitidae)

The drywood termites do not maintain subterranean nests and their moisture requirements are less demanding, so they can live in all types of wood. They inhabit the southern and western regions of the United States and occasionally are introduced into the Great Lakes region in articles of furniture or in building materials. The western drywood termite, *Incisitermes minor* (Hagen), and the short drywood termite, *Cryptotermes brevis* (Walker), have both been reported from time to time.

Identification Resources. For an identification key to families, see Arnett 1985 or Bland 1978; for keys to the termites of the United States, see Helfer 1987; for further information on the identification of termites known to occur in Canada, see Vickery and Kevan 1985.

Barklice and Booklice
(Order Psocoptera)

Introduction. Because they are small and obscure, psocids (booklice and barklice) are frequently overlooked. However, they are highly mobile and are often one of the first insect groups to invade and colonize areas of new habitat. Despite the common names "booklice" and "barklice" and their pale, soft-bodied appearance, they are in no way parasitic.

Description. Psocids are minute to small (less than 6 mm), soft-bodied, and pale (dirty white, yellowish, or grayish). Wings may be present, reduced, or absent. In winged species the two pairs of wings are membranous (with simple, reduced venation) and are held rooflike over the body. The front of the head is swollen, so the face is somewhat bulging. The head bears chewing mouthparts and slender antennae. Generally the antennae are at least half as long as the body (or longer).

Life Cycle. Metamorphosis is simple or incomplete, with egg, nymph, and adult stages. The young are similar to the adults and generally occur with them. Little is known about the details of the life cycle, but some species are known to lay eggs, while others give live birth to their young. Some species reproduce without males (parthenogenesis). Indoors many booklice have a continuous cycle, and outdoors many species are capable of completing several generations each year.

Habits. Many species of booklice are very active and run about in an erratic manner when disturbed. Many species of barklice are gregarious (living in groups) and spin protective silken webs over themselves. At least one species is capable of making a ticking noise by tapping its abdomen on the surface beneath its body.

Habitat. The indoor species (booklice) are primarily wingless and feed on glues, starches, paper, flour, mold, insect and plant specimens, and beeswax. The outdoor species (barklice) are found mostly on or under bark, on dead foliage, in nests of birds and other animals, among lichens, and in beehives.

Ecological and Economic Status. Some species of booklice can be pests that invade homes, schools, warehouses, granaries, libraries, and museums, where they damage stored products, collections, books, and foodstuffs. Their vague resemblance to the sucking lice (order Anoplura) has occasionally caused unnecessary worry in schools and institutions. Barklice are scavengers and, although numerous at times, have never been reported causing any damage to plants.

Distribution and Faunistics. The barklice and booklice of the Great Lakes region are poorly known and there are only a small number of published studies, including one on the psocids collected in Voyageurs National Park in Minnesota (Eertmoed and Eertmoed 1983), another on the psocids from southern Ontario (New and Loan 1971), and a third on the Indiana species (Mockford 1951). Altogether, at least 56 species have been reported from the Great Lakes region. Many species probably remain to be "discovered," described, and investigated.

Granary Booklice
(Family Trogidae)

These psocids are pests of granaries and warehouses, and most species are wingless or have greatly reduced wings. The granary booklouse, *Lepinotus inquillinus* Heyden, can reach extraordinary numbers from time to time. The reticulate booklouse, *L. reticulatus* Enderlein, is also a pest of stored products. The deathwatch, *Trogium pulsatorum* (L.), makes the infamous ticking sound that was once thought to be a warning of death. However, the sound is nothing more than a method of booklouse communication. The body is pale yellowish and the wings are reduced and scalelike.

Common Booklice
(Family Liposcelidae)

The wingless common booklouse, *Liposcelis bostrychophilus* Badonell, is a cosmopolitan pest of homes, libraries, warehouses, museums, and schools. It is minute (1.5–2.5 mm) and pale whitish, and the femur of the hind leg is greatly enlarged.

Fig. 43. A wingless adult booklouse of the genus *Liposcelis*. (From Michelbacher and Furman 1951.)

Gnawing Barklice
(Family Polypsocidae)

The corrupt barklouse, *Polypsocus corruptus* (Hagen), has a brown forebody and a purplish abdomen and is winged. It is commonly found on dead foliage and tree trunks.

Identification Resources. For identification keys to families of the psocids, see Arnett 1985, Bland 1978, or Borror and White 1970.

Fig. 44. A winged adult barklouse of the genus *Polypsocus*. (From Cornell University Extension 1975.)

Chewing Lice
(Order Mallophaga)

Introduction. The members of this louse family infest primarily birds but also some mammals. They do not infest moles and shrews, rabbits, or primates (including humans).

Description. Chewing lice are small (0.5–10 mm), pale whitish or yellowish (occasionally tan, brown, dark reddish, or black), flattened (dorsoventrally), wingless, and broadly oval to elongate. The head, which is at least as wide as or wider than the thorax, bears chewing mouthparts and short three- to five-segmented antennae that may be exposed or concealed. The compound eyes are small, and simple eyes (ocelli) are absent. The thorax is small with the segments fused, and the legs are short with one or two small tarsal claws at the tip.

Life Cycle. The metamorphosis is gradual. Eggs (called nits) are glued to the hairs or feathers of the host. The nymphs, which are found on the host alongside the adult lice, molt several times, changing very little in appearance as they grow. The body heat of the host provides a very stable, uniform temperature year-round, and many chewing lice have several generations each year.

Habits. Many of the chewing lice are highly host specific, but a few may occur on a variety of not more than a half dozen related host species. They feed on skin (epidermal cells), hair, feathers, body oils and secretions, dried blood (wounds), and even other ectoparasites, such as mites. Chewing lice are unable to survive for more than a day or two off the host and their dispersal is through direct contact of host animals.

Habitat. Chewing lice are highly dependent on their hosts for food, shelter, and warmth. As a result, the distribution of chewing lice in the Great Lakes region is primarily determined by the distribution of appropriate host animals. They are reported from domestic chickens, geese, ducks, coots, gallinules, pigeons, turkeys, vultures, falcons, hummingbirds, songbirds, small hawks, cattle, horses, goats, sheep, cats, dogs, coyotes, wolves, skunks, raccoons, and guinea pigs. There are even elephant and warthog chewing lice, but in the Great Lakes region these are restricted to zoo animals.

Ecological and Economic Status. Heavy infestations of lice can cause serious health problems for host animals, including general fatigue, severe skin irritations, and increased risk of infection by diseases.

Distribution and Faunistics. The chewing louse fauna of the Great Lakes region is poorly known. Even though the chewing lice of the region reside on the bodies of domestic and wild animals and should be easily obtainable, this relatively large group of insects has only received limited study. In total, 118 species of chewing lice are currently known to occur in the Great Lakes region.

Bird Chewing Lice
(Family Philopteridae)

This family is the largest in the order, with many species infesting wild and domestic birds. Two common species are the slender pigeon louse, *Columbicola columbae* (L.), which infests pigeons and squabs; and the wing louse, *Lipeurus caponis* (L.), which is common on the first wing feathers of young chickens.

Mammal Chewing Lice
(Family Trichodectidae)

Members of this family infest various mammals, including most domestic farm animals. Common species include the

Fig. 45. The cattle chewing louse, *Bovicola bovis* (L.). (Courtesy of Michigan State University Extension.)

horse chewing louse, *Bovicola equi* (Denny); the goat chewing louse, *B. limbatus* (Gervais); the cattle chewing louse, *B. bovis* (L.); and the sheep chewing louse, *B. ovis* (Schrank)—all found on their respective host animals.

Hawk Chewing Lice
(Family Laemobothriidae)

These lice infest birds of prey. The hawk louse, *Laemobothrion maximum* Scopoli, occurs on hawks and ospreys throughout North America (Arnett 1985), including the Great Lakes region.

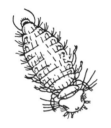

Poultry Chewing Lice
(Family Menoponidae)

Two common species of this family are the chicken shaft louse, *Menopon gallinae* L., which is yellowish and infests chickens, preferring to remain near the feather shafts; and the chicken body louse, *Menacanthus strameneus* (Nitzsch), which is pale yellow and found on the bodies of mature chickens.

Fig. 46. The chicken shaft louse, *Menopon gallinae* L. (Courtesy of Michigan State University Extension.)

Hummingbird Chewing Lice
(Family Ricinidae)

Chewing lice of the genus *Ricinus* are commonly found on many species of songbirds and hummingbirds. Very little is known about their presence and distribution in the Great Lakes region.

Identification Resources. For further information on the identification of chewing lice of the Great Lakes region, see Ewing 1929, Arnett 1985, and Bland 1978.

Sucking Lice
(Order Anoplura)

Introduction. The members of this louse family are highly specialized ectoparasites with sucking mouthparts that infest mammals, including primates (humans not excepted). Some members of this order comprise the so-called cooties that infest humans.

Description. Chewing lice are small (0.5–10 mm), pale whitish or yellowish (occasionally brownish or grayish), flattened (dorsoventrally), wingless, and broadly oval to elongate. The head, which is much narrower than the thorax, bears sucking mouthparts and short three- to five-segmented antennae. The compound eyes are small and ocelli are absent. The thorax bears short legs equipped with large grasping claws specially designed for latching onto the hairs of the host animal.

Life Cycle. The metamorphosis is gradual. Eggs (called nits) are glued to the hairs of the host, except in the case of the body louse, which lays its eggs in the seams of clothing. The nymphs molt several times until they reach maturity, changing very little in appearance as they grow. As is the case with the chewing lice, the host's body heat provides a very stable, uniform temperature year-round, and many species of sucking lice have several generations each year.

Habits. All sucking lice are host specific. They feed primarily on the blood of the host, which they obtain by piercing the skin with their beaklike mouthparts. Most sucking lice are unable to survive for more than a day or two off the host and their dispersal is generally through direct contact of the host animals. In the case of human lice, the body louse may be spread via clothing and bedding and the head louse may be spread via hats, combs, and brushes.

Habitat. Sucking lice infest the bodies of mammals. They have been found on seals, sea lions, walruses, river otters, cattle, horses, goats, sheep, peccaries, deer, cats, dogs, coyotes, wolves, moles and shrews, rodents (gophers, squirrels, flying squirrels, chipmunks, voles, and mice), and primates (monkeys, apes, and humans).

Ecological and Economic Status. Chewing lice, depending on their numbers, may be a nuisance problem or a serious health problem for their hosts. Their feeding can cause severe irritation of the skin, resulting in restlessness, weight loss, and poor health. They are also involved in the transmission of certain diseases, such as epidemic typhus, relapsing fever, and trench fever.

Distribution and Faunistics. The sucking lice are a smaller group than the chewing lice (order Mallophaga), with only about three dozen species occurring in the Great Lakes region. Species that infest humans and domestic animals have received the greatest attention; the rest of the species are in need of further study. Amin 1976 reports on some species infesting mammals in southeastern Wisconsin, and Wilson and Johnson 1971 and Lawrence et al. 1965 report on a few Michigan species. Wilson 1957 reports on the species known from Indiana. In all, 30 species of chewing lice have been reported from the Great Lakes region, but many host species are underinvestigated and many lice remain undetected and even undescribed.

Pale Sucking Lice
(Family Linognathidae)

Most of the sucking lice in this family are found on domestic
animals, including cattle, goats, sheep, and dogs. One species,
Solenopotes ferrisi (Farenholz), infests deer in Minnesota, New
York, and Ontario (and probably in all other parts of the region
as well).

Ungulate Sucking Lice
(Family Haemotopinidae)

Members of this family infest hoofed animals (ungulates),
including wild and domestic deer, cattle, sheep, goats, horses,
and pigs. The short-nosed cattle louse, *Haemotopinus erys-
trenus* Denny, the hog louse, *H. suis* (L.), and the horse sucking
louse, *H. asini* (L.), are all recorded from the Great Lakes
region.

Fig. 47. The hog louse,
Haemotopinus suis (L.).
(Courtesy of Michigan
State University
Extension.)

Armored Sucking Lice
(Family Hoplopleuridae)

Members of this family are ectoparasites of rodents, moles, and
shrews. The members of the genus *Hoplopleura* are most frequently encountered in the Great Lakes region.

Spiney Rat Lice
(Family Polyplacidae)

The rabbit louse, *Haemodipus ventricosus* (Denny), the spiny
rat louse, *Polyplax spinulosa* (Burmeister), and the deer louse, *P.
auricularis* Kellogg and Ferris, are common in the Great Lakes
region.

Human Lice
(Family Pediculidae)

The head louse, *Pediculus humanus capitis* De Geer, and the
body louse, *P. humanus humanus*, both occur in the Great
Lakes region. The head louse is generally the more common of
the two species, and localized outbreaks in schools and other
institutions are not uncommon. There are no known morphological differences between these two lice, and their habits alone
serve to differentiate them. The body louse resides in clothing

Fig. 48. The human body
louse, *Pediculus humanus
humanus*. (Courtesy of
Michigan State University
Extension.)

when not feeding on the host. The head louse remains on the host, usually about the head, rarely if ever—leaving.

Pubic Louse
(Family Pthiridae)

The pubic louse, *Pthirus pubis* (L.), is the sole member of this family. It infests humans and is generally found in the pubic region, although it has been reported from other parts of the body (chest hair, armpits, beards, and eyebrows). It is dispersed only through body contact.

Identification Resources. For further information on the identification of sucking lice of North America, including the Great Lakes region, see Ewing 1929, Kim and Ludwig 1978, and Arnett 1985.

True Bugs
(Order Hemiptera)

Introduction. The true bugs constitute the fifth largest group of insects, and they are highly modified for all types of aquatic, semiaquatic, terrestrial, and arboreal life styles.

Description. The true bugs are small to large insects (5–60 mm) with piercing/sucking mouthparts and special forewings that have a thickened, leathery basal part and a thinner, membranous tip (hemelytra). These forewings lay flat over the back of the abdomen and typically overlap at the tip when not in use. The crisscrossed forewings, along with the scutellum (a large, generally triangular part of the thorax), create a highly characteristic appearance.

The head is equipped with large compound eyes, ocelli (which may be lacking in some families), and a beaklike mouthpart that arises from the front of the face and is directed rearward between the legs (in resting position). The antennae may either be short and inconspicuous (as in most aquatic and semiaquatic bugs) or long and prominent (as in terrestrial bugs and water striders).

Other characteristics of shape and color are highly variable and are discussed under each respective family.

Life Cycle. The true bugs are the largest order with gradual metamorphosis. After hatching from the eggs, the nymphs join their adult counterparts in the appropriate habitat. The nymphal bugs resemble the adults but differ in size (being smaller) and lack functional wings. The wings develop externally as small nonfunctional wing pads that grow larger with

Quick Guide to Identification

True Bugs
(Order Hemiptera)

Terrestrial and Semiaquatic

Diagnostic Characteristics	Common Name(s)	Scientific Family Name
Small; front legs heavily spined	burrower bugs	Cydnidae
Small; scutellum large, covering abdomen	negro bugs	Thyreocoridae
Small; fingerlike processes on head; body punctate	ash-gray leaf bugs	Piesmatidae
Body distinctly flattened, oval	flat bugs	Aradidae
Body flattened; wingless, with hairs	bed bugs	Cimicidae
Body lacelike, pitted; pronotum large	lace bugs	Tingidae
Shield-shaped; five-segmented antennae	stink/shield bugs	Pentatomidae/Scutelleridae
Body angular; raptorial forelegs	ambush bugs	Phymatidae
Beak short; forelegs not raptorial	assassin bugs	Reduviidae
Sticklike (large); beak short	thread-legged bugs	Reduviidae
Sticklike (small); head elongate	marsh treaders	Hydrometridae
Forelegs tiny; claws not at tip	water striders	Gerridae
Forelegs tiny; broad-bodied	ripple bugs	Veliidae
Eyes large; membrane with four or five cells	shore bugs	Saldidae
Membrane with many veins; leaflike legs	leaf-footed bugs	Coreidae
Membrane of forewing with four or five veins	seed bugs	Lygaeidae
With cuneus; one or two closed cells in forewing	plant bugs	Miridae
Small; with cuneus; no cells in forewing	minute pirate bugs	Anthocoridae
Elongate; no veins in wing membrane	water treaders	Mesoveliidae
Beak long; body velvety	velvet water bugs	Hebridae
Slender body and broad head	broad-headed bugs	Alydidae
No scent glands; ocelli tuberculate	scentless plant bugs	Rhopalidae
Front femur enlarged; ocelli present	damsel bugs	Nabidae

Aquatic

Diagnostic Characteristics	Common Name(s)	Scientific Family Name
Large; with breathing tube (at tip of abdomen)	waterscorpions	Nepidae
Large; legs flat and fringed with hair	giant water bugs	Belostomatidae
Body keel-shaped; light color above	backswimmers	Notonectidae
Body more or less flattened; dark color above	water boatmen	Corixidae
Body humpbacked; wings elytralike	pygmy backswimmers	Pleidae
Toadlike: broadly oval and bumpy	toad bugs	Gelastocoridae
Oval and convex; no veins in membrane	creeping water bugs	Naucoridae

each passing molt. When the sexually mature adult stage is finally reached (after four or more nymphal molts), the wings become fully developed and functional.

Habits. Most terrestrial species of bugs have special "stink," or scent, glands that are capable of producing smelly (even caustic) defensive fluids. This defensive ability is especially well developed in the stinkbug family, but to one degree or another all true bugs "stink." Most of the larger bugs are also capable of giving a pretty good bite (stab, actually) when handled carelessly.

Bugs use their beak to extract fluids from both plants and animals. Some are phytophagous, feeding on plant sap, nectar, and fruit juices. Others are predacious, feeding on other insects and arthropods. A few are ectoparasites, feeding on the blood of warm-blooded animals.

Habitats. True bugs can be found in a wide variety of aquatic, semiaquatic, terrestrial, and arboreal habitats.

Ecological and Economic Status. Some true bugs are plant pests, sucking the sap from leaves, stems, and fruit. Infestation by such bugs may seriously weaken plants, making them more susceptible to some of the plant diseases spread by true bugs and other insects. Some true bugs are predatory and feed on other insects and small invertebrate animals, helping to reduce the impact of some pests. A few are specialized bloodsuckers and bite humans and other warm-blooded animals, in the process transmitting certain pathogens that cause human diseases.

Distribution and Faunistics. Many species of true bugs, representing some three dozen families, are widely distributed throughout the Great Lakes region. There are significant gaps in our knowledge about hemipterans in some states (especially Minnesota and Wisconsin) and about the species of many of the smaller families. At least 900 species of true bugs can be found in the Great Lakes region.

Burrower Bugs
(Family Cydnidae)

Burrower bugs are small (4–8 mm), elongate-oval, and blackish or reddish brown, with legs flattened and fitted with many spines for digging in the soil. One species, *Sehiris cinctus* (Beauvois), is shiny black or blue black with narrow white margins. It is the only species that spends large amounts of time above ground, and it can be found on nettles and mint plants. Another, *Amnestus spinifrons* (Say) is chestnut brown with light brown wings and yellowish legs. Burrower bugs are found on low vegetation (grasses, herbaceous plants, or shrubs), under boards and stones, or around the roots of plants. However, they are most likely to be encountered at lights.

This group is small; for example, only 13 species are known from Illinois (McPherson 1980).

Fig. 49. A negro bug of the genus *Galgupha*. (From Kansas State University Extension 1962.)

Negro Bugs
(Family Thyreocoridae; formerly Corimelaenidae)

Negro bugs are small (3–6 mm), broadly oval, strongly convex, and dark colored (shiny black, purple, or bronze), with the scutellum enlarged and covering most of the abdomen. One common species, *Galgupha atra* Amyot and Serville, is small (4–5 mm) and shiny black, with small punctures on the body. The forewing (hemelytron) has a groove along the side margin of the thickened basal part of the wing. This species is usually found near the ground, in dense vegetation or under plant debris. The negro bug, *Corimelaena pulcaria* (Germar), is also small (3–4 mm) and black. There is no groove on the wing, but the thickened basal part of the wing is paler at the base. It is usually found on flowers or berries.

This group is small; for example, only 17 species are known from Illinois (McPherson 1980).

Fig. 50. The negro bug, *Corimelaena pulcaria* (Germar). (From Kansas State University Extension 1962.)

Shield Bugs and Stink Bugs
(Families Pentatomidae and Scutelleridae)

These common bugs are small- to medium-sized (7–18 mm), greenish or brownish insects that are characteristically shield shaped. The scutellum is large and triangular, and in the shield bugs it extends all the way to the tip of the abdomen. The head is generally quite narrow and the antennae are prominent and five-segmented.

This group of bugs is moderately large. For example, 54 species of stink bugs are known from Ohio (Furth 1974) and 94 species (82 pentatomids and 12 scutellerids) are known from Illinois (McPherson 1978, 1979a, and 1979b).

There are several common species of shield bugs. The alternate shield-backed bug, *Eurygaster alternata* (Say), is small (7–9 mm) and dull yellow with numerous brownish punctures and irregular grooves on the scutellum. The "shield" (scutellum) is somewhat narrow and usually has a pale stripe down the center. The anchor shield-backed bug, *Stiretrus anchorago* (Fabricius), is small (8–11 mm) and boldly patterned in blue and white (or yellow, orange, or red). The dark blue coloration is variable but includes a large, irregular marking down the center of the body, from the head to near the tip of the scutellar shield. The pale areas on each side of the pronotum usually have two or three dark spots. The two-spotted shield-backed bug, *Tetyra bipunctata* (Herrich-Schaffer), is medium sized (12–17 mm) and brownish yellow with scattered brownish and black punctures and spots. There are usually two black spots on the base of the scutellum.

There are many common species of stink bugs. The spined soldier bug, *Podisus maculiventris* (Say), is small (6–8 mm) and dull grayish brown, with some red at the outer end of the thickened portion of the forewing and two black marks on the membranous part, especially near the tip. The sides of the pronotum are fashioned into sharply pointed spines, and there is a single spine on the underside of the abdomen. Soldier bugs are predacious and feed on many noxious insects. They are generally found on tall weeds near streams, crop fields, and woodlands. The dimidiate stink bug, *Banasa dimidiata* (Say), is bright green with reddish purple overtones on the pronotum and on parts of the forewings. It is usually found on berries, and the coloration is therefore cryptic; it often flies to lights.

Uhler's stink bug, *Chlorochroa uhleri* Stal, is medium sized (15–17 mm) and bright green with yellow margins and a yellow-tipped scutellum. It is most commonly found on junipers. Lugen's stink bug, *Mormidea lugens* (Fabricius), is small (6–7 mm) and dull bronze or grayish yellow with a white-margined scutellum. It is generally found on common mullein and other herbaceous plants, especially grasses. The one-spot stink bug, *Eustichus variolarius* (Beauvois), and the brown stink bug, *E. servus* (Say), are closely related and similar in appearance. The brown stink bug is mottled in various shades of brown and has black spots on the underside of the abdomen near the margins,

Fig. 51. The spined soldier bug, *Podisus maculiventris* (Say). (Courtesy of Michigan State University Extension.)

Fig. 52. The one-spot stink bug, *Eustichus variolarius* (Beauvois). (From Kansas State University Extension 1962.)

while the one-spot stink bug lacks these lateral spots and has one large black spot near the tip of the abdomen and a distinctive longitudinal groove on the hind tibia. The brown stink bug is a major pest of vegetable crops and fruits.

The twice-stabbed stink bug, *Cosmopepla bimaculata* (Thomas), is small (5–7 mm) and shiny black with yellow or red stripes on the front and sides of the pronotum and along the edges of the leathery part of the forewing. The tip of the scutellum has two red spots. This species is most commonly found on the foliage of herbaceous plants. The harlequin bug, *Murgantia histrionica* (Hahn), is easily recognized by its brightly colored orange-and-black markings. It is southern in distribution but occurs as far north as southern Ohio, Indiana, and Illinois. It can be a serious pest, as it feeds on the foliage of various cruciferous plants (cabbage, cauliflower, broccoli, etc.).

Fig. 53. The harlequin bug, *Murgantia histrionica* (Hahn). *Left*, nymph; *right*, adult. (From Chittenden 1925.)

Fig. 54. The green stink bug, *Acrosternum hilare* (Say). (Courtesy of the Young Entomologists' Society.)

Generally found near woodlands, the green stink bug, *Acrosternum hilare* (Say), is medium sized (12–17 mm) and bright green with a more or less elongate body. The head and pronotum are narrowly margined with yellow or orange. The two-spotted stink bug, *Perillus bioculatus* (Fabricius), is an important predator of the Colorado potato beetle, *Leptinotarsa decemlineata* (Say), and can be found in the same areas where this destructive pest occurs. This stink bug is highly variable in color, but the pronotum, entire scutellum, front edge of the forewings, and abdomen are brightly patterned with red, yellow, orange, or white. The color is determined by the temperature at which the bug developed.

For identification keys to stink bugs of Ohio, see Furth 1974; for Illinois, see Hart 1919; for Michigan, see McPherson 1970, 1979a, and 1979b.

Leaf-Footed and Squash Bugs
(Family Coreidae)

Members of this family are mostly dull brownish, grayish, or blackish, sometimes with white markings. They are medium to large bugs (10–20 mm) that are usually heavy bodied and elongate. The top surface of the abdomen is frequently concave and the wings rest in this depression. The membranous part of the forewing has many veins. The hind tibia of some species is dilated into a leaflike configuration (hence the common name "leaf-footed bugs"). The scent (stink) glands are well developed in most species. All species are plant feeders, and some are pests of agricultural crops and coniferous seeds.

The leaf-footed bug *Acanthocephala terminatus* (Dallas) is a large (20–22 mm) brown bug. The antennae are dark brown or reddish brown, except for a contrasting fourth segment, which is orange yellow. The leaflike dilations of the hind tibia are well developed. This species can be found on many types of weeds, shrubs, and trees. Three species of eastern leaf-footed bugs occur in the Great Lakes region, including *Leptoglossus phyllopus* (L.), *L. occidentalis* Heidemann, and *L. corculus* (Say). All three are reddish brown with a distinct white crossband on the forewings. They are often reported as pests of conifer seeds. The leaf-footed bug *L. occidentalis* is well distributed throughout the southwestern United States, but in recent years it has been expanding its range to the north and east, and it is now quite common in the Great Lakes region. This range expansion has been well documented because this particular species is fond of overwintering in buildings, and there have been many reports of thousands of these bugs congregating on homes, offices, and schools. For a key to species of *Leptoglossus*, see McPherson et al. 1990, Gall 1992, or Slater and Baranowski 1978.

Anyone who tries to grow pumpkins, squash, or watermelons is probably familiar with the squash bug, *Anasa tristis* (De Geer). This medium-sized bug (10–14 mm) is dull brownish yellow with coarse, gray punctures. There are also dull yellow stripes on the forebody (head and pronotum). This species has no spines or leaflike dilations on the legs.

Broad-Headed Bugs
(Family Alydidae)

These bugs have slender bodies and broad heads that are nearly as long as the pronotum. Most are yellowish brown or blackish and medium sized (10–15 mm). Scent glands are present. All are foliage feeders and can be found on grasses, weeds, and flowers, often in areas where the soil is sandy. The nymphs of some species resemble ants.

Two species are commonly encountered in the Great Lakes region: *Alydus conspersus* (Montandon) and *A. eurinus* (Say). Both are medium sized (11–15 mm) and elongate and more or less parallel sided. The first species is brownish yellow with a

Fig. 55. The leaf-footed bug *Leptoglossus phyllopus* (L.). (From Kansas State University Extension 1962.)

Fig. 56. The squash bug, *Anasa tristis* (De Geer). (From USDA 1952.)

Fig. 57. A broad-headed bug of the genus *Alydus*. (From Kansas State University Extension 1962.)

black forebody (head and front half of the pronotum) and legs.
The membranous area of the forewing is spotted with brown.
The second species is blackish and the sides of the abdomen are
spotted with yellow.

For more information on identification of broad-headed
bugs of the United States, see Fracker 1918.

Scentless Plant Bugs
(Family Rhopalidae)

These bugs resemble small leaf-footed bugs, but the scent glands
are greatly reduced or absent and the ocelli are large and raised
on little bumps (tubercles). Most species, with the exception of
the box elder bug, can be found feeding on plants in weedy
fields.

The most common species in the Great Lakes region is the
box elder bug, *Leptocoris trivittatus* (Say). The head, beak,
pronotum, and wings are black with reddish orange markings.
Body length is 11–14 millimeters. These bugs feed on the twigs
and seeds of box elder trees, and in the fall they often seek out
houses as overwintering sites (much to the displeasure of the
human occupants). The reflexed scentless plant bug, *Har-
mostes reflexulus* (Say), is found on herbaceous plants. It is pale
green to brownish with scattered brown spots on the body. The
thick, upper segment of the hind legs (femora) is slightly
enlarged and has several stout spines on the inner margin near
the tip.

For identification keys for scentless plant bugs of eastern
North America, see Hoebecke and Wheeler 1982.

Fig. 58. The box elder
bug, *Leptocoris trivittatus*
(Say). (From Knowlton
1951.)

Seed Bugs
(Family Lygaeidae)

Seed bugs are small to medium-sized bugs (2–18 mm) that are
generally brownish, although some are brightly colored. There
are only four or five distinct veins in the membranous part of
the forewing. The antennae and beak are four-segmented. Most
feed on seeds, but a few are sap feeders or even predators.

Two seed bugs are commonly associated with milkweed
plants. The small milkweed bug (*Lygaeus kalmii* Stal) is
medium sized (11–13 mm) and marked in red and black (the red
on the wings forming an irregular X). There is also a red spot on
the base of the head. The large milkweed bug, *Oncopeltus fasciatus* (Dallas), is 10–18 mm in length and black with an orange
Y-shaped marking on the front of the head, orange markings on
the sides of the pronotum, and two broad orange cross bands on

Fig. 59. The large milkweed bug, *Oncopeltus
fasciatus* (Dallas). (From
Kansas State University
Extension 1962.)

the wings. This bug is most common in the southern portions of the region, but by late summer it often establishes colonies in some northern areas. It can often be found in large numbers on the flowers and seedpods of milkweed in the late summer and early fall. Northern populations rarely survive the winter and new individuals must colonize the area again the following year. This bug is also easily reared on milkweed seed. For those who want to avoid the hassle of harvesting milkweed seeds, laboratory strains that feed on unsalted sunflower seeds have been developed.

The chinch bug, *Blissus leucopterus* (Say), is small (3–4 mm) and can be a serious pest of both lawns and grain fields. Chinch bugs are dull black and white, and covered with grayish hairs. The forewing is mostly white, but there is a large black spot on the base of each wing. The large big-eyed bug, *Geocoris bullatus* (Say), is appropriately named. The compound eyes are very large and prominent. The body is grayish and the top of the head is "dimpled" (punctate) and roughened. These bugs are generally found in areas of sparsely vegetated soils, where they feed on other insects (including the chinch bug). One of the most unusual seed bugs is the long-necked seed bug, *Myodoca serripes* Olivier. This small bug (8–10 mm) has a shiny black head with a long narrow neck, a gray pronotum, and forewings that are dark brown with pale margins. The legs are bicolored— yellowish near the tips with darker bands of brown on the femora.

Ash-Gray Leaf Bugs
(Family Piesmatidae)

Most of the members of this bug family are small (less than 4 mm) and grayish. The pronotum has five longitudinal ridges and the base of the forewing is reticulate (with lacelike punctures). The tarsi are two-segmented, and ocelli are present. Only the genus *Piesma* occurs in North America (and the Great Lakes region).

The most commonly encountered species is the ash-gray leaf bug *Piesma cinerea* (Say), which is flattened and coarsely punctate. The third antennal segment is as long as the first and second combined. The body is overall grayish yellow. There are stout, short spines between the antennae and the eyes. This species can be found on herbaceous weeds, especially pigweed. It frequently hibernates under the bark of sycamore and horse chestnut (Osborn and Drake 1916).

For identification keys to the ash-gray leaf bugs of North America, see McAtee 1919.

Fig. 60. The chinch bug, *Blissus leucopterus* (Say), a common pest of lawns. (From USDA 1951.)

Fig. 61. A big-eyed bug of the genus *Geocoris*. (From Kansas State University Extension 1962.)

Flat Bugs
(Family Aradidae)

Members of this family commonly feed on fungi located under tree bark, which explains the need for their unusual, elongate-oval, incredibly flattened bodies. The nymphs vaguely resemble bed bugs but lack the hairy bodies found in most bed bug species. Adult flat bugs are small to medium sized (3–11 mm) and dark brown, reddish brown, or black, with a granular appearance. The antennae are four-segmented and ocelli are lacking. They feed primarily on fungi and are generally found on or beneath the bark of dead trees or logs. Flat bugs are one of the few insects that care for their young. The adults of some species, usually the female, will stand guard over the eggs and nymphs to protect them.

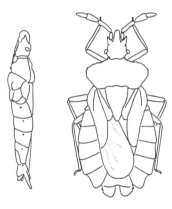

Fig. 62. A flat bug of the genus *Aradus*. *Left,* side view; *right,* top view. (Courtesy of the Young Entomologists' Society.)

This group of bugs is small, for example, only 17 species are reported from the Great Lakes region (Matsuda 1977).

One commonly encountered species is the robust flat bug, *Aradus robustus* Uhler, which is broadly oval with the body roughly sculptured with bumps and ridges. It is dark ashen brown with wing veins that are lighter in color. This species is typically found under the bark of red and black oak.

For a key to the flat bugs of Canada and the adjacent United States, see Matsuda 1977; for a key to flat bugs of Ohio, see Osborn 1903.

Lace Bugs
(Family Tingidae)

The common name for these bugs comes from the reticulate (pitted), lacelike texture of the forewing and enlarged pronotum that is typical of most of the species in this family. Lace bugs are pale whitish or grayish, sometimes patterned with darker colors. They are small bugs, less than 4 millimeters, and are not always easily seen on their host plants, because many are semi-translucent. However, their presence is often given away by the shiny black fecal droplets they produce, which are more noticeable than the bugs themselves. The antennae are four-segmented and ocelli are generally lacking.

This group of bugs is fairly small; for example, only 31 species are reported from Ohio (Osborn and Drake 1916).

Commonly encountered species include the sycamore lace bug, *Corythucha ciliata* (Say), which is found on the underside of leaves of sycamore, ash, hickory, and mulberry. It is whitish with dark spots on the prothorax and the middle of the wings. The chrysanthemum lace bug, *C. marmorata* (Uhler), is similar to the sycamore lace bug, but the wings are marked with lots of brown spots that form two cross bands near the wingtips. This species can be found on the flowers of chrysanthemum, goldenrod, aster, daisy, and other composites. Other species of lace bugs found in the Great Lakes region (and their associated host plants) are the oak lace bug, *C. arcuata* (Say) (oak and occasionally maple, apple, and roses); hawthorne lace bug, *C. crataegi* (Osborn and Drake) (hawthorne); cherry lace bug, *C. pruni* Osborn and Drake (wild cherry); elm lace bug, *C. ulmi* Osborn and Drake (elm); hackberry lace bug, *C. celtidus* Osborn and Drake (hackberry); and walnut lace bug, *C. juglandis* (Fitch) (walnut, butternut, and sometimes basswood).

For an identification key to the lace bugs of Ohio, see Osborn and Drake 1916.

Ambush Bugs
(Family Phymatidae)

These medium-sized bugs hide in flower heads, from which they ambush their prey. They are stout bodied, with the sides of the abdomen flared laterally and the top surface concave. The scutellum (triangular segment of the thorax located at the base of the forewings) is small. The front legs are raptorial with a greatly enlarged femur and sickle-shaped tibia; they are well suited for the predatory life style. These bugs are especially common in the flowers of goldenrod, where they do not hesitate to grab large insects, such as bees, wasps, and flies.

Two species are found in the Great Lakes region. The American ambush bug, *Phymata americana* Melin, is yellow or yellowish orange with dark brown or blackish markings. The dilated sides of the abdomen have an obtuse, rounded outline. The Pennsylvania ambush bug, *P. pennsylvanica* Handlirsch, is

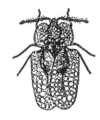

Fig. 63. The sycamore lace bug, *Corythucha ciliata* (Say). (Courtesy of Michigan State University Extension.)

Fig. 64. The chrysanthemum lace bug, *Corythucha marmorata* (Uhler). (From Kansas State University Extension 1962.)

Fig. 65. An ambush bug of the genus *Phymata*. (From Kansas State University Extension 1962.)

very similar and marked in orangish and black, although occasionally some individuals are bright green and black. However, the dilated sides of the abdomen have a sharply angled outline.

Assassin Bugs and Thread-Legged Bugs
(Family Reduviidae)

This family includes small to large bugs, either broadly elongate or very slender, that are colored in various shades of brown or blackish. The beak is short, three-segmented, and fits into a groove on the underside of the thorax between the first pair of legs. These bugs are strictly predators of other insects and ectoparasites of warm-blooded animals, including humans. Despite the predatory habits of these bugs the front legs are not raptorial.

The thread-legged bug, *Emesaya brevipennis* (Say), is long (33–36 mm), about the same diameter as a large knitting needle, and brownish with silver pubescence. The body is very slender, as are the legs (the front pair is slightly enlarged for grasping prey), and the head is ridged. The antennae are long and thread-like. The wings are short, no more than one-half the length of the abdomen. They somewhat resemble walkingsticks but can be easily distinguished by the beaklike mouthparts (walkingsticks have chewing mouthparts). Thread-legged bugs are often found in old barns and abandoned buildings and they sometimes rob spiderwebs of their catch.

The wheel bug, *Arilus cristatus* (L.) is large (28–36 mm) and dark brown with dense gray hairs. The unique pronotum is elevated and shaped like a cog wheel (a semicircular crest with 8–12 stout points). No other true bug looks like this. It is more of a southern insect but occurs as far north as New York and Illinois. The spined assassin bug, *Sinea diadema* (Fabricius), is medium sized (12–14 mm) with the outline of the body narrow at the head and widest at the middle of the abdomen (with a constriction behind the prothorax). The body is angular, rough textured, and reddish brown with the forewings patterned with yellow pubescence. The pronotum has a sharp spine on each side. It is found on grass and herbaceous plants of roadsides, old fields, etc.

The black corsair, *Melanolestes pictipes* (Herrich-Schaffer), is uniformly black, although the antennae and legs are tinged reddish. The wings of the female are often small and padlike. This medium-sized bug (15–20 mm) is an occasional visitor in houses, where it feeds on other insects and occasionally bites human occupants. Another assassin bug that frequently occurs in houses is the masked hunter, *Reduvius personatus* (L.). These medium-sized bugs (17–23 mm) are dark brown or blackish in

Fig. 66. The black corsair, *Melanolestes pictipes* (Herrich-Schaffer). (From Kansas State University Extension 1962.)

color, with a shiny forebody. The rest of the body is covered with conspicuous upright hairs. The immatures (nymphs) are spined and cover ("mask") themselves with dust, lint, and other debris. They typically feed on other insects but have occasionally been known to bite humans.

The bloodsucking conenose, *Triatoma sanguisuga* (LeConte), is not as casual about its desire for blood. It regularly feeds on the blood of small mammals (especially rodents) and will bite humans whenever it gets the chance. These medium- to large-sized bugs (16–24 mm) are broad bodied with the side of the abdomen flared laterally. The body is dark brown to blackish with the base and middle of the forewing spotted with reddish yellow and the sides of the abdomen marked with alternating spots of dark brown and reddish orange. Primarily southern in distribution, it does occur as far north as the southern parts of Illinois, Indiana, Ohio, and Pennsylvania.

For further information on the 27 species and subspecies of assassin bugs recorded from Michigan, see McPherson 1992.

Fig. 67. The masked hunter, *Reduvius personatus* (L.). (From Felt 1900.)

Damsel Bugs
(Family Nabidae)

These fiercely predatory bugs are far from damsel-like in their habits. Most are small, dull brown in color, and more or less slender and elongate. The beak is four-segmented and the antennae are long and slender. The forelegs are stout and somewhat raptorial. The nymphs of some species are ant mimics.

The common damsel bug, *Nabis americoferus* Carayon, is dull grayish yellow with an indistinct stripe down the center of the head and front of the pronotum. It is most abundant on herbaceous vegetation and grasses in fields and gardens. The beetlelike damsel bug, *Nabia subcoleoptrata* (Kirby), is a pear-shaped, shiny black bug with the antennae, legs, and sides of the abdomen bright yellow. The wings are small and padlike. The nymphs are ant mimics, and both nymphs and adults can be found in fields and meadows, where they feed on the meadow plant bug (Miridae).

Fig. 68. A damsel bug of the genus *Nabis*. (From Kansas State University Extension 1962.)

Bed Bugs
(Family Cimicidae)

This family includes not only its namesake, the bed bug, but the bat bug, swallow bug, and other flattened bloodsucking bugs as well. All are ectoparasites of warm-blooded hosts (birds and mammals, including humans) that live away from their hosts when not feeding. They are small (less than 5 mm), highly flattened, and wingless.

Fig. 69. The common bed bug, *Cimex lectularis* L. (From Back 1937.)

The common bed bug, *Cimex lectularis* L., is cosmopolitan and infests human habitations worldwide. The bed bug is reddish brown, broadly oval, flattened, and covered with short bristles. They prefer to do their feeding at night. During the day they seek refuge in cracks and spaces of bed frames, mattresses, baseboards, light switches, picture frames, and so on. When hungry, they make their way to the host animal to feed. The bat bug, *C. adjunctus* Barber, is a very closely related species with shorter hind legs that is usually associated with bat colonies but can also infest houses (with or without bats). The swallow bug, *Oeciacus vicarius* (Horvath), is an ectoparasite of the cliff swallow (and sometimes barn swallows) and is more elongate and elliptical in shape. It is covered with long, pale, silky hairs.

Minute Pirate Bugs
(Family Anthocoridae)

Since many of these bugs are commonly found on flowers, where they are predators of other insects, some species are also referred to as flower bugs. They are small (2–5 mm), elongate-oval, often flattened, and patterned in black and white. These bugs resemble plant bugs, but the membrane of the forewing has at most a few veins (generally none) and no closed cells. The last antennal segment is covered with long hairs. They can be found on flowers, under bark, in leaf litter and decay fungi, and in animal and bird nests. This group is small, with only 15 species recorded from Ontario and the Great Lakes region (Kelton 1978).

One of the most common species is the minute pirate bug *Orius insidiosus* (Say). The head, thorax, tip of the forewings, and tip of the abdomen are black. The basal part of the forewing is largely white or yellowish and strongly contrasts with the dark forebody and abdominal tip. This bug is highly beneficial, feeding on insects and spider mites, although they occasionally bite humans (with no serious effects). Another common species is *Lyctocoris campestris* (Fabricius). It is small (3–4 mm) and reddish brown with the membrane of the forewing cloudy brown with only one distinct vein. It is common in granaries, houses, barns, and animal nests, and it also occasionally bites humans.

For a key to the species of Canada and the adjacent United States, see Kelton 1978.

Fig. 70. The minute pirate bug, *Orius insidiosus* (Say), also known as the insidious flower bug. (From Kansas State University Extension 1962.)

Plant Bugs
(Family Miridae)

Miridae is the largest family of true bugs and contains almost one-half of the described species in the order. Consequently, there is an incredible amount of variation among the many genera and species. However, all members of this bug family share one common diagnostic characteristic: the membrane of the forewing has a small number of veins arranged into one or two closed cells. The forewing also has a structure known as a cuneus, a more or less triangular section of the leathery part of the forewing that is located at the tip and separated by a thin fracture. In many species of plant bugs the tip of the forewing beyond this fracture is angled downward. Ocelli are generally absent in most species of plant bugs. Most are drab or cryptic-colored plant feeders—some being serious agricultural pests—but a few are predatory. Some are wingless and mimic ants. The fleahopper plant bugs have globular bodies and jumping legs and somewhat resemble flea beetles. Most species overwinter as eggs, with the notable exceptions of species in the genera *Lygus* and *Stenodema*.

Miridae is also the largest family of true bugs in the Great Lakes region. It is estimated there are 440 species of plant bugs in Illinois and at least 316 in New York state (Knight 1941). A total of 158 species of plant bugs are reported from Ohio (Watson 1928).

Many of the species are common, but it is impossible to discuss all of them. Only a few of the more distinct or economically significant species can be mentioned here. The garden fleahopper, *Halticus bractatus* (Say), is small (3–4 mm) with a globular, teardrop-shaped body. It is black with tiny white spots; parts of the legs and antennae are pale. It has jumping hind legs, and its appearance and habits are remarkably similar to those of flea beetles. The garden fleahopper is generally found on garden plants and many types of herbaceous weeds. The rapid plant bug, *Adelphocoris rapidus* (Say), is small (7–8 mm) and dark brown with a yellow margin on the pronotum and wings. The antennae are also alternating black and yellow.

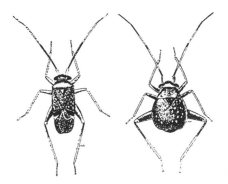

Fig. 71. The garden fleahopper, *Halticus bractatus* (Say), an unusually shaped plant bug. *Left,* adult male; *right,* female. (From Knight 1941.)

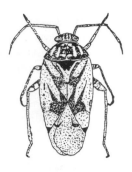

Fig. 72. The tarnished plant bug, *Lygus lineolaris* (Beauvois), a highly variable insect. *Left,* dark individual; *right,* light individual. (From Felt 1900.)

Fig 73. The four-lined plant bug, *Poecilocapsus lineatus* (Fabricius). (From Knight 1941.)

One of the most common and widespread plant bugs is the tarnished plant bug, *Lygus lineolaris* (Beauvois). It is often abundant on ornamental plants, fruit trees, crop plants, and many types of herbaceous plants and weeds. The tarnished plant bug is yellowish brown to olive with brown mottling. The scutellum has a broad, Y-shaped marking in yellow. The clouded plant bug, *Neurocolpus nubilis* (Say), is bright reddish tan to yellowish brown with a yellow spot on the wings. The basal segment of the antenna is covered with flattened hairs, and the second segment is swollen at the base. Four broad, black, longitudinal stripes on the pronotum and forewings contrasting with a background of bright green or yellow green is characteristic of the four-lined plant bug, *Poecilocapsus lineatus* (Fabricius). The antennae are black and the membrane of the forewing is brownish or yellowish. They are common on many herbaceous plants, including garden plants and flowering annuals and perennials.

Fig. 74. The meadow plant bug, *Leptopterna dolobrata* (L.). (From Knight 1941.)

The meadow plant bug, *Leptopterna dolobrata* (L.), is introduced from Europe and is now common throughout most of the northern United States. It is especially fond of bluegrass and timothy and often can be found in large swarms in the spring and summer. The males are reddish brown or yellowish orange, while the females are pale greenish white with black stripes on the pronotum. The wings of the female are often short. The three-spined plant bug, *Stenodema trispinosa* (Reuter), has three spines on the underside of the hind femora. They are small (6–8 mm) and greenish. They inhabit wet meadows, where they are generally found on grasses. The red plant bug, *Coccobaphes sanguinareus* Uhler, is bright red with only the wing membrane, base of the antennae, and tibia black. It is found on sugar and red maples. The military plant bug, *Hadronema militaris* Uhler, is a handsome grayish blue insect with white lateral wing margins. It is a common inhabitant of western prairies but also occurs in Michigan, Indiana, and New York.

Fig. 75. The red plant bug, *Coccobaphes san-guinareus* Uhler. (From Knight 1941.)

Many of the plant bugs are difficult to tell apart. For identification keys to *Lygus* plant bugs of North America, see Knight

1917; for keys to *Deraeocoris* plant bugs of North America, see Knight 1921; for keys and information on plant bugs of Illinois, see Knight 1941; for plant bugs of Ohio see Watson 1928.

Marsh Treaders
(Family Hydrometridae)

The marsh treaders (also referred to as water measurers) are small (8 mm) and elongate (sticklike) with a long head and thin, threadlike legs. Wings are usually absent. Marsh treaders walk on the surface of the water and climb plants looking for other insects to eat. Only members of the genus *Hydrometra* occur in the Great Lakes region.

One common species is Martin's marsh treader, *Hydrometra martini* Kirkaldy. It is 8 to 11 millimeters long, brown, and slender. The head is very long (almost as long as the entire thorax), and the eyes are situated near the middle, just behind the midpoint between the front and rear. The antennae are thin and threadlike and less than half the length of the body. Wings are present and they are whitish with dark veins. At first glance marsh treaders appear similar to the thread-legged bug, but there are several important differences. Thread-legged bugs are much bigger, their head is short, and their front legs are slightly raptorial.

For a key and checklist of the North American marsh treaders, see Drake and Lauck 1959.

Water Striders
(Family Gerridae)

Also known as pond skaters and water spiders, no other insect can be mistaken for a water strider. Their body is elongate and narrow, and the wings may be fully developed, reduced, or even absent. Winged forms seldom fly, however. The front legs are short and held up in front of the body, where they are used to capture prey. The four remaining legs (middle and hind pairs) are used to "skate" about on the surface of the water. The legs and body are clothed with waterproof hairs that enable water striders to walk on the surface film of the water. The claws are located above the tip of the tarsus to avoid puncturing the surface film. The weight of the water strider depresses the surface film of the water and you can easily see four dimples on the water. These dimples often cast a shadow on the bottom of the stream or pond, and sometimes these are easier to see than the water strider itself. Water striders feed on small insects, dead or alive, that fall onto the surface of the water. Since they hiber-

Fig. 76. A water strider of the genus *Gerris.* (From Kansas State University Extension 1962.)

nate as adults they can often be seen skating about on warm, sunny days in late winter and early spring.

This group of bugs is small, with 14 species (in six genera) known from Wisconsin (Hilsenhoff 1986), for example.

The most common species of water strider is *Gerris remigis* Say. It is also one of the largest species, up to 16 millimeters in length. It is dark brown to black with a yellowish or orange stripe down the center of the pronotum.

For an identification key to water striders of Indiana, see Deay and Gould 1936; for water striders of Wisconsin, see Hilsenhoff 1986.

Small Water Striders
(Family Veliidae)

The small water striders, also known as broad-shouldered water bugs or ripple bugs, are somewhat similar in appearance and habits to their larger cousins. However, they are much smaller (rarely exceeding 5 mm) and the thorax is much broader, giving them a wide-shouldered appearance rather than a narrow-bodied appearance. They are predators, feeding on small insects. Two genera, *Rhagovelia* and *Microvelia*, occur in the Great Lakes region, and Hilsenhoff 1986 reports 14 species (only 2 of them *Rhagovelia* species) from Wisconsin.

The obese ripple bug, *Rhagovelia obesa* Uhler, occurs in groups in the ripple areas of fast-flowing streams. They are black with two orange spots on the front edge of the pronotum. The American broad-shouldered water strider, *Microvelia americana* (Uhler), is also small (2–2.5 mm), but the abdomen has large patches of silvery hairs. They are found on the surface of pools, eddies, and backwaters of streams and small ponds.

For further information on identification of small water striders of North America, see Smith and Polhemus 1978; and for Wisconsin, see Hilsenhoff 1986.

Water Treaders
(Family Mesoveliidae)

The water treaders are small (2–4 mm), greenish or yellowish green bugs that are more or less elongate and winged. There are no veins in the membrane of the forewing. They are semi-aquatic, living on ponds and lakes with thick, luxuriant vegetation (especially duckweed). They are predatory and spend their time walking about the surface vegetation seeking prey. Only the genus *Mesovelia* occurs in the Great Lakes region, with only three species being reported.

The most common species is Mulsant's water treader, *Mesovelia mulsanti* White. It is small (3–4 mm) and brownish, greenish, or greenish yellow with a yellow stripe down the center of the back. There are both winged and wingless individuals. The hemelytra on winged individuals are largely white with contrasting, brownish areas and veins. They look a little bit like small plant bugs with long legs. They are predatory and feed on other small insects.

For a key to the species of water treaders found in Wisconsin, see Hilsenhoff 1986.

Velvet Water Bugs
(Family Hebridae)

These small bugs (generally less than 4 mm) are covered with minute hairs ("velvet") and are typically darkly colored. The beak is long, reaching to the middle pair of legs. The wings may be well developed, short, or lacking. The membrane of the wing, when present, lacks distinct veins. They are predators and can be found crawling on the floating vegetation of ponds. Only the genera *Merragata* and *Hebrus* occur in the Great Lakes region, with at least two species in each genus.

The most commonly encountered species is the common velvet water bug, *Merragata hebroides* White. This velvet water bug is small (less than 2 mm) with a reddish brown and black forebody. The forewings are white at the base and there are four white spots on the membranous area as well. The legs are yellowish.

For an identification key to the velvet water bugs of Wisconsin, see Hilsenhoff 1986; for a key to the *Merragata* of North America, see Drake 1917.

Shore Bugs
(Family Saldidae)

These bugs, as you might guess, are commonly encountered on the sandy and muddy shores of streams, ponds, and lakes. They are very active and can be seen scurrying about in search of their prey. The membranous area of the forewing has four or five elongate closed cells. The eyes are very large and flylike and the body is often widest at or beyond the middle. The antennae are four-segmented and the beak is three-segmented. Most species are patterned in black and white or brown and white.

Twenty species of shore bugs are presently found in the Great Lakes region, and six more species are very likely to occur here (Schuh 1967).

Fig. 77. The turbarian shore bug, *Saldoida turbaria* Schuh. (From Schuh 1967.)

One of the most common species is the pallipes shore bug, *Saldula pallipes* (Fabricius). It is small (5–6 mm) and black bodied, except for yellow or pale white markings on the wings and the pale membrane of the forewing. They are generally found along streams, where they are active and quick to fly if approached. The turbarian shore bug, *Saldoida turbaria* Schuh, is an inhabitant of sphagnum bogs in Michigan (and probably other areas as well). It has the front of the pronotum swollen and enlarged into two thick, upright tubercles.

For an identification key to the shore bugs of the Great Lakes region, see Schuh 1967; for a key to the shore bugs of North America, see Drake and Hottes 1950.

Backswimmers
(Family Notonectidae)

These medium-sized (less than 16 mm), elongate, keel-shaped aquatic bugs are unique in their habit of swimming upside down. This behavior provides protection in two ways. First, since the back is light colored any potential predator beneath the backswimmmer would have a difficult time seeing it against the lighted background above. Second, any reflection of the backswimmer on the surface of the water would result in the reflective image being right side up—a tempting, but erroneous, choice for any predator above the water.

Backswimmers are accomplished scuba divers and carry a supply of air with them beneath the water, trapped in the hairs of the undersurface of the body. Occasionally they must return to the surface to replenish their supply, at which time they will be seen with the tip of the abdomen at the surface and the head directed downward at an angle. The hind legs are very long and fringed with hairs, forming "oars" for swimming through the water. They catch and eat other insects, crustaceans, tadpoles, and even small fish, and they will administer a painful bite if handled carelessly. They are also accomplished aerialists and frequently turn up at lights at night. They occasionally invade swimming pools, causing nuisance problems.

Only the genera *Buenoa* (antennae three-segmented) and *Notonecta* (antennae four-segmented) occur in the Great Lakes region, with a total of nine species known from the region (four and five species, respectively).

The undulated backswimmer, *Notonecta undulata* Say, is one commonly encountered species. It is 10–12 millimeters long and yellowish white with a black scutellum and cross band on the forewings. It inhabits weedy ponds and streams. The lunar backswimmer, *N. lunata* Hungerford, is similar to the *N. undulata*, but can be distinguished by the white scutellum and

Fig. 78. The undulated backswimmer, *Notonecta undulata* Say. (From Kansas State University Extension 1962.)

lack of cross bands (occasionally present). The insulated back-swimmer, *N. insulata* Kirby, is similar but larger (12–16 mm) and the forebody is pale greenish yellow. It inhabits cold-water pools. The Margarita backswimmer, *Buenoa margaritacea* Torre Bueno, is pearly white or straw yellow with an orange scutellum (at least on live specimens). The forewings lack pubescence.

For an identification key to the backswimmers of Wisconsin, see Hilsenhoff 1984.

Pygmy Backswimmers
(Family Pleidae)

These small (less than 3 mm) water bugs are similar to back-swimmers in behavior but not in appearance. They are convex (but lack a keeled midline), have a hard shell (elytra) over the back, and lack the fringe of hairs on the hind legs. Pygmy back-swimmers are predacious, feeding on small insects and crustaceans. They are found in permanent, weedy ponds, especially those with duckweed. A single, tan-colored species, *Neoplea striola* Fieber, is found in the Great Lakes region and occurs in ponds with heavy plant growth.

Water Boatmen
(Family Corixidae)

Water boatmen have a parallel-sided, more or less flattened and elongate body with a broad, blunt head and a tapered abdomen. The head is slightly wider than the pronotum and the compound eyes are large. The scutellum and forewings are patterned with many thin, dark, herringbonelike cross bands.

The beak is short, wide, and unsegmented. The front tarsi are scoop-shaped and used to strain food. Unlike other aquatic bugs these cannot bite humans. Water boatmen inhabit the shallow margins of ponds and lakes and sometimes slow-moving streams. They are frequently attracted to lights at night.

This family of aquatic bugs is relatively large; for example, 49 species are reported from Wisconsin (Hilsenhoff 1984).

Most species are very similar in appearance and difficult to tell apart. Common species include the interrupted water boatmen, *Hesperocorixa interrupta* (Say), which is yellowish brown with 8 to 10 broad, dark bands across the pronotum and a regular zigzag pattern of cross bands on the forewings; and the alternate water boatmen, *Sigara alternata* (Say), which is brown or reddish brown with eight or nine narrow cross bands on the pronotum and an irregular, scattered pattern of cross bands on the forewings.

Fig. 79. The alternate water boatman, *Sigara alternata* (Say). (From Kansas State University Extension 1962.)

For an identification key to the water boatmen of Wisconsin, see Hilsenhoff 1970 and 1984.

Waterscorpions
(Family Nepidae)

These aquatic bugs are easily recognized by the elongate breathing tube at the tip of the abdomen. The body can be either elongate (sticklike) or broadly elongate-oval (water bug–like). The front legs are raptorial. They inhabit ponds and lakes, where they crawl about on the submerged vegetation hunting for insects and small fish. Sometimes they sit in ambush, with the breathing tube at the surface and the head directed downward. They are capable of biting humans if handled carelessly. Only two genera, *Ranatra* (with three species) and *Nepa* (with one species), occur in the Great Lakes region.

The fuscated waterscorpion, *Ranatra fusca* P. Beauvois, is dark reddish brown to dull yellowish brown with a wide, shallow groove on the underside of the front of the thorax. It has an elongate, sticklike body. There is a distinct tooth in the middle of the front tibia. The American waterscorpion, *R. americana* Mondanton, is similarly colored (brownish) but has conspicuously banded legs. The broad waterscorpion, *Nepa apiculata* Uhler, is stout and resembles a small giant water bug (Belostomatidae). However, it can be easily separated from these other aquatic bugs by the long breathing tube and the lack of flattened swimming legs with a fringe of hairs.

For more information on the waterscorpions that occur in the region, see Hungerford 1922; for keys to Wisconsin water scorpions, see Hilsenhoff 1984.

Fig. 80. The fuscated waterscorpion, *Ranatra fusca* P. Beauvois. (Courtesy of Michigan State University Extension.)

Giant Water Bugs
(Family Belostomatidae)

Giant water bugs are large to very large (25 to 60 mm), elongate-oval, more or less flattened, and brownish or greenish. The middle and hind pairs of legs are flattened and fringed with hairs (for swimming). The front legs are large and raptorial. Giant water bugs inhabit ponds, lakes, and slow-moving streams, where they lurk in the vegetation or muck, awaiting a chance to ambush their prey (snails, tadpoles, small fish, and other insects). They are capable of inflicting a painful bite if handled carelessly, and they have occasionally bitten swimmers (hence the common name "toe biters"). In the summer they are frequently seen at lights, and the name "electric light bug" is commonly used to refer to these insects in some areas. Only two genera, *Letho-*

cerus (with two species) and *Belostoma* (with two species), occur in the Great Lakes region.

The American giant water bug, *Lethocerus americanus* (Leidy), is very large (50–60 mm) and brownish with dark brown markings on the legs. The first beak segment is wider than it is long and is shorter than the second segment. The toe biter, *Belostoma flumineum* Say, is large (20–25 mm) and light brown with darker brown markings on the legs. The first beak segment is longer than it is wide and is as long as the second segment. The female water bug glues the eggs to the back of the male, an interesting form of bug-world "child care."

For more information on identification of species in the genus *Lethocerus*, see Menke 1963; for species of *Belostoma*, see Lauck 1964. For keys to the species from Wisconsin, see Hilsenhoff 1984.

Fig. 81. The toe biter, *Belostoma flumineum* Say, a giant water bug. (From Kansas State University Extension 1962.)

Creeping Water Bugs
(Family Naucoridae)

Creeping water bugs are broadly oval, 9–13 millimeters long, and fitted with raptorial front legs. There are no veins in the membranous part of the forewing. Ocelli are lacking. They inhabit ponds and streams, living among dense vegetation and debris. They feed on other insects and will bite humans if handled carelessly.

The only species from this small family of bugs found in the Great Lakes region is the creeping water bug, *Pelocoris femoratus* (Beauvois). It is small (9–12 mm) with a greenish yellow or yellowish brown forebody with dark olive spots. The forewings are brownish with pale shoulder marks.

Toad Bugs
(Family Gelastocoridae)

These semiaquatic bugs resemble tiny toads in both behavior (hopping and crouching) and appearance. The body is broadly oval with an irregular, bumpy surface and large bulging eyes. They are found along the sandy or muddy shores of ponds and streams, where they hunt for their prey. They are generally colored like the substrate of their habitat and may be difficult to see unless they are moving about. They also burrow in the moist soil but can be easily flushed from their burrows by splashing water on the beach.

A single common species, the eyed toad bug, *Gelastocoris oculatus* (Fabricius), occurs in the Great Lakes region, and it varies in color from yellowish to blackish (according to the color

Fig. 82. The eyed toad bug, *Gelastocoris oculatus* (Fabricius). (Courtesy of Michigan State University Extension.)

of their habitat). The legs are usually ringed with alternating darker and lighter bands.

Identification Resources. For further information on the identification of true bugs (Hemiptera) of North America, see Slater and Baranowski 1978; for bugs of the eastern United States, see Blatchley 1926; for Heteroptera of Illinois, see McPherson 1989; for families of aquatic bugs in North America, see Lehmkuhl 1979. For species of aquatic bugs of northern Michigan, see Hussey 1919; for Wisconsin, see Hilsenhoff 1984; for Minnesota see Bennett and Cook 1981. For species of semiaquatic bugs of Wisconsin, see Hilsenhoff 1986. Sources for identification keys and information on other families, when available, are mentioned under each respective family.

For identification keys to families and subfamilies of immature bugs (nymphs), see DeCoursey 1971. Keys to families of immature bugs can also be found in Herring and Ashcock 1971 and Lawton 1959.

Cicadas, Hoppers, Aphids, Scales, and Whiteflies
(Order Homoptera)

Introduction. At first glance cicadas, hoppers, aphids, scale insects, and whiteflies would not appear to have much in common. However, despite tremendous diversity in size and appearance, they all share several important characteristics and indeed are closely related.

Description. The homopterans are minute to very large insects (1–50 mm, but most are small) with piercing/sucking mouthparts and uniformly textured wings. The wings may be thin and membranous, thickened and parchmentlike, or even absent in some cases. The forewings usually lay tentlike over the back of the abdomen. The head is equipped with compound eyes, ocelli (which may be lacking in some families), and a beaklike mouthpart that arises from the rear of the face and is directed rearward between the legs (in resting position). The antennae are generally short and often more or less inconspicuous. Homopterans may be soft bodied or hard bodied, and they may be naked, pubescent, or covered with waxy secretions. Other characteristics of shape and color are highly variable and are discussed under each respective family.

Life Cycle. Most homopterans develop through gradual metamorphosis. After hatching from the eggs, the nymphs join their adult counterparts in the appropriate habitat. Usually the nymphal homopterans resemble the adults, though the nymphs are smaller and lack wings. However in some cases the nymphs

---

Quick Guide to Identification

Cicadas, Hoppers, Aphids, Scales, and Whiteflies
(Order Homoptera)

Winged

Diagnostic Characteristics	Common Name(s)	Scientific Family/ Superfamily Name
Large, heavy-bodied; membranous wings	cicadas	Cicadidae
Pronotum enlarged, covering abdomen	treehoppers	Membracidae
Froglike; circle of spines on tibia	froghoppers	Cercopidae
Elongate; antennae in front of eyes	leafhoppers	Cicadellidae
Body more or less elongate; antennae below eyes	planthoppers	Fulgoroidea*
Small; powdery white wings/body	whiteflies	Aleyrodidae
Small; cornicles on abdomen	aphids	Aphididae
Small; wings rooflike over body	psyllids	Psyllidae
Small; single pair of wings	scales (male)	Coccoidea

Wingless

Diagnostic Characteristics	Common Name(s)	Scientific Family Name
Small; cornicles on abdomen	aphids	Aphididae
Flattened; disklike or domelike	scale insects	Coccoidea
Flattened; with waxy threads	mealybugs	Pseudococcidae

*Includes the following difficult to separate families: Cixiidae, Delphacidae, Acanaloniidae, Flatidae, and Dictyopharidae.

---

are distinctly different, such as in the scale insects. Some homopterans have special life stages—for example, some active, first-stage nymphs are known as crawlers—and some even have pupalike stages. The life cycles of some homopterans are incredibly complex, involving parthenogenesis and alternate hosts. Some homopterans (such as aphids and leafhoppers) are capable of completing many generations each season, while others (such as cicadas) may require many years to complete a single generation.

Habits. Homopterans use their beak to extract fluids from plants. All are phytophagous, feeding on plant sap, nectar, or fruit juices. One group of homopterans, the cicadas, are noted for their ability to sing.

Habitats. Homopterans can be found in a wide variety of subterranean, terrestrial, and arboreal habitats.

Ecological and Economic Status. Many homopterans are plant pests, sucking the sap from leaves, stems, and fruit. This action may seriously weaken plants, making them more susceptible to some of the plant diseases spread by homopterans and other insects.

Distribution and Faunistics. Many species of homopterans are widely distributed throughout the Great Lakes region. Since most families and species of homopterans are significant plant pests, many groups have received attention over the years. Some of the homopteran groups that have been studied in the Great Lakes region are the fulgorid planthoppers, the jumping plant lice, and the armored scale insects of Ohio; the spittlebugs of Canada and the adjacent United States; the leafhoppers of Illinois, Minnesota, Ohio, Wisconsin, and Michigan; the plant lice (aphids) of Illinois, Indiana, and Minnesota; and the lecanium scales of Michigan.

Cixiid Planthoppers
(Family Cixiidae)

This relatively large group of planthoppers is best recognized by the broad body and long legs with only a few spines at the base of the tibia. The head is more or less normal and is not swollen or extended forward (although there may be two or three keel-like ridges). There are usually three ocelli present on the head. The forewings are membranous and often have dark spots along the veins.

The huckleberry planthopper, *Oliarus cinnamonensis* Provancher, is small (6–7 mm) and blackish with a narrow, pale front margin of the forewings. They are bog inhabitants said to feed on blueberries (huckleberries). The humble planthopper, *O. humilis* (Say), is smaller (5 mm) with a dark brown body. The wings are smoky brown or blackish on the outer one-third. The middle of the thorax has five or six longitudinal ridges.

More information on planthoppers of the Great Lakes region is available in Osborn 1938.

Delphacid Planthoppers
(Family Delphacidae)

This group comprises the largest planthopper family, containing small (2–9 mm), more or less elongate planthoppers that are usually found on grasses in moist habitats (wet meadows, stream-sides, etc.). They are recognized by the large spur at the tip of the tibia. More information on planthoppers of the Great Lakes region is available in Osborn 1938.

The three-ridged planthopper, *Stobaera tricarinata* (Say), is small (4 mm) and brown or brownish yellow. The face, legs, and forewings are banded with white and dark markings. The corn planthopper, *Peregrinus maidus* Ashmead, occurs on corn worldwide. It is small (4–5 mm) and light colored with fuscous markings.

Fulgorid Planthoppers
(Family Fulgoridae)

Members of this family are among some of the largest of the planthoppers (1–8 mm and larger). Most are brownish or reddish and the front of the head may be prolonged forward. The forewings have many reticulate veins at the base of the wing near the tip of the scutellum. Two rare species, *Cyrpoptus belfragei* Stal and *Poblicia fuliginosa* (Olivier), are recorded from Ohio (Osborn 1938).

The planthoppers (superfamily Fulgoroidea) are well distributed in the region, with 107 species known from southern Illinois, for example (Wilson and McPherson 1980).

For identification keys and information on planthoppers of the Great Lakes region, especially Ohio, see Osborn 1938.

Acanaloniid Planthoppers
(Family Acanaloniidae)

Acanaloniid planthoppers are medium sized (10–12 mm) and greenish or yellowish with greenish markings. The forewings are broad, rounded at the tip, and folded down beside the body. The front edge of the forewing has many netlike veins.

The two-striped planthopper, *Acanalonia bivittata* (Say), is greenish or greenish yellow with brown markings forming two parallel stripes on the face, pronotum, and leading edge of the wings. The legs are also brown. This species can be found on cranberry, goldenrod, and a variety of other herbaceous and weedy plants.

More information on planthoppers of the Great Lakes region, especially in Ohio, is available in Osborn 1938.

Flatid Planthoppers
(Family Flatidae)

Flatid planthoppers are mothlike and wedge shaped (more elongate than the previous family). They are generally pale green or

Fig. 83. The three-ridged planthopper, *Stobaera tricarinata* (Say). (From Osborn 1938.)

Fig. 84. The fulgorid plant hopper *Cyrpoptus belfragei* Stal. (From Osborn 1938.)

Fig. 85. The fulgorid planthopper *Poblicia fuliginosa* (Olivier). (From Kansas State University Extension 1962.)

Fig. 86. The two-striped planthopper, *Acanalonia bivittata* (Say). (From Osborn 1938.)

brownish. The elongate, lower section (tibia) of the hind leg has many spines on it and there are generally only two ocelli on the head. The front edge of the wing has many regularly spaced short cross veins and the wing tips are squarely truncated. The nymphs produce white waxy threads.

Two common species are the New World planthopper, *Anormenis septentrionalis* Spinola, which is pale green to whitish with a length of 9–11 millimeters; and the citrus planthopper, *Metcalfa pruinosa* (Say), which is bluish gray above and pale gray below with dark spots on the forewings. The body is covered with a whitish powder. The New World planthopper is generally found on bittersweet, dogwood, plum, grape, prickly ash, red oak, and black ash. The citrus planthopper is found on many woody plants, including citrus.

More information on planthoppers of the Great Lakes region, especially Ohio, is available in Osborn 1938.

Dictyopharid Planthoppers
(Family Dictyopharidae)

These planthoppers are generally small to medium sized (8–12 mm) and greenish or brownish. The forewings are long and narrow and many species have the head prolonged forward into a slender, beaklike structure.

Fig. 87. The meadow planthopper, *Scolops sulcipes* Say. (From Osborn 1938.)

The meadow planthopper, *Scolops sulcipes* Say, is very common. It is 6–10 millimeters in length and brownish. The forewings have many meshlike cross veins (reticulate) and there are both long-winged and short-winged forms. The head is prolonged forward as a slender conical spine. They are most commonly found in wet meadows (and sometimes lawns), where they feed on grasses.

More information on planthoppers of the Great Lakes region is available in Osborn 1938.

Spittlebugs and Froghoppers
(Family Cercopidae)

The adults are small (less than 13 mm) and froglike. Their bodies are elliptical and greenish or brownish, and they actively hop about. The forewings are thickened and leathery, without conspicuous veins. The hind leg has one or two stout spines. The nymphs, referred to as spittlebugs, reside within frothy, bubblelike masses of "spittle." These spittle masses provide the young with protection against desiccation and predation. The bubbles are formed by forcing air from the abdomen into plant juices eliminated from the anal opening of the spittlebug. Cercopids feed on the sap of grasses, herbaceous plants, shrubs, or trees. Some species are involved in the transmission of plant diseases (viruses and fungi).

A little less than two dozen species are found in the Great Lakes region. For example, 21 species are known from Michigan (Hanna and Moore 1966) and 22 species are known from the Canadian part of the Great Lakes region (Hamilton 1982).

The Saratoga spittlebug, *Aphrophora saratogensis* (Fitch), is small (9–11 mm) and brownish with a pale scutellum and a light middorsal stripe on the head and pronotum. The forewing is mottled with tan and silver cross bands. Adults are most commonly found on red pine. The pine spittlebug, *Aphrophora parallelus* Say is a similar species but the wings flare broadly to the sides near the middle of the body and the front margin of the head is angular (differences apparent only when viewed from above). The diamond-backed spittlebug, *Lepyronia quadrangularis* (Say) is brownish above and blackish below and densely clothed with fine hairs. There are V-shaped dark brown markings on the wings that form a diamond shape when the wings are closed. The base of the wing, near the "shoulder," is also dark in color. This species is one of the most commonly collected spittlebugs in meadows, fields, pastures, and croplands of the Great Lakes region.

The meadow spittlebug, *Philaenius spumarius* (L.), is also a very common species that occurs in meadows and grasslands. It was an early introduction into the region from Europe and is now widespread. This species has a light-colored head (occasionally the pronotum is also light colored) and a dark oval spot on the underside of the thorax in the center. This species is highly variable in color, ranging from tan to brownish to grayish to blackish, either uniformly colored or patterned with darker and/or lighter colors. The dogwood spittlebug, *Clastoptera proteus* Fitch, is a short, stout spittlebug (4–5 mm) that is blackish and marked with yellow cross bands on the head, thorax, and forewings. The scutellum is also yellow.

Fig. 88. The pine spittlebug, *Aphrophora parallelus* Say. (From State University of New York 1964.)

Fig. 89. The meadow spittlebug, *Philaenius spumarius* (L.). *Right,* adult froghopper on leaf; *left,* young spittlebug and spittle mass. (From University of Maryland Extension 1951.)

For identification keys to the spittlebugs and froghoppers of Canada and the adjacent United States, see Hamilton 1982; for a key to Michigan species, see Hanna and Moore 1966; for a key to Illinois species, see Moore 1955.

Cicadas
(Family Cicadidae)

Cicadas are the largest homopterans, measuring up to two inches (50 mm) in length. They have heavy, robust bodies and four membranous wings. The beak is prominent and the compound eyes are large. Three ocelli are present on the forehead. The antennae are short and bristlelike. The males produce loud, sometimes shrill, monotonous buzzing noises during daylight hours. For a key to the songs of Michigan cicadas, see Alexander et al. 1972. Cicadas are also known as harvestflies, because of their sudden appearance at the end of the summer at harvest time. They are often improperly referred to as "locusts" because of their plaguelike mass emergences, but they are not really related or similar to the migratory locusts of the Old World.

The cicada fauna of the Great Lakes region is small, with a total of 10 species known from Michigan, for example (Moore 1966).

The dog-day cicada, *Tibicen canicularis* (Harris), is large (32 mm) and blackish with the front of the pronotum and the leading edge of the wings marked in dark green. The underside of the abdomen has a black stripe down the center. Wing length is less than 40 millimeters. Nymphs require two to five years to complete their development. This cicada is one of the most abundant in many areas of the Great Lakes region. Linne's cicada, *T. linnei* (Smith and Grossbeck), is similar to the dog-day cicada, but the wing length is greater than 40 millimeters.

There has been much confusion about the status of periodi-

Fig. 90. The dog-day cicada, *Tibicen canicularis* (Harris). (From USDA 1952.)

cal cicadas in North America, but the work of Alexander and Moore (1962) has provided answers to many of the old questions. They have shown that there are not two species of periodical cicadas but six distinct species occurring in all of the 20–25 recognizable broods of periodical cicadas. Linnaeus' 17-year cicada, *Magicicada septendecim* (L.), and Walsh and Riley's 13-year cicada, *M. tredecim* (Walsh and Riley), are indistinguishable, except for the length of their life cycle. They are both 27–33 millimeters in length. The body is black with red eyes and major wing veins. There is a distinct W-shaped black marking near the wing tip. The lateral extension between the eyes and wing bases are reddish and the underside of the abdomen is also primarily reddish or yellowish.

Fig. 91. The 17-year (periodical) cicada, *Magicicada septendecim* (L.), with left wing extended to show W-shaped marking near wing tip. (From Felt 1900.)

Leafhoppers
(Family Cicadellidae)

Not only is the Cicadellidae the largest family of homopteran insects, but it is also one of the largest insect families known to science. DeLong 1948 estimates that at least 700 species inhabit Illinois. A total of 227 species are found in Michigan (Taboada 1979), 123 species in Wisconsin, and 105 species in New York (Sanders and DeLong 1917). Leafhoppers are mostly small (less than 13 mm), elongate, and greenish, brownish, or brightly colored (striped). The hind tibia has many rows of small spines and are suited for jumping. Only two ocelli are present on the head. Many species are involved in the transmission of plant diseases (viruses, mycoplasms, etc.). Some leafhoppers migrate into the Great Lakes region each spring and then migrate south in the fall.

The candy-stripe leafhopper, *Graphocephala coccinea* (Forster), is easily recognized by the red-and-blue or red-and-green stripes on its body. The head and pronotum are yellowish or reddish. These small leafhoppers (8–9 mm) are generally found on shrubs and blackberries. The potato leafhopper, *Empoasca fabae* (Harris), is a significant agricultural pest. It is uniformly pale green with six to eight white spots on the front of the pronotum and a white H-shaped mark on the scutellum. It feeds on potato, alfalfa, beans, clover, and apple and is responsible for causing "hopper burn"—a browning of the leaves caused by this leafhopper's excessive feeding. DeLong and Sanders 1930 revises the list of *Empoasca* of North America and provides information on the many other species in this genus (most of which are of economic importance). The painted leafhopper, *Endria inimica* (Say), is small (4mm) and grayish yellow with two black spots on the head, front of the pronotum, and base of the scutellum. They are found on grasses in lawns and meadows.

Fig. 92. The potato leafhopper, *Empoasca fabae* (Harris). (From DeLong 1948.)

For identification keys to leafhoppers of Illinois, see DeLong 1948 and Moore 1955; for keys to leafhoppers of Ohio, see Osborn 1928 and Johnson 1935; for leafhoppers of Minnesota, see Medler 1942.

Treehoppers
(Family Membracidae)

Fig. 93. The two-spotted treehopper, *Enchenopa binotata* (Say). (From Kansas State University Extension 1962.)

Treehoppers are easily recognized by the enlarged pronotum, which covers the thorax and abdomen (and usually projects forward over the head as well). Treehoppers may appear thornlike, humpbacked, or lobed. Both pairs of wings are membranous and they are usually covered by the pronotum. There are two ocelli on the head, but they are also usually covered by the pronotum. These jumping insects are small (less than 12 mm) and usually greenish, brownish, or blackish. The nymphs (and occasionally the adults) of many species are tended by ants because of the sweet secretions (honeydew) they produce.

The Great Lakes treehopper fauna is quite large, with a total of 101 species of treehoppers known from Ohio, for example (Osborn 1940).

The thorn bug, *Umbonia crassicornis* Amyot and Serville, is small (5 mm), yellowish, and distinctly thorn shaped. The tip of the "thorn" (pronotum) is dark. This species occurs primarily in the southeast but ranges as far north as southern Ohio and possibly Indiana. The two-spotted treehopper, *Enchenopa binotata* (Say), is small (5–6 mm) and dark brown with two yellow spots near the center of the pronotum. The pronotum projects forward and is slightly expanded at the tip. The buffalo treehopper, *Stictocephala bubalus* (Fabricius), is small (9–10 mm) and easily recognized by its bright green color and pronotum with two distinct lateral "horns," giving a buffalo-like appearance. It is usually found on herbaceous plants and trees, especially elm and apple. The locust treehopper, *Thelia bimaculata* Fabricius,

 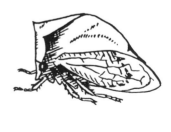

Fig. 94. An adult buffalo treehopper, *Stictocephala bubalus* (Fabricius). *Left,* top view; *right,* side view. (From Kansas State University Extension 1962 and Courtesy of Michigan State University Extension.)

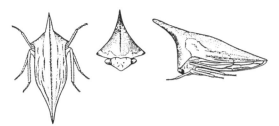

Fig. 95. The locust treehopper, *Thelia bimaculata* Fabricius. (From Osborn 1940.)

deserves a brief mention because it is often quite common on locust trees, where it is tended and protected by ants.

For an identification key to the treehoppers of Ohio, see Osborn 1940.

Jumping Plant Lice
(Family Psyllidae; formerly Chermidae)

Jumping plant lice, or psyllids, are small (less than 5 mm) and have four membranous wings, three ocelli, and a body shaped like a miniature cicada. The antennae are short (10-segmented) and stout. Jumping plant lice are found on a wide variety of trees, and some species cause economic damage.

A total of 33 species of jumping plant lice are reported from Ohio (Caldwell 1938).

The hackberry nipple gall maker, *Pachypsylla celtidis-mamma* Riley, produces nipplelike galls on the leaves of hackberry. The top of the head is yellowish brown and the thorax is reddish brown to black, often marked with longitudinal light stripes. The forewing is whitish and densely spotted with brownish. The adults emerge from the galls in the fall and can often be found trying to gain entry into buildings to hibernate. The pear psylla, *Psylla pyricola* (Foerster), is a major pest of pears. They are very small (2–3.5 mm) and reddish with brownish black markings. The scutellum is marked with four narrow, longitudinal light stripes. The eyes are bronze colored. The adults overwinter in bark crevices.

For an identification key to the jumping plant lice of Ohio, see Caldwell 1938.

Whiteflies
(Family Aleyrodidae)

Adult whiteflies are small (less than 2–3 mm), with four wings that are transparent, clouded whitish, or mottled with spots or bands. There are two ocelli on the head. The body is generally covered with a white, powdery substance. The nymphs are flattened, scalelike, and covered with white, waxy secretions.

Whiteflies have an unusual type of metamorphosis that is midway between gradual and complete. When whiteflies hatch from the egg stage they develop into an active stage known as a crawler. This stage is followed by an immobile nymphal stage (two molts), then a pupalike stage, and finally the adult stage.

The greenhouse whitefly, *Trialeurodes vaporariorum* (Westwood), occurs both in greenhouses (year round) and outdoors in vegetable and flower gardens (in summer months). They feed on many species of herbaceous and ornamental plants and can be found on the underside of the leaves (the small, whitish adults fly away from the plants when approached). The nymphs are pale green.

Aphids or Plantlice
(Family Aphididae)

Aphids are minute to small (1–5 mm), pale greenish or yellowish (occasionally pinkish) insects with a pair of tubular appendages (cornicles) located near the tip of the abdomen. These cornicles produce a special alarm scent or pheromone that aphids use to warn one another of danger. The antennae are distinct and four- to six-segmented. Wings, when present, are membranous and transparent. Aphids are generally gregarious, with nymphs and adults living together in groups on their host plants, where they extract copious amounts of sap with their needlelike mouthparts. They may feed on the roots, stems, twigs, leaves, flowers, or fruits of plants. They produce large amounts of honeydew and therefore many species are tended and protected by ants.

Fig. 96. Adult aphids. *Left,* winged individual; *right,* nonwinged individual. Note the presence of tubelike cornicles near the tip of the abdomen in both individuals. (From USDA 1952 and Courtesy of Michigan State University Extension.)

Aphids have complex life cycles involving parthenogenetic generations mixed with biparental generations and/or alternating between different host plants (often with different body forms). This complex cycle, in combination with their small size, makes aphids difficult to identify. Some species, however, are very common. It is a relatively large family; for example, Knapp 1972 records 218 species in 64 genera from Indiana.

Only a few of the many species of aphids found in the Great Lakes region can be discussed here. The corn leaf aphid, *Rhopalosiphum maidis* Fitch, is minute (1.5–2.5 mm) and dark blue green with a green head. There are both winged and wingless forms and they can be found on the tassels, ear tips, and leaf whorls of corn plants. The melon aphid, *Aphis gossypii* Glover, is greenish yellow with dark spots on top of the abdomen. The cornicles are also black. They are found on melons, cucumbers, vegetable plants, ornamental plants, and even weeds. The rose aphid, *Macrosiphum rosae* (L.), is generally found on roses. These small aphids (2–3.5 mm) are usually green or pinkish. The winged adults have blackish antennae, head, thorax, and cornicles. The English grain aphid, *M. granarium* (Kirby), occurs on grasses, cereals, and small grains. Wingless females are pale green with long black antennae. Winged adults are also marked with brownish or blackish on the thorax. The green peach aphid, *Myzus persicae* (Sulzer), is pale to dark green with stripes in the summer and pinkish in the fall. The winged individuals are dark brown with a brown patch on an otherwise yellow abdomen. There are two prominent pimplelike bumps (tubercles) on the front of the head (in all forms). They are found on all types of fruit, vegetable, and ornamental plants.

For identification keys to aphids of Illinois, see Hottes and Frison 1931; for keys to aphids of Minnesota, see Oestlund 1887.

Spruce and Pine Adelgids
(Family Adelgidae)

This family of aphidlike insects is a small group that often cause gall formations on various species of spruce and pine. They differ from true aphids by lacking cornicles. The wingless females are usually covered with a white, waxy substance. Commonly encountered adelgids include the Cooley spruce gall adelgid, *Adelges cooleyi* (Gillette), on Colorado blue spruce and Douglas fir; eastern spruce gall adelgid, *Adelges abietis* (L.), on Norway and white spruce; balsam woolly adelgid, *Adelges piceae* (Ratzenburg), on balsam fir; pine bark adelgid, *Pineus strobi* (Hartig), on white pine; and pine leaf adelgid, *Pineus pinifoliae* (Fitch), on pines.

Phylloxerids
(Family Phylloxeridae)

This family of aphidlike insects is a small group, feeding on the roots and leaves of plants (and sometimes causing galls). The adults are wingless and covered with a white powder. There are no cornicles on the abdomen. The most commonly encountered species is the grape phylloxera, *Phylloxera vitifoliae* Fitch, which feeds on the roots and leaves of both cultivated and wild grapes. The rootstocks of most North American varieties are resistant to grape phylloxera damage, but leaf feeding is usually quite evident by the presence of small galls on the undersides of the leaves.

Scale Insects
(Superfamily Coccoidea)

Fig. 97. An adult male (winged) scale insect. (From Comstock 1880.)

Scale insects are perhaps some of the most unusual insects imaginable. The immatures and females are wingless, often legless, and disk or scale shaped. The body segments are highly fused. Only the mouthparts and the scalelike covering are well developed. Scale insects produce various types of waxy body coverings that may be either soft or hard and either flattened or domelike. The nymphs that hatch from the eggs are known as crawlers. They have functional legs that enable them to migrate onto new plant shoots and foliage. After finding a suitable feeding site they settle down, insert their mouthparts, and molt into a legless form. The nymphs molt anywhere from two to four more times, depending on the species and the sex of the individuals. Males molt fewer times and then form special cocoonlike structures in which they pupate into a minute, two-winged, flylike adult. (They differ from true flies, which also have a single pair of wings, by having only a single tarsal segment, one or two filaments at the tip of the abdomen, and nonfunctional mouthparts.) Females, when mature, deposit their eggs beneath the protective scale and die soon thereafter.

For an annotated key to the families of scale insects of North America, see Howell and Williams 1976. For identification information on scale insects of Ohio, see Sanders 1904; for Indiana, see Dietz and Morrison 1916.

Mealybugs
(Family Pseudococcidae)

The mealybugs are the second largest family of scale insects. The body is covered with a whitish, waxy material, giving these bugs a powdery, or "mealy," appearance by which they can be recognized. Unlike most other scale insects, however, the legs of the female are well developed. The outer edge of the elongate-oval body bears many long, waxy filaments. Mealybugs are primarily tropical insects, although several species occur in the temperate region (especially in greenhouses), and they infest all plant parts (roots, stems, foliage, etc.), where they extract plant fluids with their piercing/sucking mouthparts.

The citrus mealybug, *Planococcus citri* (Risso), is small (3 mm) and whitish with a slightly darker median stripe. The waxy filaments at the tip of the abdomen, as well as those surrounding the body, are short and less than one-fourth the length of the body. The long-tailed mealybug, *Pseudococcus longispinus* (Targioni-Tozzetti), is also whitish, but the waxy filaments are much longer. The lateral filaments are as long as one-half the width of the body and several of the terminal filaments are longer than the length of the body.

Soft, Wax, and Tortoise Scales
(Family Coccidae)

These scales are characterized by a soft and often more or less flattened body covering. They are found on trees (both fruit and ornamental) and other ornamental plants.

Fig. 98. Adult females of the cottony maple scale, *Pulvinaria innumerabilis* (Rathvon). (From Felt 1900.)

The female brown soft scale, *Coccus hesperidum* L., is small (3–5 mm), flat, oval, and brownish (sometimes marbled with pale brown or greenish). They are generally found on greenhouse plants and houseplants. The female cottony maple scale, *Pulvinaria innumerabilis* (Rathvon), is brown or reddish brown with a large white, cottony mass of wax lifting the rear of the scale. It occurs on silver maple and other deciduous trees. The lecanium scale group contains a large number of closely related species, many of which are difficult to tell apart with any high degree of certainty. A total of 12 species of lecanium scales are reported from Michigan (Pettit and McDaniel 1920). Female terrapin scales, *Lecanium nigrofasciatum* Pergande, are hemispherical and reddish brown with tortoiselike black markings around the edge of the scale. They can be found on a variety of hardwood trees.

For an identification key to the Michigan lecanium scales, see Pettit and McDaniel 1920.

Pit Scales
(Family Asterolecaniidae)

These scale insects are found on the twigs and shoots of deciduous trees, causing pitlike depressions in the bark. The golden oak scale, *Asterolecanium variolosum* (Ratzenburg), is relatively common in the Great Lakes region and occurs on the branches and twigs of oaks. This small, oval scale is covered with a greenish gold secretion.

Fig. 99. The oyster shell scale, *Lepidosaphes ulmi* L., on twig. (Courtesy of Michigan State University Extension.)

Armored Scales
(Family Diaspidae)

Diaspidae is the largest family of scale insects. For example, 52 species of armored scales are known from Ohio (Kosztarab 1963). These insects are covered with a tentlike covering of wax, excretions, and shed skins that is independent of the body. These scale insects are small, flattened, and disklike. They feed on trees and shrubs and may often be so abundant that they encrust entire twigs and branches with their bodies.

The oyster shell scale, *Lepidosaphes ulmi* L., has a distinctive shape like an oyster shell—narrowly pointed at one end, slightly curved, and broader at the other end. They are grayish brown and can be found on a variety of trees and shrubs. The San Jose scale, *Quadraspidiotus perniciosus* Comstock, was introduced into California from the Orient in the late 1800s. It is now widespread throughout North America. These small, gray, circular scales have a nipplelike center and can be found on a wide variety of trees and shrubs. Pine needle scale, *Chionaspis pinifoliae* (Fitch), looks like small flecks of paint on pine needles. The scales are whitish and elongate-oval (shaped slightly like an oyster shell). They can be found on the needles of pine and other conifers.

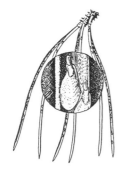

Fig. 100. Pine needle scales, *Chionaspis pinifoliae* (Fitch), on Austrian pine. (From Knowlton 1936.)

Identification Resources. Sources for identification keys and other detailed information, when available, are mentioned under the respective family.

Thrips
(Order Thysanoptera)

Description. Thrips are minute to small (1–4 mm), elongate, soft-bodied insects. They may be wingless or may have two pairs of wings. The wings, when present, are long and narrow with the front and rear margins fringed and featherlike. They are folded flat over the back when not in use. Thrips may be black, gray, brown, tan, or yellowish, and the wings and/or body may be spotted or blotched with darker colors. The head bears short antennae (six to nine segments), compound eyes, ocelli (in winged forms), and a conical mouth with piercing stylets.

Fig. 101. An adult thrips of the genus *Euthrips;* note unusual fringed wings. (From Foster and Jones 1911.)

Life Cycle. Thrips have a type of metamorphosis that has some of the attributes of gradual metamorphosis and some of the attributes of complete metamorphosis; however, it is generally described as gradual metamorphosis, even though there are five stages of development (egg, larva, prepupa, pupa, and adult). The eggs, which are quite large for the size of the female, are deposited in soil, bark crevices, or plant tissue. Some thrips give birth to live young (the eggs being retained in the body of the female). The early immature stages are referred to as larvae and they resemble the adults in appearance and habits (although smaller and with developing external wing pads). Later, they change into pupae, although in many instances they may remain relatively mobile and active. Some pupae, however, do form protective cocoons. The pupae develop into adults, completing the cycle. Some species totally lack males, and parthenogenesis is not uncommon among thrips. There are generally several generations per year, up to five in some areas.

Habits. Many species of thrips are plant feeders, feeding on plant juices and sometimes on pollen. Some species are scavengers, feeding on dead plant materials, fungi, or algae. A few species are predatory, feeding on mites, small insect larvae, and even other species of thrips. Thrips crawl about plants with a slow, steady pace but run swiftly and leap suddenly when disturbed. Their feet (tarsi) are equipped with retractile bladderlike organs that, when inflated, provide adhesion to smooth surfaces.

Habitat. Thrips can be found in flower blossoms, on the undersides of leaves, in leaf whorls and axils, under bark, in mosses, in leaf litter and soil, and on fruits and flower bulbs.

Ecological and Economic Status. The predatory species and scavengers are beneficial in their activity. The plant-feeding species, however, are frequent pests of vegetables, fruits, flowering plants, and grasses. About two dozen plant-feeding thrips that have been introduced into the Great Lakes region have become well established and widespread pests. Some are involved in the transmission of plant diseases. Their populations may reach extraordinary numbers, and under such conditions they may cause respiratory and dermal irritation to agricultural workers. They have even been reported to bite under these conditions.

A note on the common name "thrips" should be interjected at this point. The word is always used in its plural form, regardless of the number of thrips being referred to. Therefore, there is no such thing as a "thrip."

Distribution and Faunistics. Despite the large number of species that are plant pests, these minute insects are still in need of study in many parts of the Great Lakes region. Many species of thrips have yet to reinvade the previously glaciated (northern) parts of the region, hence the greatest number of species are found south of the Wisconsinan glacial maxima. A total of 74 species are reported from New York state, while a total of 223 species are reported from Illinois, a state where many southern areas were not glaciated (Stannard 1968). With more than 400 species occurring in North America, it seems likely that most states in the Great Lakes region are inhabited by more than 100 species (more in the south, fewer in the north).

Common Thrips
(Family Thripidae)

Members of this large family are collected on leaves and flowers, and many species are of economic importance. The greenhouse thrips, *Heliothrips haemorrhoidalis* (Bouche), is minute (1 mm) and dark brown or black. The ridged body has a netlike pattern of wrinkles. The antennae are eight-segmented. Wings are usually present. They are found on many plants in greenhouse situations throughout the region, and may occasionally be found outdoors on flowers during the summer months. The grass thrips, *Anaphothrips obscurus* (Müller), is commonly found on bluegrass and corn. They are minute (1.5 mm) and yellowish with dark blotches (no blotches occur on the last two abdominal segments). Wings are usually present. The antennae have nine segments, and segments three and four are brownish. The flower thrips, *Frankliniella tritici* (Fitch), are minute (1 mm) and yellowish or brownish. The antennae are eight-segmented, and the fourth segment bears a forked sense cone. The front angles of the pronotum are each armed with a long, stout bristle. Wings are usually present. Flower thrips are found on the flowers of many flowering plants, including grasses, herbaceous plants, shrubs, and trees. The onion thrips, *Thrips tabaci* Lindeman, is a serious pest of onions, carrots, melons, cucumbers, peas, beans, and roses. It is minute (1 mm) and pale yellowish or brownish. The wings are a uniform dusky gray. The head is wider than it is long and the antennae are seven-segmented. The gladiolus thrips, *Thrips simplex* (Morrison), is minute (1 mm) and dark brown or black. The base of the wings and the third antennal segment are white. This thrips is generally found on gladiolus and iris plants. The mullein thrips, *Haplothrips verbasci* (Osborn), is a very common species that is associated with the mullein plant. It can be found in the whorls of the plant even in the early and late season. It is black, except for light areas adjacent to the abdominal sutures, and larger than many other thrips in the region (up to 3 mm).

Identification Resources. For identification keys to the thrips of Illinois, see Stannard 1968.

Dobsonflies, Lacewings, Antlions, and Relatives
(Order Neuroptera)

Introduction. Some specialists divide the order Neuroptera into three separate orders (Megaloptera, Raphidiodea, and Planipennia), which is an indication of the wide variation within this group of insects. Of these three groups, which are treated as suborders here, only the Megaloptera and Planipennia occur in the Great Lakes region. The snakeflies (Raphidiodea) are restricted to the western United States.

Description. All neuropterans have chewing mouthparts and two pair of membranous wings, with many longitudinal veins and cross veins. The body is usually elongate, more or less soft, and dull brown, green, gray, or black in color. The antennae are generally long and slender, although in a few cases they are short and even clubbed at the tip. The dobsonflies and alderflies (Megaloptera) have large, membranous wings that fold flat over the back, while the lacewings and antlions (Planipennia) have smaller, many-veined, lacelike, transparent wings that are held rooflike over the back when not in use.

Life Cycle. The neuropterans are the most primitive group of

Quick Guide to Identification

Dobsonflies, Lacewings, Ant Lions, and Relatives
(Order Neuroptera)

Diagnostic Characteristics	Common Name(s)	Scientific Family Name
Large; male has sicklelike mandibles	dobsonflies	Corydalidae
Large; with ocelli	fishflies	Corydalidae
Medium; no ocelli; dark wings	alderflies	Sialidae
Small and mantislike	mantidflies	Mantispidae
Small; covered with white powder	dusty-wings	Coniopterygidae
Slender body; knobbed antennae	antlions	Myrmeleontidae
Long, knobbed antennae	owlflies	Ascalaphidae
Membranous wings; greenish body	green lacewings	Chrysopidae
Membranous wings; brownish body	brown lacewings	Hemerobiidae
Lacewinglike; few wing veins	spongillaflies	Sisyridae
Lacewinglike; wings hairy	pleasing lacewings	Dilaridae
Lacewinglike; very large	giant lacewings	Polystoechotidae
Lacewinglike; with large scales	beaded lacewings	Berothidae

insects with complete metamorphosis (egg, larva, pupa, and adult stages). Eggs may be laid in the water, in soil, or on plants. The larvae, which are either aquatic or terrestrial (sometimes arboreal), are predaceous. Pupation occurs in earthen cells or silken cocoons. Adults emerge one to two weeks later. The life cycle of the dobsonflies and alderflies generally requires more than one year for completion, while the lacewings and antlions have an annual cycle.

Habits. Adult neuropterans are generally terrestrial, although they spend large amounts of time in the air. The flying ability of adult neuropterans varies considerably, but typically their flight is fluttering and feeble. The larvae are generally aquatic (Megaloptera), terrestrial, or arboreal (Planipennia). Almost all neuropterans are predators of other insects, both as adults and larvae, but some are also known to feed on nectar and pollen. The predaceous larvae are equipped with special hollow mandibles for feeding. They grab their prey, puncture them with the sharply pointed mandibles, and suck out the body fluids. When done they typically fling the empty carcass over their back with a flick of the head. Adult megalopterans are nocturnal and are frequently attracted to lights.

Ecological and Economic Importance. As predators, all neuropterans are highly beneficial to the interests of humans. Some lacewings, for example, are even raised commercially for release in yards and gardens. The larvae of the dobsonfly, which is known as the hellgrammite, is prized as fish bait. Most other larval megalopterans are eaten by fish and are an important component of the aquatic food chain.

Habitat. Habitat preferences vary with the habits and life cycles of the various members of the order. Larvae of the Megaloptera are found in all types of aquatic environments, including streams, pools, ponds, and lakes. The adults are generally found in nearby terrestrial habitats. The larvae of the Planipennia may be found on the ground (antlions) or on plants (aphidlions). Adults are generally found in nearby habitats.

Distribution and Faunistics. Our knowledge on the distribution of neuropteran insects is weak and there is still much to be learned. The neuropteran fauna is not large—less than 100 species. For example, 65 species are known from Indiana (Lawson and McCafferty 1984), 60 taxa (species and subspecies) are known from Minnesota (Parfin 1952), 63 species are known from New York (Leonard 1928), and 47 species (Planipennia only) are known from Wisconsin (Throne 1971a, 1971b, and 1972).

Alderflies
(Family Sialidae)

All the alderflies of the Great Lakes region are members of the genus *Sialis*. The smoky alderfly, *Sialis infumata* Newman, is a commonly encountered species with a dusky brown body and smoky wings. They are usually found sitting on alders next to streams. When sitting on the vegetation they are easily approached and picked up with the fingers. The larvae occur in aquatic habitats with areas of thick aquatic plants and debris.

Dobsonflies and Fishflies
(Family Corydalidae)

This family is represented by three genera, *Corydalus*, *Chauliodes*, and *Nigronia*. The dobsonfly, *Corydalus cornutus* L., is the sole member of its genus in the region. It is easily recognized by its very large size (greater than 100 mm) and dull gray color. The males have long, sickle-shaped mandibles, longer than half the length of the body. The females have much smaller

Fig. 102. The smoky alderfly, *Sialis infumata* Newman. (From Kansas State University Extension 1962.)

Fig. 103. The dobsonfly, *Corydalus cornutus* L. *Top,* adult male; *bottom,* head of adult female. (From Kansas State University Extension 1962.)

mandibles. The fishfly, *Chauliodes rastiicornis* (Rambur), is smaller (30–50 mm) and has a mottled gray body and comblike antennae. The wings may be clear or with dark spots. The saw-combed fishfly, *Nigronia serricornis* (Say), has a black body and comblike antennae. Fishflies range much further north than dobsonflies.

Dusty-Wings
(Family Coniopterygidae)

The dusty-wings are small (about 3 mm) and the body and wings are covered with a whitish, waxy powder. The head bears chewing mouthparts, antennae that are long and slender, and prominent compound eyes. Ocelli, however, are lacking. The wing venation is simple and somewhat reduced. Only a handful of species occur in the Great Lakes region, almost half of them in the genus *Coniopteryx*. They are often considered to be rare or uncommon, but it is more likely that their small size causes them to be easily overlooked.

Mantidflies
(Family Mantispidae)

Fig. 104. The brown mantidfly, *Climaciella brunnea* (Say). (From Kansas State University Extension 1962.)

These mantidlike neuropterans are often a source of puzzlement for beginning collectors who are lucky enough to obtain one. The prothorax is long and narrow and the front legs are raptorial. They look very much like a tiny little praying mantid, but the size is too small (20–35 mm) and the wings are transparently membranous. The brown mantidfly, *Climaciella brunnea* (Say), is brownish or blackish mixed with some yellow. The wings are uniformly dark (at least in the front half). The interrupted mantidfly, *Mantispa interrupta* Say, is greenish brown or brown and the wings have a dark leading edge and three dark spots near the tip. Uhler's mantidfly, *M. uhleri* Banks, is similar but lacks the three dark spots. The larvae of mantidflies are parasites of spider egg sacs, and the adults are predators. They are generally collected in weedy fields, brushy areas, and woodland openings.

For a key to the species of northern Michigan, see Hungerford 1936.

Brown Lacewings
(Family Hemerobiidae)

These lacewings are medium sized (12–20 mm), brownish in color, and hairy. The antennae are long and slender. The wings only have a small number of cross veins. The larvae are known as aphidwolves and feed primarily on aphids. They also consume mealybugs and mites when available. The spotted brown lacewing, *Hemerobius stigma* Stephens, is the most common species in the region. It is brown or reddish brown and the wings are spotted with darker brown on the forewing, especially along the cross veins. It is frequently found on conifer trees but may also be taken by sweeping vegetation. The striped brown lacewing, *H. humulinus* L., is similar but has a distinct yellowish stripe on the prothorax and a yellowish head. The wing margin has alternating dark and light spots. Barber's brown lacewing, *Sympherobius barberi* (Banks), is pale light brown and occurs in the southern portions of the region. The wings are evenly clouded with paler brown, but some of the longitudinal veins are lightly spotted. It can be taken from oaks and is attracted to lights.

Green Lacewings
(Family Chrysopidae)

Fig. 105. The goldeneye lacewing, *Chrysopa oculata* Say. (From Felt 1900.)

These insects are well known because of their highly beneficial nature. They are raised commercially and sold as biological control agents throughout the United States. Green lacewings are medium sized (10–20 mm) with greenish bodies and wings. The wings are transparent and have many veins and cross veins. Several species have golden eyes. Several species are commonly encountered, including the goldeneye lacewing, *Chrysopa oculata* Say. It is medium sized (15–22 mm) and green with pale green wings, golden eyes, and a black banding on the head. As many as eight color varieties, each with a subspecific name, have been recognized. The weeping goldeneye, *Chrysoperla carneia* Stephens, is similar to the goldeneye lacewing but lacks the black banding on the head. Another species found in the Great Lakes region is the black-horned green lacewing, *Chrysopa nigricornis* Burmeister. It is the largest species, with a length of greater than one inch (25 mm), and can be easily recognized by the antennae, which are black at the base. It inhabits wooded and brushy areas and frequently flies to lights. One of the rarest lacewings in the Great Lakes region is the Canadian lacewing, *Eremochrysa canadensis* Banks, which is known only from the type locality at Go Home Bay in Ontario and central Wisconsin (and a few places in New England). It is mostly brown or reddish brown in color and the wing veins are spotted with brown.

For an identification key to the genera of green lacewings, see Bickley and MacLeod 1956; for a key to green lacewings of the nearctic region, see Banks 1903; for a key to the green lacewings of Canada, see Smith 1932.

Pleasing Lacewings
(Family Dilaridae)

A single, rare species of pleasing lacewing, *Nallachius americanus* (MacLachlan), occurs in the southern parts of the Great Lakes region. The antennae of the males are pectinate (comblike) and the female has a protruding ovipositor that is as long as the body. The wings are rather hairy and the forewing is more or less triangular.

Beaded Lacewings
(Family Berothidae)

These poorly known, and possibly rare, lacewings are represented in the Great Lakes region by at least two species, *Lomamyia banksi* Carpenter and *L. flavicornis* (Walker). The females have enlarged scales on the thorax and hind wings. The top of the head is flattened and some of the cross veins near the front margin of the wing are forked. In general appearance they somewhat resemble caddisflies.

Giant Lacewings
(Family Polystoechotidae)

The punctured giant lacewing, *Polystoechotes punctatus* (Fabricius), the largest lacewing (greater than 35 mm) to inhabit the Great Lakes region, is quite rare. It reaches the eastern limit of its distribution in the western part of the Great Lakes region (Ontario, Minnesota, and Indiana). In addition to the large size, the wing venation is peculiar, with forked cross veins along the margins of the wing and simple cross veins in the center of the wing.

Spongillaflies
(Family Sisyridae)

These small insects (6–8 mm) resemble brown lacewings but have unbranched cross veins along the front margin of the wing. They are generally found near water, because their larvae live in and feed on the freshwater sponge (*Spongilla fragilis*). The common spongefly, *Sisyra vicaria* Walker, is one species in the Great Lakes region that is commonly encountered near lakes. Another species that is likely to be encountered is the fuscated spongefly, *S. fuscata* Fabricius. The former species has a yellowish or yellowish brown head and yellowish antennae (except for the basal segments), while the latter species is dark brown or blackish with dark antennae.

For more information on identification of spongillaflies of North America, see Parfin and Gurney 1956.

Antlions
(Family Myrmeleontidae)

This family is named for the habits of its larvae, which dig conical pits in the sand to catch ants and other small terrestrial insects. The adults resemble damselflies (order Odonata) but can be told apart by the prominent antennae (longer than the head) that are clubbed at the tip. Adults are most likely to be active in the early morning and late afternoon or on cloudy days. They are taken at lights and must also be active at night.

The spotted-wing antlion, *Dendroleon obsoletus* (Say), is the only antlion in the Great Lakes region with dark spots, blotches, and bars on the wings. The front of the body is brown and the abdomen is dark brown. The Say's antlion, *Brachynemurus abdominalis* (Say), occurs throughout the eastern United

Fig. 106. The spotted-wing antlion, *Dendroleon obsoletus* (Say). (From Kansas State University Extension 1962.)

Fig. 107. The common antlion, *Myrmeleon immaculatus* De Geer. *Left,* adult; *center,* larva; *right,* pit trap in sand. (From Kansas State University Extension 1962 and Courtesy of Michigan State University Extension.)

States and most of the Great Lakes region. The adults are brownish gray to gray with a darker body. The larvae of this species do not build pits. Another widespread and common species is the common antlion, Myrmeleon immaculatus De Geer. It is mostly dark gray, marked with dark brown and orange. There are four orange spots on the thorax and the abdomen is banded in gray and brown. The wing membrane is unspotted, but the cross veins are alternately colored brown and gray.

For identification information on the antlions of North America, see Banks 1927.

Owlflies
(Family Ascalaphidae)

These conspicuous insects are primarily western and southwestern in distribution and therefore are not common in the Great Lakes region. They are readily recognized by the elongate body, very large size (40–50 mm), and long antennae (nearly as long as the body) that are clubbed at the tip. The wings are clear and membranous. The larvae do not dig pits but frequently cover themselves with dust and dirt as they lie in wait for passing prey. A single species, the four-spotted owlfly, Ululodes quadrimaculatus (Say), has been recorded from the region (Indiana, Wisconsin, and New York).

Identification Resources. For identification keys to the Megaloptera and Neuroptera of Minnesota, see Parfin 1952; for keys to the Planipennia of Wisconsin, see Throne 1971a, 1971b, and 1972. Sources for identification keys and information for other families, when available, are mentioned under each respective family.

Beetles and Weevils
(Order Coleoptera)

Introduction. This order is the largest in the entire animal kingdom. Some of the individual families contain more members than whole orders in other insect and animal groups. While many families of beetles are similar in appearance, the immense diversity of beetle species, a reflection of their adaptations to all types of environments and habitats, is incredible. Needless to say, the challenges of beetle identification can be considerable at times.

Description. Despite their variation, most beetles share a few common characteristics. The forewings are generally modified

Quick Guide to Identification

Beetles and Weevils
(Order Coleoptera)

Aquatic

Diagnostic Characteristics	Common Name(s)	Scientific Family Name
Small; oval and tapered body	crawling water beetles	Haliplidae
Small; curved spine on front tibia	burrowing water beetles	Noteridae
Two pair of compound eyes; large forelegs	whirligig beetles	Gyrinidae
Antennae long and slender	predaceous diving beetles	Dytiscidae
Antennae short; palps long	water scavenger beetles	Hydrophilidae

Terrestrial (Tip of Abdomen Exposed)

Diagnostic Characteristics	Common Name(s)	Scientific Family Name
Small; rounded or oval body	clown beetles	Histeridae
Body elongate and narrow	rove beetles	Staphylinidae
More or less elongate-oval; clubbed antennae	sap beetles	Nitidulidae
Soft-bodied; head wider than pronotum	blister beetles	Meloidae
Large; body wedge shaped or rounded; often marked with orange or red	carrion beetles	Silphidae
Short broad snout; clubbed antennae	seed weevils	Bruchidae

Terrestrial (Soft, Flexible Elytra)

Diagnostic Characteristics	Common Name(s)	Scientific Family Name
Head hidden from above; bioluminescent	fireflies	Lampyridae
Head hidden; netlike elytra	net-winged beetles	Lycidae
Head visible; antennae long	soldier beetles	Cantharidae
Head visible; antennae short; wedge-shaped	soft-winged flower beetles	Melyridae
Head wider than pronotum; tarsi 5-5-4	blister beetles	Meloidae
Fire-colored; pectinate antennae	fire-colored beetles	Pyrochroidae

Terrestrial (Clubbed Antennae)

Diagnostic Characteristics	Common Name(s)	Scientific Family Name
Prominent snout	weevils	Curculionidae
Short broad snout; tip of abdomen exposed	seed weevils	Bruchidae
Short broad snout; abdomen concealed	fungus weevils	Anthribidae
Hind tarsi four-segmented; humpbacked shape	tumbling flower beetles	Mordellidae
Hind tarsi four-segmented; notched eyes; 11-segmented antennae	darkling beetles	Tenebrionidae
All tarsi appear three-segmented; more or less rounded, spotted	ladybird beetles	Coccinellidae
All tarsi five-segmented; body hairy	checkered beetles	Cleridae
Antennae "elbowed"; large jaws	stag beetles	Lucanidae
Lamellate antenna; "peg" on head	bess beetles	Passalidae
Lamellate antennae; robust body	scarab beetles	Scarabaeidae
Body with hairs or scales	dermestid beetles	Dermestidae
Head partially covered; pronotum with two longitudinal grooves	handsome fungus beetles	Endomychidae
Head concealed; tip of antennae enlarged	deathwatch beetles	Anobiidae
Head hidden; antennae club paddlelike; tip of elytra often concave	bark beetles	Scolytidae
Head hidden; teeth on pronotum	horned powder-post beetles	Bostrichidae
Head visible; parallel-sided; small	powder-post beetles	Lyctidae
Elongate-oval; three-segmented club; often with red or orange	pleasing fungus beetles	Erotylidae
Broadly rounded; shiny	shining flower beetles	Phalacridae

Terrestrial (Serrate or Pectinate Antennae)

Diagnostic Characteristics	Common Name(s)	Scientific Family Name
Oval, wedge-shaped; pubescent	waterpenny beetles	Psephenidae
Bullet-shaped body; metallic sheen	metallic wood-boring beetles	Buprestidae
Body elongate; clicking mechanism	click beetles	Elateridae
Fire-colored; pectinate antennae	fire-colored beetles	Pyrochroidae

Terrestrial (Nonclubbed Antennae)

Diagnostic Characteristics	Common Name(s)	Scientific Family Name
With prominent, curved snout	weevils	Curculionidae
With prominent, straight snout	straight-snouted weevils	Brentidae
Small to large; antennae on side of face between mandible and eye	ground beetles	Carabidae
Medium to large; antennae above base of mandible	tiger beetles	Cicindelidae
Hind tarsi four-segmented; humpbacked shape	tumbling flower beetles	Mordellidae
Hind tarsi four-segmented; notched eyes; eleven-segmented antennae	darkling beetles	Tenebrionidae
Bullet-shaped body; metallic sheen	metallic wood-boring beetles	Buprestidae
Body elongate; clicking mechanism	click beetles	Elateridae
Bilobed tarsus; antennae long	long-horned beetles	Cerambycidae
Body spiderlike; long legs	spider beetles	Ptinidae
Body flattened, narrow	flat bark beetles	Cucujidae
Small and convex (pill-like)	pill beetles	Byrrhidae
Small; grooved or wrinkled pronotum	wrinkled bark beetles	Rhysodidae
Small; elytra with squared punctures	reticulated beetles	Cupedidae
Small; head wider than pronotum, which is often prolonged and hornlike	antlike flower beetles	Anthicidae
Small; spined forelegs; hairy	variegated mud-loving beetles	Heteroceridae
Long-legged; long claws; elongate-oval	riffle beetles	Elmidae
Tarsal claws toothed; antennae long	comb-clawed bark beetles	Alleculidae
Bilobed tarsus; antennae short	leaf beetles	Chrysomelidae

into hardened, armorlike wing covers, or elytra. The elytra meet in a straight line down the back. Beetles have chewing mouthparts as adults and larvae. The mandibles of some predatory aquatic beetle larvae are hollow, which enables them to suck out the body fluids of their prey.

The antennae are generally well developed and conspicuous, though variable (threadlike, beadlike, comblike, featherlike, or clubbed). The legs may be modified for running, digging, climbing, swimming, or grasping. Body shape is similarly variable, from round to oval to elongate, and may be flattened, cylindrical, or even hemispherical. While a large number of species are a basic brown or black, many species are boldly striped, spotted (with red, yellow, or orange), or brilliantly colored (in metallic and iridescent blues, greens, purple, or bronze). The surface of the body, including the elytra, may be smooth, grooved, or irregularly roughened. Beetles range in size from minute (less than 2 mm) to very large (greater than 20 mm).

Life Cycle. This order is the largest to undergo complete metamorphosis, with separate egg, larva, pupa, and adult stages. Overwintering may occur in any stage of development but most commonly takes place during the adult or egg stage. The larva (often called a grub) passes through a series of four to six molts and then develops into a pupa. The adult emerges from the pupa in a week or so, unless the pupa is the overwintering stage. The complete life cycle of beetles may be as short as several months for some species and as long as nine years for others.

Habits. The habits of beetles are as variable as beetles themselves. They occupy all types of aquatic and terrestrial habitats. Every ecological niche and microhabitat has at least one beetle occupant. They are predators, ectoparasites, endoparasites, and plant feeders. The hardened elytra provide beetles with a remarkable degree of protection. Beetles also have a second pair of wings that fold up beneath the elytra. These hind wings are used for flying, and this mobility is part of the beetle success story. Since the elytra are not used for flying, they must be held up and out of the way for movement of the hind wings. Arnett (1985) describes the beetle's mobility as follows:

> Besides this feature [the wings], most species have developed the ability to crawl into almost all known habitats. Actually, the wings are used much less often than those of most other flying insects. Although their primary function is to provide rapid transit from place to place as with other species, they have the added feature of combining distance with powers of penetration. Wingless species cannot move rapidly over great distances; other winged insects are hampered by the delicate nature of their wings. One might compare the airplane and automobile; combine the two and call the results a beetle.

Some beetles can make sounds by rubbing various parts of their bodies together or by expelling air from the spiracles. Others communicate with bioluminescence, especially the fireflies (family Lampyridae) and glowworms (Phengodidae).

Habitats. Most species of beetles are associated with plants, feeding on the roots, stems, twigs, flowers, or fruits. Some are scavengers and decomposers and are found in soil, leaf litter, animal excrement, carrion, fungi, or rotting fruit. Some inhabit the nests of birds and other animals; some are common in the dwellings of humans. Others are predators and parasites and are found searching for their prey or hosts. Some beetles are aquatic and semiaquatic and are found in and around water.

Ecological and Economical Status. Most beetles are beneficial as scavengers, decomposers, pollinators, predators, and parasites. As the largest insect group, they are an important component of the aquatic and terrestrial food chains. A few species are pests, consuming valuable plants (crops, ornamentals, and forest trees) and infesting stored products.

Distribution and Faunistics. Many species of beetles, in more than 100 families, inhabit the Great Lakes region. Any area in the region is likely to have more than a thousand species. For example, Pettit 1869, 1870, 1871, and 1872 reports 1,143 species from Grimsby; Harrington 1884b reports 1,003 from Ottawa; and Brimley 1930 lists 489 species collected from the Rainey River District—all in Ontario. Similarly, a total of 1,424 beetle species are listed for Buffalo, New York (Zesch and Reineche 1881). A total of 886 species of beetles are reported from Whitefish Point, Michigan, from just two seasons of collecting (Andrews 1923). Wickham 1897 reports 691 species taken in six weeks of collecting in Bayfield, Wisconsin; and Adams 1909 reports 206 species from Isle Royale, Michigan, collected over a similarly short time period. Andrews 1916 reports collecting 623 species of beetles from Charity Island (650 acres) in a nine-day period in 1910. A total of 580 species were recorded from Charlevoix County, (including the Beaver Islands) in Michigan during the early 1920s (Hatch 1924).

As one moves further south in the region, the number of species increases significantly (almost doubles). For example, Blatchley 1910 reports 2,535 species from Indiana, and Downie 1956 and 1958 reports an additional 143 species not listed by Blatchley. Klages 1901 lists 2,500 for southwestern Pennsylvania (Pittsburgh area), and Dury 1879b, 1884, 1902, 1906, and 1909 list 2,290 species of beetles for the Cincinnati area in Ohio. Many of these published lists are quite old and represent only a portion of the actual beetle fauna of those respective areas. There are many prominent families and thousands of common species of beetles and weevils in the Great Lakes region.

Reticulated Beetles
(Family Cupedidae)

These small beetles (7–11 mm) are elongate, slightly flattened, and covered with broad "scales." The elytra have rows of closely spaced, square punctures. Only one species of this small beetle family is common in the Great Lakes region.

The mottled cupes, *Cupes concolor* (Westwood), is known from Indiana, Michigan, New York, and Pennsylvania.

Wrinkled Bark Beetles
(Family Rhysodidae)

These small beetles (5–8 mm) are elongate, reddish brown, and narrow bodied, with beaded antennae. The head and pronotum are deeply grooved. They live under the bark of hardwood trees (ash, beech, or elm) and pine. Two species likely to be encountered are *Omoglymmius americanus* (Laporte), which have deeply punctured elytra, and *Clinidium sculptile* Newman, which have longitudinally grooved (striate) elytra.

Tiger Beetles
(Family Cicindelidae)

These beetles are closely allied to the ground beetles and are often included as a subfamily of that group. The brightly patterned adult tiger beetles are active mostly on warm, sunny days. They inhabit a variety of habitats, including beaches, sand dunes, sand pits, trails, woodland openings, and barren hillsides. The larvae construct special burrows in the ground and ambush unsuspecting prey that passes within reach. Adult tiger beetles are often wary and difficult to catch.

Fig. 108. The bronzed tiger beetle, *Cicindela repanda* Dejean. (From Kansas State University Extension 1962.)

About two dozen tiger beetle species inhabit the Great Lakes region. One of the most common is the bronzed tiger beetle, *Cicindela repanda* Dejean, which is bronzy brown with white markings on the elytra and a metallic green or blue green underside. It can be found in a variety of sandy habitats, from lake beaches to stream banks to sand pits. The beach tiger beetle, *C. hirticollis* Say, is similar, but it is slightly larger and more robust, has the underside of the pronotum covered in white hairs, and inhabits clean, sandy beaches of the Great Lakes. The six-spotted tiger beetle, *C. sexguttata* Fabricius, is brilliant green or blue green with four to eight white spots (some individuals may even lack spots) and inhabits woodland trails, barren hillsides, and sparsely vegetated fields. The ghost tiger beetle, *C. lepida* Dejean, occurs in areas of bleached white sand with sparse vegetation. The white elytral markings are greatly expanded, leaving only small areas of pale brown. Against the white sand this camouflage works so well that these beetles cannot be seen until they move. Fortunately, they fly to lights at night and can be easily captured by hand. The punctured tiger beetle, *C. punctulata* Olivier, is smaller than the preceding species and is dark gray or blackish with a purplish or greenish tinge and a row of green punctures on each elytron. The white elytral markings are thin or virtually invisible.

The most common tiger beetle on trails and similar habitats in boreal forests is the long-lipped tiger beetle, *C. longilabris* Say. It is dark slate gray with sparse white markings on the elytra. The upper lip (labrum) is long and white. The noble tiger beetle, *C. formosa generosa* Dejean, is the largest beetle of the genus *Cicindela* found in the Great Lakes region. It has a bronzy brown or greenish brown dorsal surface with very broad white markings on the elytra and a greenish copper underside. It inhabits sand "blowouts"—areas of loose sand and sparse vegetation. A single representative of the big-headed tiger beetles, the Virginia big-headed tiger beetle, *Megacephala virginica* (L.), occurs in the southern parts of the region (Ohio, Indiana, and Illinois). It is dark green with a yellowish crescent-shaped marking near the tip of each elytron. Big-headed tiger beetles are active at night and hide beneath logs, stones, and debris during the day. They are flightless and can be collected at night around lights.

For further information on identification and life histories of tiger beetles of Indiana, see Knisley et al. 1987; for Michigan, see Graves 1963; for Ontario, see Graves 1965; for Ohio, see Graves and Brzoska 1991; for Minnesota, see Horn 1928; for Canada, see Wallis 1961.

Ground Beetles
(Family Carabidae)

These beetles are variable in form, but most are more or less broadly elongate and at least partly black (some are brilliantly colored). Antennae are threadlike and attached to the side of the head between the eye and the mandible. The head is generally much narrower than the pronotum (except in the mud-loving marsh ground beetles, *Elaphrus* species). The legs are typically long and the second leg segment (trochanter) of the hind legs is usually large and kidney shaped. The elytra are generally grooved longitudinally (or with rows of punctures). Most species are omnivorous, feeding on both animal and plant material. Some are strictly predaceous, feeding on insects, slugs, and snails, while others feed exclusively on seeds of grasses and herbaceous plants.

The ground beetles are a large group. There are at least 450 species found in Michigan alone (Dunn, unpublished data). Schrock 1985 reports a total of 465 species from Indiana.

The American round sand beetle, *Omophron americanum* Dejean, inhabits sandy banks of streams, ponds, and lakes. These oval, dome-shaped beetles are active at night. The body

Fig. 109. The northern fiery hunter, *Calosoma frigidum* Kirby. (From Lugger 1897.)

Fig. 110. The slender seedcorn beetle, *Clivina impressifrons* LeConte. (From Kansas State University Extension 1962.)

Fig. 111. The burrowing
ground beetle, *Geopinus
incrassatus Dejean.* (From
Kansas State University
Extension 1962.)

Fig. 112. The common
stenolophus beetle,
Stenolophus comma
Fabricius. (From Kansas
State University Extension
1962.)

Fig. 113. A ground beetle
of the genus *Harpalus.*
(From Kansas State
University Extension
1962.)

and elytra are light brown with irregular dark cross bands. The bombardier beetles, *Brachinus* species, have orange forebodies (head and pronotum) and dark bluish black elytra that are truncate at the tip, exposing a small part of the abdomen. The body and elytra are covered with pubescence. These ground beetles are capable of spraying an odorous, hot liquid at would-be predators. Picking up one of these beetles results in an audible pop, followed by a puff of smoke and a warm sensation on the fingertips (which are temporarily stained dark brown).

The caterpillar hunter, *Calosoma scrutator* Fabricius, is a large (25–36 mm) shiny green or blue green beetle that loves to climb trees and shrubs in search of caterpillars. The margin of the shiny black head and pronotum is gold, green, or red, and the margins of the elytra are generally purple. The fiery hunter, *C. calidum* (Fabricius), is also large (20–27 mm) but is black with greenish elytral margins and a row of red punctures on each elytron. The mud-loving ground beetle, *Elaphrus ruscarius* Say, looks like a tiny tiger beetle (7–8 mm), because the head is wider than the pronotum. It is brassy greenish gray with shiny, mirrored punctures on the elytra. These beetles inhabit the muddy banks of streams, ponds, and bogs. The slender seedcorn beetle, *Clivina impressifrons* LeConte, is small (6–7 mm long), elongate and parallel sided, and uniformly reddish brown in color. The body has a narrow, stalked constriction between the pronotum and the elytra.

The minute ground beetles, *Bembidion* species, are some of the smallest beetles to be found anywhere. They are usually quite common, but because of their small size and difficult identification, they are generally passed over by most collectors. Many of the species are brassy with irregular cross bands or shiny black with spots. The green ground beetle, *Chlaenius sericeus* Forster, is medium sized (13–17 mm), elongate-oval, and uniformly green above and black below. The body is covered with dense pubescence. The burrowing ground beetle, *Geopinus incrassatus* Dejean, is easily recognized by its pale coloration and stout body. It is cream or light tan and patterned with some areas of slightly darker brown. Typically these beetles burrow in sandy soils and only come to the surface at night or on cloudy days.

The common stenolophus beetle, *Stenolophus comma* Fabricius, is small (5–8 mm long) and reddish brown, although the center of the pronotum is paler and the center of the elytra is dark brown or blackish. It is one of the most commonly encountered species of ground beetles at lights. The seedcorn beetle, *S. lecontei* Chaudoir, is very similar, but the dark marking on the center of the elytra extends all the way to the scutellum. The Pennsylvania dingy ground beetle, *Harpalus pennsylvanicus* De Geer, is very common in grassy areas and fields, where it hides beneath logs, stones, and debris during the day. It readily flies to lights. It is medium sized (15–17 mm) and uniformly black above and orangish brown below (including the legs). The Pennsylvania long-necked ground beetle, *Colliuris pennsylvanicus* (L.), is easily recognized by its shape and coloration. The head is slightly elongate and diamond shaped and the pronotum is slender and cylindrical. The reddish orange elytra are marked with several black spots on the rear half. It lives on vegetation in wet areas and is most commonly seen at lights and in beach drift. The large foliage ground beetle, *Lebia grandis* Hentz, has a rusty orange forebody (head and pronotum) and shiny dark blue elytra, which are short, leaving the tip of the abdomen exposed. This small beetle (8–10 mm) feeds on the eggs of other insects and is especially important in the control of the Colorado potato beetle (family Chrysomelidae). The false bombardier beetle, *Galerita janus* Fabricius, is large (up to 25 mm) and distinctly orange-and-black. Its larger size and bicolored pattern (black head, orange pronotum, and black elytra) are distinctive, and it can only be confused with less common members of the same genus. It inhabits open woodlands.

Further information on the identification and life histories of ground beetles of Canada and the adjacent United States can be found in Lindroth 1961.

Crawling Water Beetles
(Family Haliplidae)

These minute aquatic beetles are seldom seen or collected. They are broadly oval, with a rounded head and pointed elytra. The basal segments (coxae) of the hind legs are enlarged and cover much of the abdomen. The space beneath these enlarged coxae is used to store oxygen for underwater use. These beetles are brownish or reddish yellow with dark blotches on the elytra. Although their legs are fringed with hairs for swimming, they prefer to crawl along the bottom in search of food. The adults occur in weedy areas of ponds and streams and feed on algae.

The immaculate crawling water beetle, *Haliplus immaculatus* Harris, is reddish yellow with black spots. There are two impressions on the rear margin of the pronotum and deep grooves on the underside of the prothorax. Hungerford's crawling water beetle, *Brychius hungerfordi* Spangler, was described from specimens collected in 1952 from the Maple River in northern lower Michigan (the Douglas Lake area). This tiny water beetle is thought to be rare (it is listed as an endangered species in Michigan), but its actual status is unknown since few people know how to collect these beetles. It was collected from among plant roots under approximately two feet of water. It is

Fig. 114. The Pennsylvania long-necked ground beetle, *Colliuris pennsylvanicus* (L.). (From Kansas State University Extension 1962.)

Fig. 115. The large foliage ground beetle, *Lebia grandis* Hentz. (From Kansas State University Extension 1962.)

Fig. 116. The body and markings of *Brychius hungerfordi* Spangler, an endangered species of crawling water beetle from Michigan (legs and antennae omitted). (From Spangler 1954.)

light yellowish tan with rows of punctures on the elytra and dark longitudinal markings. The pronotum has two dark spots near the center of the front edge and there are two impressed folds parallel to the sides. The twelve-spotted crawling water beetle, *Peltodytes duodecempunctatus* Say, is dull yellow with six black spots on each wing cover and two black dots on the pronotum near the base.

For identification keys to the crawling water beetles of Wisconsin, see Hilsenhoff and Brigham 1978.

Predaceous Diving Beetles
(Family Dytiscidae)

These oval, streamlined beetles inhabit all types of aquatic habitats. They prefer quiet waters with lots of vegetation and prey. They fly from habitat to habitat and sometimes mistake the shiny surfaces of cars as water. They also sometimes appear at lights. The body is convex above and below, and most species are brownish or blackish with yellowish markings on the pronotum and/or elytra. There are spiracles located at the tip of the abdomen, and oxygen is stored beneath the wing covers for use underwater. The hind legs are long, flattened, and fringed with hairs. The antennae are threadlike and more or less visible. The larvae, frequently referred to as water tigers, are also predaceous. Their large, curved mandibles are hollowed and used to suck the body juices out of their prey. Hilsenhoff 1992 reports 148 species of predaceous diving beetles from Wisconsin.

The fimbriolate predaceous diving beetle, *Cybister fimbriolatus* (Say), is large (30–33 mm) and dark brown or blackish with a greenish tinge. The front of the head, the sides of the pronotum, and the sides of the elytra are margined in yellow. The wing covers have two rows of punctures and the body is widest beyond the middle. The females have five longitudinal furrows on the pronotum. The giant predaceous diving beetle, *Dytiscus harrisii* Kirby, is our largest dytiscid beetle (greater than 38 mm). It is black with the front, sides, and rear of the pronotum and the sides of the elytra yellowish. The body is widest at the middle. The wing covers are unpunctured. The undulated predaceous diving beetle, *Hydroporus undulatus* Say, is small (4–5 mm), very convex, and pale yellowish to yellowish brown. The front and rear edges of the pronotum are black and the elytra is black with irregular yellowish patches. The disintegrated predaceous diving beetle, *Agabus disintegratus* (Crotch), is slightly larger than the *Hydroporus undulatus* (7–9 mm). The forebody is reddish yellow and the pronotum has two reddish cross bands.

The elytra are brownish yellow with three or four longitudinal stripes on each wing cover.

For an identification key to the predaceous diving beetles of Wisconsin, see Hilsenhoff 1992, 1993a, 1993b, and 1993c.

Burrowing Water Beetles
(Family Noteridae)

This small family of water beetles (Hilsenhoff 1992 reports two species from Wisconsin) are similar in appearance to predaceous diving beetles, except the front legs bear a curved or hooked spine. All the tarsal segments are parallel sided. Most burrowing water beetles are uniformly reddish brown and the scutellum of the thorax is hidden. They inhabit weedy ponds and lakes and the adults are free-swimming beetles (the larvae burrow in mud).

For an identification key to the burrowing water beetles of Wisconsin, see Hilsenhoff 1992.

Whirligig Beetles
(Family Gyrinidae)

Whirligig beetles are oval, flattened, and shiny black or dark metallic green. Adult whirligigs are easily recognized by their peculiar compound eyes (divided into two halves, one above the waterline and one below the waterline) and by their habit of swimming about in a circular motion on the surface of the water. The front legs are most prominent, with the middle and hind legs noticeably shorter and flatter. The antennae are short and clubbed. Adults are able to swim under water, and they also fly to lights at night. Both adults and larvae are predaceous.

The common whirligig beetle, *Dineutus americanus* (Fabricius), is shiny black with a bronze sheen. The edge of the elytra, near the tip, are curved inward (emarginate). The discolored whirligig beetle, *D. discolor* Aube, is shiny black above and brown or reddish orange below. The boreal whirligig, *Gyrinis borealis* Aube, is not restricted to the boreal regions of North America. It is distributed from Maine to Virginia and west to Michigan and Indiana. It is small (6.5–7.5 mm) and shiny black with bronzed elytral margins.

For a key to the whirligig beetles of Wisconsin, see Hilsenhoff 1990; for a key to Minnesota whirligigs, see Ferkinhoff and Gunderson 1983.

Fig. 117. The common whirligig beetle, *Dineutus americanus* (Fabricius). (From Kansas State University Extension 1962.)

Water Scavenger Beetles
(Family Hydrophilidae)

Most water scavenger beetles are uniformly black or brown, but
some are patterned. In general appearance they resemble predaceous diving beetles, but they have long, four-segmented maxillary palpi and short, clubbed antennae (which are often concealed from above). The palpi are as long as or longer than the
antennae, and at first glance it sometimes looks like these
beetles have two pair of antennae. A few of the large, black
species have a characteristic keel or spine on the underside of
the thorax between the legs. The adults are true scavengers,
feeding on decaying plant and animal matter, but the larvae are
predaceous. Many species fly to lights at night.

The giant water scavenger beetle, *Hydrophilus triangularis*
Say, is large (34–37 mm) and shiny black with an olive green
reflection. There is a sternal keel present between the legs on
the underside of the thorax. The obtuse water scavenger beetle,
Hydrochara obtusatus Say, is medium sized (13–16 mm) and
shiny black with a greenish reflection. The abdomen is marked
with yellowish or reddish triangular marks along the outer
edges. A sternal keel is present between the legs. The striated
water scavenger beetle, *Berosus striatus* (Say), is small (4–5 mm)
and greenish yellow with a black head. There are two black
stripes and spots on the pronotum and scattered indistinct spots
on the elytra. The elytra are grooved longitudinally, with the
grooves finely punctured. The ridges between the grooves are
flattened and coarsely punctured. The dung-feeding water scavenger beetle, *Sphaeridium scarabaeoides* (L.), is the "black
sheep" of the water scavenger beetle family. It is terrestrial and
feeds on cow manure. It is small (5–7 mm), broadly oval, and
blackish with two large red spots near base of the elytra. The
tips of the elytra are yellowish.

Fig. 118. The giant water
scavenger beetle,
Hydrophilus triangularis
Say. (From Kansas State
University Extension
1962.)

Sexton, Burying, and Carrion Beetles
(Family Silphidae)

These medium- to large-sized beetles are usually black and
boldly marked with splotches of red, orange, or yellow. The
body may be broadly oval or wedge shaped (widening toward
the rear) and the tip of the abdomen is usually exposed beyond
the tip of the shortened elytra. The antennae are clubbed. These
beetles are associated with carrion and dung and sometimes
with decay fungi. Some species fly to lights.

The American carrion beetle, *Silpha americana* L., is
broadly oval with a large, mostly yellow pronotum (the center is
black). When in flight it makes a loud buzzing noise and looks

and sounds very much like a large bumble bee. They are primarily associated with rotting vegetation and decay fungi. The round-necked burying beetle, *Necrophorus orbicollis* Say, is large (20–25 mm) and shiny black with four large orange red or yellow markings on the elytra. The hind angles of the pronotum are broadly rounded. This species is generally associated with dead animals, including small mammals, snakes, fish, and frogs.

The American burying beetle, *N. americanus* (Olivier), once common throughout the forests of eastern North America (including the Great Lakes region), has now all but vanished (and is accordingly listed as an endangered insect in the United States). It is now known from only a few widely separated sites in Rhode Island, Arkansas, Nebraska, and Oklahoma. The most plausible hypothesis so far put forth to explain the disappearance of this beetle is that its primary food, the fallen nestlings of the passenger pigeon, are no longer available. With its host extinct the beetle may soon follow. The hairy burying beetle, *N. tomentosus* Weber, is medium sized (15–20 mm) and shiny black. The elytra are short with two orange red bands. The pronotum is covered with dense, yellow hairs. The Surinam burying beetle, *Necrodes surinamensis* Fabricius, is similar in shape and color to the other burying beetles, but the elytral bands are reduced to three orangish or reddish dots on each wing cover, and the tip of the elytra is rounded, not truncate.

For further information on the life histories of the carrion and burying beetles in Minnesota, see Hatch 1927; for Canada, see Anderson 1982. For identification keys to the carrion and burying beetles of Canada and the adjacent United States, see Anderson and Peck 1985.

Rove Beetles
(Family Staphylinidae)

The rove beetle family is one of the largest groups of beetles, but all the species are easily recognized by the narrow, elongate body and the short elytra that leaves virtually all the abdomen exposed. In addition most species have the habit of running with the tip of the abdomen curled up and over the body in a "threatening" manner. Rove beetles are both predators and scavengers and are found in association with other insects, carrion, dung, rotting plant materials, and the nests of other animals. Identification is difficult, but a few species are easily recognized.

The blue-winged rove beetle, *Philonthus cyanipennis* (Fabricius), is medium sized (12–15 mm) and shiny black with metallic blue, green, or purple elytra. The elytra are densely punctured and the abdomen only sparsely so. The brown and gold rove beetle, *Ontholestes cingulatus* (Gravenhorst), is small

Fig. 119. A burying beetle of the genus *Necrophorus*. (From Kansas State University Extension 1962.)

Fig. 120. The Surinam burying beetle, *Necrodes surinamensis* Fabricius. (From Kansas State University Extension 1962.)

Fig. 121. A rove beetle of the genus *Philonthus*. (From USDA 1952.)

(4–6 mm) and brownish with golden pubescence. It is found on carrion in wooded areas. The spotted rove beetle, *Staphylinus maculosus* Gravenhorst, is medium sized (17–21 mm) and velvety yellowish brown. The head is marked with reddish brown and the abdomen is spotted in brown. The body is covered with dense hairs and the elytra are densely punctured. The hairy rove beetle, *S. maxillosus* (L.), is medium sized (10–21 mm) and shiny black with a band of yellowish gray or yellowish brown hairs on the elytra. The base of the abdomen is also covered with yellowish hairs. Both *S. maculosus* and *S. maxillosus* are associated with carrion.

Clown, or Hister, Beetles
(Family Histeridae)

The clown, or hister, beetles are recognized by their rounded or broadly oval shape, concealed head, and one or two exposed abdominal segments. They are generally minute to small (1–10 mm) and shiny black (occasionally greenish or bronze) with clubbed antennae. The upper leg segments (femora) are generally enlarged and flattened. These beetles are usually associated with carrion, dung, and decaying plant materials (as scavengers), but some species are predaceous.

The short hister beetle, *Hister abbreviatus* Fabricius, is small and black with two parallel grooves near the side margins of the pronotum. The elytra are deeply grooved. The Pennsylvania hister beetle, *Saprinus pennsylvanicus* (Paykull), is easily recognized by its metallic green or bronze green coloration. The elytra have coarse punctures near the tip. The Carolina hister beetle, *Platysoma carolinum* (Paykull), is also small (3–4 mm) and highly flattened. It is shiny black with five shallow grooves on the elytra. It is found under bark.

Fig. 122. A clown (hister) beetle of the genus *Saprinus*. (From Kansas State University Extension 1962.)

Stag Beetles
(Family Lucanidae)

Stag beetles are another easily recognized family. The males have large, antlerlike mandibles and both sexes have asymmetrically clubbed antennae that are "elbowed" (lamellate). The segments of the club are fixed in place and cannot be moved. Stag beetles are robust, more or less elongate beetles that reach lengths of up to 40 millimeters. They are generally encountered in wooded areas, especially those with sandy soils. The larvae often feed inside rotting logs.

The pinching beetle, *Pseudolucanus capreolus* (L.), is large to very large (22–35 mm) and shiny, dark reddish brown. The mandibles of the males end in a single point. The pinching bug ranges north into Michigan and southeastern Canada (Ontario). The elephant or American stag beetle, *Lucanus elephas* Fabricius, is also very large (28–40 mm) and shiny reddish brown. The mandibles of the male are two pronged at the tip. The elephant stag is more southern in distribution but is known from the southern portions of Ohio, Indiana, and Illinois. The antelope stag beetle, *Dorcus parallelus* Say, is medium sized (15–26 mm) and dark brown to blackish. The mandibles of the male are less pronounced than most other species, and there is a prominent tooth on the inner surface of the mandible near the center.

Bess Beetles
(Family Passalidae)

The bess or peg beetle, *Popilius disjunctus* (Illiger), is distinctively shaped—elongate, parallel sided, and with an hourglass waist between the pronotum and elytra. The bess beetle is large (30–35 mm) and shiny black with an unusual saddlehorn-shaped "peg" on the front of the head. These beetles are gregarious and adults and larvae can be found together inside rotting logs and stumps of oak.

Scarab, or Lamellicorn, Beetles
(Family Scarabaeidae)

Despite the very large size of this beetle family, a single characteristic instantly separates them from all other beetles. The antenna has a lopsided club composed of three to seven movable plates (lamellae) that can be manipulated into a tight ball or a comb-toothed arrangement. Aside from this antennal characteristic, the scarab beetles are extremely variable in size, shape, and coloration. They range from 4 to more than 25 millimeters in length. Most are robust and heavy bodied, but some are not. Many are plain black, brown, or gray, but some are quite colorful (even metallic). The C-shaped larvae (called white grubs) of many species feed on the roots of a wide range of plants, so this family has a large number of economically important pest species in it. Other species live in dung, carrion, animal nests, rotting wood and vegetation, and ant nests. The adults of many species are leaf feeders and cause damage to crops, ornamental plants, and forest trees.

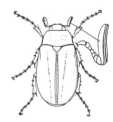

Fig. 123. A Junebug (family Scarabaeidae) showing the characteristic lamellate antennae. (From USDA 1952.)

    The eastern tumblebug, *Canthon pilularius* (L.), has the delightful habit of forming animal dung into a neat little ball that is rolled to a suitable site and buried—that is, after an egg

Fig. 124. The lesser dung
beetle, *Aphodius fimetarius* (L.). (From Kansas
State University Extension
1962.)

Fig. 125. The rose chafer,
*Macrodactylus sub-
spinosus* (Fabricius).
(From Felt 1900.)

Fig. 126. The spotted
grape beetle, *Pelidnota
punctata* L. (From Felt
1900.)

has been deposited in the dung ball. These medium-sized dung beetles (11–19 mm) are uniformly black with a greenish, bluish, or bronze sheen. The legs are slightly curved and the upper long segment of the front leg (femora) is enlarged for digging and pushing. The small black dung beetle, *Copris minutus* (Drury), is medium sized (8–12 mm) and shiny black. Both the males and females have a well-developed horn on the head. The splendid dung beetle, *Phaenaeus vindex* MacLeay, is medium sized (14–22 mm) and brilliantly colored. The head is bronze, the pronotum is coppery bronze, and the elytra are green, sometimes with blue reflections. The males are horned; the females have a short pimplelike bump (tubercle). The splendid earth-boring dung beetle, *Geotrupes splendidus* (Fabricius), is broadly oval and shiny metallic green, bronze, or purple. The front legs are broad and toothed for excavating in the soil. The top of the head is roughened and there is a short horn in the middle.

The hecate dung beetle, *Onthophagus hecate* Panzer, is small (5–9 mm) and black with bronze or purplish reflections. The body is covered with short gray hairs and the pronotum is bumpy and projects forward in a scoop-shaped manner. This species is usually associated with animal dung. The lesser dung beetle, *Aphodius fimetarius* (L.), is one of the smallest dung beetles of the Great Lakes region. It is 6–8 millimeters long with a black forebody (reddish spots occur on the front corners of the pronotum) and reddish elytra. It can be found in horse and cow dung, even during the winter months.

The skin beetles, *Trox* species, are recognized by their grayish color and wrinkled and bumpy appearance. They are found on carrion (especially old and dry carrion), feathers, skin, and animal nests.

The iridescent chafer, *Serica sericea* Illiger, is small (8–10 mm), dull black or purplish brown (usually with iridescent reflections), and frequently taken at lights. June (or May) beetles, *Phyllophaga* species, are familiar to most of us because of their summertime attraction to lights at night. Their approach is signified by a distinct buzzing sound that ends in an equally discernible thud when they land on a screen. All species are uniform brown and difficult to distinguish from one another. There may be a dozen or more common species in any given area; for example, 18 species were reported from Wisconsin (Chamberlain et al. 1943) and 21 species were reported from southeastern Michigan (Morofsky 1943). The rose chafer, *Macrodactylus subspinosus* (Fabricius), is a serious pest of ornamental and garden plants. It is yellowish brown and densely covered with yellowish scales or hairs, and it has long, orangish legs.

The goldsmith beetle, *Cotalpa lanigera* L., is large (20–26 mm) and golden yellow or greenish yellow. The color fades to a dull yellow in dead specimens. The underside is densely hairy.

They can be found on the foliage of poplar, willow, and oak. The spotted grape beetle, *Pelidnota punctata* L., is also large (17–25 mm), but it is brown or brownish orange with two black spots on the pronotum and six black spots on the elytra. The underside is dark green or blackish. As indicated by its common name this beetle is associated with grape plants and feeds on the foliage. It also comes to lights at night.

The Japanese beetle, *Popillia japonica* Newman, is a well established introduced species that occurs in the eastern and southern portions of the Great Lakes region. It is a very attractive beetle, with a shiny green forebody, coppery-colored elytra with green margins, and six to eight white tufts of hair on the sides of the abdomen near the tip. The white grubs of this species are root feeders and the adults feed on the foliage of many types of plants; in sufficient numbers these beetles can cause serious damage to plants.

The odor-of-leather beetle, *Osmoderma eremicola* Knoch, is found in rotten wood of deciduous trees. The beetles have a peculiar odor, similar to the aroma of wet leather. They are large (25–30 mm) and shiny brown, with the head depressed and scooped out between the eyes. The bumble flower beetle, *Euphoria inda* (L.), feeds on fruits and vegetables and flies very quickly and evasively. These medium-sized beetles (13–16 mm) have a triangular pronotum that produces a conical appearance to the front of the body. The front of the body is bronze and mottled with yellowish spots. The elytra are brownish yellow and mottled with black spots that form irregular cross bands. The ant nest scarab beetles, *Cremastocheilus* species, are unique among the scarabs. Their larvae develop within the nests of ants, feeding on organic debris. The adults are very hard bodied and have tufts of golden hairs (trichomes) at the hind angles of the pronotum that produce tasty secretions loved by ants.

For an identification key to the *Onthophagus* dung beetles of the United States, including the Great Lakes region, see Howden and Cartwright 1963; for a key to the species of *Ataenius* in Ohio, see Wegner and Niemczyk 1979.

Fig. 127. The bumble flower beetle, *Euphoria inda* (L.). (From Kansas State University Extension 1962.)

Pill Beetles
(Family Byrrhidae)

These unusual, mostly small (5–10 mm) beetles are broadly oval with a convex upper and lower surface. The head is directed downward and cannot be seen from above. When these beetles are disturbed they pull their legs and antennae into special grooves on the underside of the body, forming a compact "pill." Most species are dull brown, gray, or black. Most species live among the roots of grasses in sandy soils. The American pill beetle, *Byrrhus americanus* LeConte, is small (7–9 mm) and black with dense gray pubescence. The elytra are marked with three or four black lines.

Waterpenny Beetles
(Family Psephenidae)

Few members of this small family of beetles are found in the Great Lakes region; for example, only three species are known from Wisconsin (Hilsenhoff and Schmunde 1992). Herrick's waterpenny beetle, *Psephenus herricki* (DeKay), is small (4–6 mm), broadly oval and flattened (widest toward the tip of the elytra), and brownish black or dull black in color. This species is found crawling about stones and vegetation near streams. The larvae are flattened and disklike (like a penny) and cling to stones in swiftly moving water, including the wave-swept shores of the Great Lakes.

For a key to the waterpenny beetles of Wisconsin, see Hilsenhoff and Schmunde 1992.

Variegated Mud-Loving Beetles
(Family Heteroceridae)

These beetles inhabit burrows in the sandy or muddy shorelines of streams and ponds. They are small (6–7 mm), elongate, and parallel sided, with flat mandibles that protrude forward in front of the head. The outer margins of the front and middle tibia are equipped with comblike spines.

The pale variegated mud-loving beetle, *Neoheterocerus pallidus* (Say), is black, though covered with irregular patches of brownish and yellowish hairs (the yellow hairs forming three indistinct cross bands). They can be extremely abundant at lights, especially if there is water nearby.

Riffle Beetles
(Family Elmidae)

Riffle beetles are associated with rapidly running, cold streams. The body of a riffle beetle is more or less elongate and cylindrical. The legs are long and the tarsal claws are large. The antennae are long and slender and often slightly clubbed. Most species are dull gray or black with some lighter markings. They inhabit the riffle area of streams and may be semiaquatic or fully aquatic. This family is relatively small; for example, only 28 species are known from Wisconsin (Hilsenhoff and Schmunde 1992).

The four-spotted riffle beetle, *Stenelmis quadrimaculata* Horn, is small (2.5–3.5 mm) and dark reddish brown to black with four oblong white markings on the elytra. The pronotum is generally pale gray. The body is covered with a white, waxy pubescence.

For a key to the riffle beetles of Wisconsin, see Hilsenhoff and Schmunde 1992.

Metallic Wood-Boring Beetles
(Family Buprestidae)

These slender, somewhat flattened, hard-bodied beetles are generally blue, green, bronze, or red in color with a metallic sheen (especially on the underside). The head is partially hidden beneath the pronotum, giving them a bullet-shaped appearance. The antennae are usually sawlike (sometimes threadlike) and more or less short. The adults are found on flowers, leaves, tree trunks, and stumps and logs in full sunlight. Dead or dying trees and freshly cut wood will attract these beetles; so will a pan of turpentine. The larvae, called flat-headed borers, look like tiny cobras and live in frass-packed galleries under the bark of trees and logs. A few species are leaf miners and gall makers.

The large flat-headed pine heartwood borer, *Chalcophora virginiensis* (Drury), is long (20–38 mm) and shiny black with bronze reflections. The pronotum and elytra are sculptured with irregularly shaped, raised, smooth, shiny areas. The depressed areas between these elevated areas are filled with brassy-colored punctures. The hemlock borer, *Melanophila fulvoguttata* Harris, is small (8–10 mm) and dark grayish or black with six yellowish spots on the elytra. It attacks unhealthy hemlock and fir trees. The flat-headed apple tree borer, *Chrysobothris femorata* (Olivier), is small (7–14 mm) and dark bronze. There are usually two irregular, light-colored cross bands on the elytra. The body

Fig. 128. An adult flat-headed apple tree borer, *Chrysobothris femorata* (Olivier). (From Kansas State University Extension 1962.)

Fig. 129. Two metallic wood-boring beetles of the genus *Agrilus. Left,* the bronze birch borer, *A. anxius* Gory; *right,* the red-necked cane borer, *A. ruficollis* (Fabricius). (From Felt 1900 and Chittenden 1922.)

is coarsely and densely punctured. The bronze birch borer, *Agrilus anxius* Gory, is small (8–10 mm), very slender, and bronze. It is associated with white-barked birches and can be a major pest of landscape birches. The red-necked cane borer, *A. ruficollis* (Fabricius), is similar to the bronze birch borer in size and shape but is slightly smaller and has a distinctive red pronotum. It lives in the canes of raspberries and blackberries, and adults can be found sunning themselves on the uppermost leaves of these plants.

For an identification key to the metallic wood-boring beetles of Michigan, see Wellso et al. 1976.

Click Beetles
(Family Elateridae)

Click beetles are elongate and narrow bodied with a rounded head and pointed elytra. The hind angles of the pronotum are pointed and directed toward the rear. The antennae are sawlike, beadlike, or sometimes threadlike. The pronotum is loosely hinged and the underside of the prosternum has a spine that fits into a socket on the mesosternum. When a click beetle falls on its back the beetle is able to quickly flex its body and pop up into thhe air, the spine and groove mechanism producing a distinct clicking sound. This behavior startles would-be predators and amuses children. Adult click beetles may be found on the ground or on foliage; they are also commonly found under loose bark. The larvae, known as wireworms, are long, cylindrical, and hard bodied. Wireworms live in the soil and feed on the roots of plants, occasionally causing economic damage.

The eyed click beetle, *Alaus oculatus* (L.), is the largest click beetle found in the Great Lakes region, measuring up to 45 millimeters in length. The body is black and mottled with patches of whitish. The pronotum has two large eye spots (black surrounded with white), which are quite distinctive and obvi-

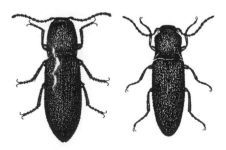

Fig. 130. Two brown click beetles, members of the genus *Melanotus*. (From Knowlton 1951.)

ously serve a protective role. The wheat wireworm, *Agriotes mancus* (Say), is small (7–9 mm) and yellowish brown to dark brown with the corners of the pronotum and the sides of the elytra dull yellow. The pronotum is wider than it is long. The body is covered in short yellow pubescence and the elytral grooves are coarsely punctured. The brown click beetles, *Melanotus* species, are uniformly brown or blackish, with sparse pubescence. Many species, all very similar in appearance, occur in the region.

Firefly Beetles
(Family Lampyridae)

The firefly beetles, also known as lightningbugs, are renowned for their ability to visually communicate with one another by flashing lights. The light is produced in special body organs located near the tip of the abdomen, in the greenish or yellowish areas. The flashing is part of the mate-attracting process and each species has its own unique pattern of flashes. This bioluminescence is especially fascinating since it produces no heat as a by-product. These beetles are generally soft bodied and the large pronotum covers most or all of the head, which cannot be seen from above. The elytra are generally greenish, brownish, or blackish. The antennae are threadlike or occasionally sawlike (serrate). Both the adults and the larvae are predaceous.

Slightly more than two dozen species of fireflies, 24 confirmed and 7 probable, are recorded from Ohio (Marvin 1965).

The common eastern firefly, *Photinus pyralis* L., is common throughout much of the Great Lakes region. It is a dull brown, medium-sized insect (10–14 mm). The pronotum has a central black spot and dull yellow margins. The edges and sutural area of the elytra are also pale. The Pennsylvania or woods firefly, *Photuris pennsylvanicus* (De Geer), is medium sized (11–15 mm) with dark brown elytra marked with yellowish stripes and margins. The head and pronotum are yellowish; the pronotum has a reddish spot in the middle that is divided in two by a black stripe. This insect is the officially designated Pennsylvania state insect.

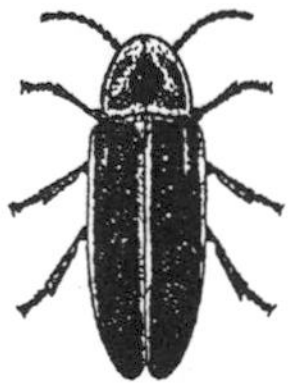

Fig. 131. The common eastern firefly, or lightningbug, *Photinus pyralis* L. (From Kansas State University Extension 1962.)

Soldier Beetles
(Family Cantharidae)

These soft-bodied beetles look somewhat like firefly beetles, but the head is not covered by the pronotum and is plainly visible. Their bodies are generally either dark yellowish with black markings or dark gray, brown, or blackish with yellow or red on the pronotum. They are commonly seen on flowers.

Fig. 132. The
Pennsylvania soldier
beetle, *Chauliognathus
pennsylvanicus* (De
Geer). (From Felt 1900.)

This group is fairly large, with 72 species of soldier beetles found in Ohio, for example (Miskimen 1956).

The Pennsylvania, or goldenrod, soldier beetle, *Chauliognathus pennsylvanicus* (De Geer), is a medium-sized beetle (9–12 mm) that is commonly found on the flowers of goldenrod, Queen Anne's lace, and other composites in the late summer. The body is dull yellow orange with a transverse black spot on the pronotum and an elongate black spot on each wingcover. The underside of the body is blackish with some yellow markings. The two-lined soldier beetle, *Cantharis bilineatus* Say, is small (6–8 mm) with a reddish yellow body. The antennae (except for the base) and the legs are black. The pronotum has two oblique black bands and the elytra are black (except for the pale side margins). The hairy soldier beetle, *Podabrus tomentosus* (Say), is also small (9–12 mm) with a reddish yellow forebody and blackish elytra. The base of the antennae and the legs are also reddish yellow.

Net-Winged Beetles
(Family Lycidae)

These beetles are not always common, but when you see one you will know what it is. The elytra are soft, wedge or fan shaped (widest beyond the middle), and sculptured with ridges and lacelike depressions. The head is concealed beneath the pronotum. The antennae are more or less flattened and usually serrate. Most net-winged beetles are black with reddish or orange markings.

The reticulated net-winged beetle, *Calopteron reticulatum* (Fabricius), is medium sized (13–19 mm) and orange yellow with two broad cross bands on the elytra, one of which includes the tip of the elytra. The pronotum is also black at the center.

Skin and Larder Beetles
(Family Dermestidae)

Fig 133. The larder beetle,
Dermestes lardarius L.
(Courtesy of Michigan
State University
Extension.)

These small- to medium-sized beetles are convex and oval or elongate-oval. The body is covered either with a combination of flattened white, yellow, red, or brown scales, or black hairs. The antennae are clubbed. Most dermestid beetles are scavengers, feeding on carrion, fur, feathers, and hair; but they will also feed on fabrics, leather, museum specimens, and stored food products.

The larder beetle, *Dermestes lardarius* L., is 7–8 millimeters long and black. The front half of the elytra is yellowish tan with three or four black spots. This species is associated with stored foods that are high in animal fat, such as meat, cheese, and pet foods, and with taxidermic specimens. The hide beetle, *D. maculatus* De Geer, is similar, although slightly larger, at 6–10 millimeters, and black with white hairs. The sides of the pronotum and the undersides of the abdomen are clothed in thick, flattened white hairs. This beetle is also fond of animal products and is frequently used to prepare skeletal specimens in the laboratory. The black carpet beetle, *Attagenus megatoma* (Fabricius), is small (3–5 mm), elongate-oval, and dark reddish black. The body is sparsely covered with pubescence.

Carpet beetles are small (3–4 mm), oval, and covered with flattened, multicolored scales. The furniture carpet beetle, *Anthrenus flavipes* LeConte, is covered with a mottled arrangement of white, yellow, orange, and black scales with no suggestion of cross bands. The varied carpet beetle, *A. verbasci* (L.), also has white, yellow, and gray scales, but there are more or less distinct transverse cross bands of white bordered by yellow. A third species, the buffalo carpet beetle, *A. scrophulariae* (L.), is similar, but the orange scales are concentrated along the suture, forming a longitudinal pattern of orange from the scutellum to the tip of the abdomen.

Fig. 134. The varied carpet beetle, *Anthrenus verbasci* (L.). (From Michelbacher and Furman 1951.)

Fig. 135. The buffalo carpet beetle, *Anthrenus scrophulariae* (L.). (From Felt 1900.)

Spider Beetles
(Family Ptinidae)

These beetles are easily recognized: the body is spiderlike, with a narrow head and pronotum, globular elytra, and long legs. The head is concealed and not (or barely) visible from above. Spider beetles are small, rarely larger than 5 millimeters in length. They infest many types of stored products in granaries, warehouses, and homes.

The American spider beetle, *Mezium americanum* (Laporte), is globular with shiny blackish elytra and light-colored hairs on the forebody and legs. The white-marked spider beetle, *Ptinus fur* (L.), is dull brownish with four irregular patches of white scales on the elytra. The body may be oval (females) or elongate-oval (males).

Fig. 136. The American spider beetle, *Mezium americanum* (Laporte). (From USDA 1953.)

Deathwatch and Drugstore Beetles
(Family Anobiidae)

Anobiid beetles are small (1–9 mm), cylindrical, and oval or elongate-oval. An enlarged, hoodlike pronotum covers the head, which cannot be seen from above. The antennae are generally serrate or clubbed, with the last few segments greatly enlarged.

Fig. 137. An adult female white-marked spider beetle, *Ptinus fur* (L.). (From USDA 1953.)

Fig. 138. The drugstore beetle, *Stegobium paniceum* (L.). (From White 1962.)

Fig. 139. The cigarette beetle, *Lasioderma serricorne* (Fabricius). (From White 1962.)

Adults and larvae infest stored products (seeds, cereals, foodstuffs, spices, tobacco, and museum specimens) or wood (twigs, logs, lumber, and furniture). The name deathwatch comes from the ticking noise made by some wood-boring species that sometimes inhabit bedposts. This noise was once believed to be an omen of death.

The anobiid fauna of Ohio includes at least 52 confirmed species, and another 19 species, known from adjacent states, are likely to be found there (White 1962).

Two commonly encountered anobiid beetles are pests of stored products. The drugstore beetle, *Stegobium paniceum* (L.), is a common pest of many types of stored products. It is small, reddish brown, and covered with fine yellowish pubescence. The elytra are finely grooved and punctured and the sides of the pronotum are serrate along the front edge. The cigarette beetle, *Lasioderma serricorne* (Fabricius), is similar, but the elytral grooves are indistinct and the pronotum lacks serrations.

For an identification key to the anobiid beetles of Ohio, see White 1962.

Horned Powder-Post Beetles
(Family Bostrichidae)

These beetles, also known as false powder-post beetles, are small (3–12 mm), cylindrical, and elongate with the head bent downward and hidden by the pronotum. The front of the pronotum is covered with tubercles having the appearance of rasplike teeth. The antennae are short, straight, and clubbed.

One common species is the apple twig borer, *Amphicerus bicaudatus* (Say), which is reddish brown to brownish black. The pronotum is rounded with many tubercles and the tip of the elytra slope downward with two inwardly curved spines.

Powder-Post Beetles
(Family Lyctidae)

Powder-post beetles are small (2–7mm), elongate and narrow, somewhat flattened, and plain brown or reddish brown. The head is visible from above and the two-segmented antennal club is distinctive. These beetles and their larvae bore in seasoned wood, reducing the interior to a fine flourlike powder.

The brown lyctus beetle, *Lyctus brunneus* Stephens, occurs throughout the Great Lakes region and infests hardwood lumber. It is uniformly colored light brown.

Checkered Beetles
(Family Cleridae)

Checkered beetles are easily recognized by the combination of brightly colored, hairy bodies, and large heads (as wide as or wider than the thorax). Both the larvae and the adults are predaceous on wood-boring insects, and they are generally found on or around woody plants, often under loose bark.

The checkered beetle fauna of the Great Lakes region includes 50 species known for Ohio (Knull 1951) and 35 species reported from Michigan (Gosling 1980).

The dubious checkered beetle, *Thanasimus dubius* (Fabricius), has a reddish brown forebody (head, pronotum, and wing bases) and black elytra with two irregular cross bands of light-colored pubescence. The black-legged checkered beetle, *Enoclerus nigripes* (Say), has a red or reddish brown forebody and a pair of black cross bands (one narrow and one wide) near the tip of the elytra. The black bands are separated by a narrow white cross band. Nuttall's checkered beetle, *Trichodes nuttalli* (Kirby), is frequently found on the flowers of wild roses, dandelions, daisies, and other plants. It is metallic blue (sometimes greenish blue or purple) with three reddish yellow cross bands on the elytra.

For an identification key to the checkered beetles of Ohio, see Knull 1951.

Soft-Winged Flower Beetles
(Family Melyridae)

The soft-winged flower beetles are small (4–8 mm), soft bodied, and hairy, with serrate antennae. The head is partially concealed from above and the body is generally widest behind the middle (they are somewhat wedge shaped). These beetles are commonly found on flowers, where they feed on pollen, insect eggs, and other insects. They are poorly known.

The two-lined soft-winged flower beetle, *Collops vittatus* Say, is dark blue black. The side of the pronotum, as well as the margin and suture of the elytra, are orange red or yellowish red. The basal segments of the antennae are greatly enlarged in the males. The four-spotted egg eater, *C. quadrimaculatus* Fabricius, has a blue head, an orange pronotum, and orange elytra with four large blue spots. The metallic green soft-winged flower beetle, *Malachius aeneus* (L.), is distinctly wedge shaped. Most of the forebody and the base and center of the elytra are metallic green. The front of the head, the front corners of the

Fig. 140. The black-legged checkered beetle, *Enoclerus nigripes* (Say). (From Knull 1951.)

Fig. 141. Nuttall's checkered beetle, *Trichodes nuttalli* (Kirby). (From Knull 1951.)

Fig. 142. The four-spotted egg eater, *Collops quadrimaculatus* Fabricius, a soft-winged flower beetle. (From Kansas State University Extension 1962.)

pronotum, and the edges and tip of the elytra are reddish brown or orange.

Sap Beetles
(Family Nitidulidae)

Sap beetles are most easily recognized by their body shape (flattened and either oval or nearly rectangular) and by their clubbed antennae. Also, the elytra in many species are short, leaving part of the abdomen exposed. Sap beetles are mostly small, ranging in size from 3 to 7 millimeters. Adults and larvae feed on decaying plant materials, including damaged and overripe fruits and vegetables, plant sap, picnic foods and beverages, nectar, pollen, and fungi.

This group of beetles is small; only about two dozen species of sap beetles are known from the Great Lakes region.

The corn sap beetle, *Carpophilus dimidiatus* (Fabricius), is brownish yellow to blackish with an orangish spot on each wing cover. It is commonly found on ears of corn that have been damaged by caterpillars or vertebrate animals. Most people are acquainted with the picnic beetle, *Glischrochilus quadrisignatus* (Say), a frequent uninvited guest to outdoor meals. It is shiny black with four yellowish, orangish, or reddish markings on the elytra (the front spots are usually moon shaped). The red-spotted sap beetle, *G. fasciatus* (Olivier), is similar, except the basal spots are not moon shaped. The two-spotted sap beetle, *G. obtusus* (Say), is also similar but has only two reddish or yellowish spots on the elytra.

Fig. 143. The corn sap beetle, *Carpophilus dimidiatus* (Fabricius). (From Kansas State University Extension 1962.)

Flat Bark Beetles
(Family Cucujidae)

These small beetles (mostly 2–5 mm, but occasionally up to 14 mm) are narrowly elongate and distinctively flattened. Most species are found underneath the loose bark of dead trees (where they feed on other insects), but some infest stored grains and foodstuffs.

The saw-toothed grain beetle, *Oryzaephilus surinamensis* (L.), is dark reddish brown with a three-ridged pronotum bearing six teeth on each side. It is commonly found in breakfast cereal, pancake mix, and other foodstuffs. The red cucujid, *Cucujus clavipes* Fabricius, is one of the largest species in the family and resides under the bark of dead trees. The bright red color makes it unmistakable. The rufus flat bark beetle, *Catogenus rufus* Fabricius, often hibernates in groups. It is dark reddish brown with a transverse groove on the head behind the eyes. The center of the pronotum has a linear depression.

Fig. 144. The saw-toothed grain beetle, *Oryzaephilus surinamensis* (L.). (From Felt 1900.)

Pleasing Fungus Beetles
(Family Erotylidae)

These shiny, smooth, elongate-oval beetles are generally black with red, orange, or yellow markings. They range in size from 3 to 20 millimeters, with most species in the Great Lakes region in the 3 to 8 millimeter range. The 11-segmented antennae have a three- or four-segmented club. Most species feed on various types of fungi and can be found on fungi, in decaying wood, under bark, or in moldy debris.

The four-spotted pleasing fungus beetle, *Ischyrus quadripunctatus* (Olivier), is elongate-oval and seven to eight millimeters long. The head is black, but the pronotum and elytra are yellowish or reddish yellow with black markings—four black spots on the pronotum and three more or less transverse groups of spots on the elytra. The heroic pleasing fungus beetle, *Megalodacne heros* (Say), is shiny black and 18–21 millimeters long. There are four red patches on the elytra—two near the base and two near the tip.

Shining Flower Beetles
(Family Phalacridae)

These minute beetles (1–3 mm) are oval or nearly spherical, distinctly convex, and generally shiny brown. They are commonly found in the flower heads of plants (goldenrod, daisies, wild carrot, etc.).

The smut beetle, *Phalacrus politus* Melsheimer, is minute and shiny black. It can be found in the leaves of skunk cabbage, sedges, and arum. Larvae (and adults) feed on the smut-infested seed heads of grains and grasses. The shining flower beetle *Stilbus apicalis* (Melsheimer) is also minute. It is shiny black with the tips of the elytra reddish. This species can be found on grasses and often flies to lights in large numbers.

Ladybird Beetles
(Family Coccinellidae)

These familiar beetles are easily recognized by their hemispherical shape, bright colors, and spotted markings (although some are without markings and others have linear markings). The head is partially or completely concealed by the pronotum. Some leaf beetles appear similar but can be separated by the longer, four-segmented tarsi (ladybird beetles only have three visible tarsal segments). The adults and larvae are predaceous, with the exception of two species of bean beetles (genus *Epi-*

Fig. 145. The two-spotted
ladybird beetle, *Adalia
bipunctata* (L.). (From Felt
1900.)

Fig. 146. The convergent
ladybird beetle,
Hippodamia convergens
Guerin. (Courtesy of
Michigan State University
Extension.)

Fig. 147. The maculated
ladybird beetle,
Coleomegilla maculata
(De Geer). (From Kansas
State University Extension
1962.)

lachna) that are leaf feeders. Because of their beneficial nature, the ladybird beetle (ladybug) has been made the official state insect of both New York and Ohio.

Many species of the tiny dusky ladybird beetles, *Scymnus* species, occur in the Great Lakes region. Most are black and pubescent, but some have light markings. The twice-stabbed ladybird beetle, *Chilocorus stigma* (Say), is shiny black with two red spots on the elytra. The abdomen is also red. The two-spotted ladybird beetle, *Adalia bipunctata* (L.), is the reverse: it is orange with two black spots on the elytra. It frequently hibernates in houses. The ash-gray ladybird beetle, *Olla sayi* (Crotch), is ashy gray to straw colored with black spots on the pronotum and elytra.

The nine-spotted ladybird beetle, *Coccinella novemnotata* Herbst, is orange or reddish yellow with four black spots on each wing cover. The black spot on the scutellum and wing bases makes the ninth spot. The transverse ladybird beetle, *C. transversoguttata richardsoni* Brown, is reddish or yellowish with a black pronotum. The black markings on the elytra consist of a single transverse bar at the base of the wings, a spot in the middle of each elytron, and two transverse bars near the tip of each elytron. The convergent ladybird beetle, *Hippodamia convergens* Guerin, is frequently used as a biological control agent. The pronotum is black-and-white and the elytra are orange with 12 black spots of various sizes (the ones near the tip being larger than those near the base of the wing covers). The 13-spotted ladybird beetle, *H. tredecimpunctata tibialis* (Say), has a black-and-yellow pronotum and orange or reddish elytra with 13 spots. The maculated ladybird beetle, *Coleomegilla maculata* (De Geer), is pinkish red with a dozen black spots on the pronotum and elytra. This species is more elongate than most of the other ladybird beetles found in the Great Lakes region and is extremely common at times.

Handsome Fungus Beetles
(Family Endomychidae)

The two longitudinal grooves in the base of the pronotum are characteristic of this family. The antennae are clubbed and most species are small (3–8 mm) and black with red or orange markings. They somewhat resemble ladybird beetles, but the head is partially exposed and the front corners of the pronotum usually project forward. Most species feed on fungi and can be found on rotting wood, bark, or fruit.

The striped handsome fungus beetle, *Aphorista vittata* (Fabricius), is brownish red or orange with three broad, longitudinal stripes (one along the suture and one near the outer margin of the two wing covers). Sometimes the pronotum is edged in black and has two faint dark brown spots. The four-spotted handsome fungus beetle, *Endomychus biguttatus* Say, is similar in shape, but the head, pronotum, and legs are black and the elytra are reddish or orange with four black spots (the hind ones being larger).

Darkling Beetles
(Family Tenebrionidae)

After the ground beetles, this family is probably the second most variable group of beetles. They range in size from 3 to 35 millimeters in length and may be oval, elongate-oval, rectangular, or elongate. Most are plainly colored, usually black (or other dark colors). Distinguishing characteristics for the group include notched (bean-shaped) eyes, 11-segmented antennae (often clubbed), and hind legs with only four tarsi (front and middle legs each with five). Most of the members of this group are found under rocks, logs, debris, and loose bark. A few are pests of stored foods. This family reaches its zenith in the arid regions of the western states, but many species are represented in the Great Lakes region; for example, 83 species are known from Michigan (Spillman 1973).

The horned fungus beetle, *Bolitotherus cornutus* (Panzer), is quite distinct. The dark grayish brown, bumpy body bears two large, blunt horns on the pronotum that project forward over the head. The horns of the female are reduced to short tubercles. The cuticle of these beetles is very hard, and many an insect pin has been bent by someone trying to pin one of these beetles. They inhabit bracket (shelf) fungi, from which they often emerge given sufficient time to complete their development.

The false mealworm, *Alobates pennsylvanica* (De Geer), is commonly found beneath loose bark. It is large (20–23 mm) and uniformly black, with a pronotum that is narrower than the elytra. The head is narrower than the pronotum. The elytra are marked with five rows of punctures. The yellow mealworm (*Tenebrio molitor* L.) and the dark mealworm (*T. obscurus* Fabricius) are common pests of stored products. The larvae of the yellow mealworm are commonly raised for fish bait, pet food, and school projects. The adult beetles are very similar (the colors mentioned in the common names referring to the larvae, not the adults). In the adult stage both species are dark reddish brown and 13–16 millimeters long. The pronotum is as wide as the base of the elytra and the head is broad (wider than long). The yellow mealworm has distinct, uniformly spaced punctures on the pronotum; the pronotum of the dark mealworm is densely punctate and many of the punctures are confluent (run together).

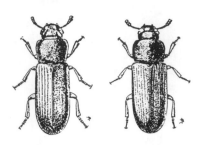

Fig. 148. The flour beetles, *Tribolium* species. *Left,* the confused flour beetle, *T. confusum* J. Duval; *right,* the red flour beetle, *T. castaneum* (Herbst). (From USDA 1953.)

The confused flour beetle, *Tribolium confusum* J. Duval, is small (4–5 mm), more or less flattened, elongate and parallel sided, and reddish brown. The tip of the antennae is gradually enlarged into a club. The red flour beetle, *T. castaneum* (Herbst), is nearly identical except that the antenna is abruptly clubbed at the tip. Both species infest flour, cake mix, pancake mix, and similar foodstuffs.

Comb-Clawed Bark Beetles
(Family Alleculidae)

These beetles resemble darkling beetles, except they are soft bodied and the antennae are threadlike or sawlike. Most species are covered with silky pubescence. The tarsal claws are equipped with blunt teeth and look like miniature garden rakes. These beetles inhabit leaf litter, loose bark, and fungi. They are not particularly common and are most frequently seen at black lights.

The black comb-clawed bark beetle, *Hymenorus niger* (Melsheimer), is black except for the brownish base of the elytra and the pale legs. The elytral grooves, which are distinctly visible at the base, disappear before reaching the tip.

Fire-Colored Beetles
(Family Pyrochroidae)

Superficially resembling soldier beetles, these soft-bodied, elongate beetles can be recognized by the narrow neck behind the head and by the large eyes. The antennae are serrate or spectacularly featherlike. They are generally found on vegetation or flowers.

The unicolorous fire-colored beetle, *Dendroides concolor* (Newman), is fire red except for the eyes and antennae. The dark-winged fire-colored beetle, *Dendroides cyanipennis* Latreille, has a black head, antennae, and elytra and a reddish yellow pronotum.

Tumbling Flower Beetles
(Family Mordellidae)

These beetles are characterized by the arched body, the shortened elytra, and the pointed abdomen that protrudes beyond the tips of the elytra. They are commonly found on flowers and jump or tumble when disturbed.

The tumbling flower beetle *Mordella atrata* Melsheimer is small (3–6 mm) and dull black with gray pubescence. The edges of the abdomen are gray.

Blister Beetles
(Family Meloidae)

These soft-bodied beetles can be recognized by the elongate body, narrow pronotum (narrower than both the head and the elytra), and hind tarsi with only four segments (the other legs have five tarsal segments). Furthermore, many species of blister beetles have short elytra, leaving part of the abdomen exposed. Most species are medium sized (10–15 mm). The adults consume plant foliage (some are significant agricultural pests), but larvae are parasites of grasshopper egg pods. All blister beetles contain the blistering agent cantharidin, and when the beetles are disturbed this chemical exudes from their leg joints. This caustic fluid helps repel enemies.

The black blister beetle, *Epicauta pennsylvanica* De Geer, is sparsely pubescent and completely dull black in color. It is generally found on flowers. The striped blister beetle, *E. vittata* (Fabricius), is dull yellow with two or three black stripes. Adults can be found feeding on potatoes, tomatoes, and a variety of weedy plants. The short-winged blister beetles (genus *Meloe*) differ in having the elytra short and wrinkled, overlapping at the base. Most species—for example, *Meloe angusticollis* Say (12–15 mm)—are blue black or violet.

Fig. 149. The black blister beetle, *Epicauta pennsylvanica* De Geer. (From Gilbertson and Horsfall 1940.)

Antlike Flower Beetles
(Family Anthicidae)

These unusual beetles are generally small (2–5 mm) with a head as wide as the pronotum but connected by a narrow neck. The pronotum of some species projects forward as a hornlike process, causing the head to be angled downward. The hind tarsus is four-segmented (the other legs have five tarsal segments). These beetles occur on flowers and foliage, and on barren soil. Several species frequent sand dunes and can be encountered in large numbers beneath driftwood on sandy beaches of the Great Lakes.

The antlike flower beetle *Notoxus monodon* Fabricius is small and dull yellowish brown with gray pubescence. The elytra have two black spots on the base and a broad cross band near the tip. A distinct pronotal horn is present. The antlike flower beetle *Anthicus cervinus* LaFerte-Senectere is also small but lacks a pronotal horn. The forebody is reddish brown and the elytra are yellow with four black markings (giving the appearance of a broad yellow cross on the elytra).

Long-Horned Beetles
(Family Cerambycidae)

These small to large beetles are generally elongate and narrow, with long antennae (at least half as long as the body, but often much longer than the length of the body). The antennae are attached to the head at the edge of the compound eyes, resulting in eyes that are somewhat notched, or bean shaped. The next to the last tarsal segment is distinctively split, or bilobed. The coloration is variable, but most species have a unique, distinctive pattern of markings that makes identification relatively easy. The adults feed on plant materials (wood, roots, leaves, and pollen) and are frequently found on flowers, tree trunks, and logs. They are strong fliers and can often be seen flying toward woodpiles and cutover areas in forests. The larvae, commonly referred to as roundheaded wood borers, bore into the bark and wood of both hardwood and coniferous trees, and in few cases they bore into seasoned lumber. This group of beetles is relatively large, with 262 species known (or very likely to occur) from Ohio (Knull 1946) and 227 species known from Michigan (Gosling 1973, 1977, and 1983).

Representatives of the Great Lakes long-horned beetle fauna include the following. The broad-necked root borer, *Prionus laticollis* (Drury), is large to very large (22–44 mm) and shiny black with three prominent toothlike projections on each side of the pronotum. The tile-horned root borer, *P. imbricornis* L., is similar, but it is reddish brown in color and the male has flattened, tilelike antennal segments. It is the largest long-horn beetle in the Great Lakes region, measuring up to 50 millimeters in length. The black-horned pine borer, *Callidium antennatum hesperum* Casey, is medium sized and metallic blue or

Fig. 150. The tile-horned root borer, *Prionus imbricornis* L. (From Knull 1946.)

bluish black and the rounded pronotum has two circular depressions. The locust borer, *Megacyllene robiniae* (Forster), is a common sight on goldenrod in the fall. It is black with bright yellow, chevron-shaped cross bands on the elytra and transverse stripes on the pronotum. The rustic borer, *Xylotrechus colonus* Fabricius, is patterned in rusty red and gray with two, broad, yellowish gray cross bars. The red-headed ash borer, *Neoclytis acuminatus* (Fabricius), has a distinctive red head and pronotum. The elytra are light brown with four narrow cross bands of yellowish hairs and have a dark tip. Both the rustic borer and the red-headed ash borer may be common in homes, emerging from firewood stored indoors.

The elderberry longhorn beetle, *Dendrobius palliatus* Forster, is found around elderberry shrubs, where they feed on the flowers. The larvae bore in the roots and pithy stems of the plant. The adult beetle is large and metallic blue except for the bases of the elytra, which are yellowish with two black spots. The red oak borer, *Enaphalodes rufulus* (Haldeman), is reddish brown and covered with brownish orange pubescence. The pronotum has two small bumps near the center and the elytra are spined at the tip. The velvet flower longhorn, *Typocerus velutinus* (Olivier), has a black forebody (including the antennae) and reddish brown elytra with four broad cross bands. The body, which is widest at the shoulders and tapered toward the rear, is covered with yellowish pubescence.

Several species of pine sawyers occur in the Great Lakes region and they are characterized by their very long antennae and large body size (greater than 13 mm). The most common is the Carolina sawyer, *Monochamus carolinensis* Olivier, which is 17 millimeters long and mottled with various shades of brown. The white-spotted pine sawyer, *M. scutellaris scutellaris* Say, is darker, (almost blackish) with a distinctive patch of white hairs on the scutellum. The southern pine sawyer, *M. titillator* Fabricius, also occurs in the region and is brown and mottled with patches of gray, brown, and black hairs (like the Carolina sawyer), but the body is larger (25 mm). The red milkweed beetle, *Tetraopes tetrophthalmus* (Forster), is a common sight on the stems and flowers of its host plant. These beetles are bright red orange with four black spots on the pronotum and six black spots on the elytra. The large red milkweed beetle, *T. femoratus* LeConte, is slightly larger and is red orange with four black spots on the elytra and gray rings on the antennae. Milkweed beetles have a unique compound eye that is actually split into an upper and lower part with the antennae rooted in the space between.

For keys to the genera and species of long-horned beetles known from Ohio, see Knull 1946.

Fig. 151. The locust borer, *Megacyllene robiniae* (Forster). (From Kansas State University Extension 1962.)

Fig. 152. The red-headed ash borer, *Neoclytis acuminatus* (Fabricius). (From Kansas State University Extension 1962.)

Fig. 153. The elderberry longhorn beetle, *Dendrobius palliatus* Forster. (From Knull 1946.)

Fig. 154. The red milk-weed beetle, *Tetraopes tetrophthalmus* (Forster). (From Kansas State University Extension 1962.)

Fig. 155. The asparagus beetle, *Crioceris asparagi* (L.). (From Felt 1900.)

Fig. 156. The Colorado potato beetle, *Leptinotarsa decemlineata* (Say). (From Knowlton 1935.)

Leaf Beetles
(Family Chrysomelidae)

Most leaf beetles are brightly colored and range from 2 to 20 millimeters in length. The body is generally broadly oval or elongate-oval. The next to the last segment of the tarsi are split, or bilobed. Some species resemble ladybird beetles (but they have four tarsal segments, not three as do ladybird beetles) or long-horned beetles (but the antennae are always shorter than one-half the length of the body and the compound eyes are generally not notched). Both the adults and the larvae are plant feeders and may consume foliage (on the surface or as leaf miners), flowers, roots, and occasionally stems. Many species damage agricultural crops or ornamental plants.

The leaf beetle fauna of the Great Lakes region is large. There are 298 taxa (species and subspecies) known from Ohio, and at least 112 more taxa are probable because they occur in adjacent states. Similarly large faunas are known for New York, with 332 taxa, and for Indiana, with 266 taxa (Hughes 1944).

The waterlily beetles (*Donacia* species) have the longest antennae of any of the leaf beetles (and for this reason are sometimes called long-horned leaf beetles). These medium-sized beetles are elongate and generally green, blue, violet, or copper in coloration with a metallic sheen. The adults are found crawling on the exposed parts of aquatic vegetation. The larvae have two spines equipped with spiracles at the tip, which they use to obtain oxygen from the roots and lower stems of aquatic plants. The asparagus beetle, *Crioceris asparagi* (L.), feeds on asparagus plants. It is metallic dark blue with a red pronotum. The elytra are marked with three pairs of yellowish spots that are connected along the margin. The dogbane beetle, *Chrysochus auratus* Fabricius, is very common on dogbane (Indian hemp) and milkweed during the summer months. It is spectacularly metallic, with a blue pronotum and with elytra that are shiny metallic green with a coppery or purplish luster.

The Colorado potato beetle, *Leptinotarsa decemlineata* (Say), is a serious pest of potatoes but can also be found feeding on horse nettles. The robust, elongate-oval body (6–11 mm) is reddish orange and the yellowish tan elytra are longitudinally striped in black. The dogwood or rowena calligrapha, *Calligrapha rowena* Knab, is found on the leaves of dogwood and is beautifully colored in metallic blue green and orange red or gold. The pronotal and elytral markings are metallic blue green. The markings on the elytra look like fancy script and include a sutural stripe with three short spurs on each side, marginal stripes, and spots between. The red orange or gold background color of the elytra often fades to a dull straw color in dead specimens.

The corn rootworm beetles are common in many parts of the region where corn is grown. The southern corn rootworm (also known as the spotted cucumber beetle), *Diabrotica undecimpunctata howardi* Barber, is greenish yellow and has 12 distinct black spots on the elytra. The western corn rootworm, *D. virgifera* (LeConte), is yellowish with three broad longitudinal black stripes on the elytra (often coalescing into one large black marking). The northern corn rootworm, *D. longicornis* Say, is uniformly pale green and the elytra have more or less distinct longitudinal grooves.

The corn flea beetle, *Chaetocnema pulicaria* (Say), is minute and shiny black with bronze or blue green reflections. The bases of the antennae are orange and the lower parts of the legs are yellowish. It is especially common on corn but can also be found on many types of grasses. The three-spotted flea beetle, *Disonycha triangularis* (Say), is larger (4–5 mm) with shiny blue or blue black elytra and a yellowish pronotum decorated with three black spots in a triangular pattern. All flea beetles possess enlarged hind legs that allow them to jump.

The locust leaf miner, *Odontota dorsalis* (Thornberg), is small and somewhat wedge shaped. The head and legs are black and the pronotum and elytra are dirty yellow or orangish. The elytra are sculptured with raised longitudinal ridges and deep punctures. The larvae are leaf miners that tunnel between the upper and lower surfaces of black locust foliage, and by late summer whole trees may take on a browned appearance from the feeding of these beetles.

The tortoise beetles are a distinct and unusual group of leaf beetles. The body is broadly rounded, flattened, and tortoiselike, and the head cannot be seen from above. This species' brilliant coloration fades to dull hues once the beetles are dead. The argus tortoise beetle, *Chelymorpha cassidea* (Fabricius), is reddish or yellowish with 6 black dots on the pronotum and 14 black spots on the elytra. The mottled tortoise beetle, *Deloyala guttata* (Olivier), has the margins of the pronotum and elytra translucent pale yellow. The elytra are blackish with irregular patches of yellow. The goldbug, *Metriona bicolor* (Fabricius), is brilliant brassy gold with the margins of the pronotum and elytra thin and flattened. Once this species dies, its gold color changes to a dull reddish yellow.

For a key to the aquatic leaf beetles of Wisconsin, see Bayer and Brockman 1975; for a key to the leaf beetles of Ohio, see Wilcox 1954.

Fig. 157. The southern corn rootworm (spotted cucumber beetle), *Diabrotica undecimpunctata howardi* Barber. (Courtesy of Michigan State University Extension.)

Fig. 158. The western corn rootworm, *Diabrotica virgifera* (LeConte). (Courtesy of Michigan State University Extension.)

Fig. 159. The northern corn rootworm, *Diabrotica longicornis* Say. (From Kansas State University Extension 1962.)

Fig. 160. The corn flea beetle, *Chaetocnema pulicaria* (Say). (From Gentner 1926.)

Fig. 161. The goldbug, or gold tortoise beetle, *Metriona bicolor* (Fabricius). (Courtesy of Michigan State University Extension.)

Fig. 162. The bean weevil, *Acanthoscelides obtectus* (Say). (Courtesy of Michigan State University Extension.)

Seed Weevils
(Family Bruchidae)

These small weevils (2–5 mm) have a short, broad snout. The body is often robust and egg shaped (widest at or beyond the middle) with short elytra, leaving the tip of the abdomen exposed. Most species are dull, mottled in various shades of gray, brown, and yellow, and often partially covered with hairs or scales. The antennae are clubbed or serrate and the femur of the hind leg is enlarged. The adult beetles are found on leaves, flowers, and seeds. Both the larvae and the adults feed inside seeds, especially peas and beans.

The cowpea weevil, *Callosobruchus maculatus* (Fabricius), has a blackish forebody and brown elytra with scattered brown and white pubescence. The elytra are marked with two large lateral spots and a black tip. The bean weevil, *Acanthoscelides obtectus* (Say), is black with yellowish or grayish pubescence. The base and the very tip of the antennae is orange-brown. The elytra have short cross bands of brownish hairs. The pea weevil, *Bruchus pisorum* (L.), is black with dense reddish brown, gray, yellowish, and white pubescence arranged in two indistinct cross bands. The exposed tip of the abdomen is white.

Fungus Weevils
(Family Anthribidae)

Fungus weevils are similar to the seed weevils, but they are more elongate and the tip of the abdomen is not exposed. The antennae are slender, often with a club at the tip. These beetles are generally found on dead branches, beneath bark, and on flowers, where they feed on wood, fungi, seeds, and pollen. At least one species infests stored food products and is spread by commerce.

The coffee bean weevil, *Araecerus fasciculatus* (De Geer), is small (2–5 mm) and dark brown with various patterns of yellowish brown hairs. The antennae and legs are reddish brown. Adults infest seeds, berries, dried fruits, and moldy foodstuffs.

Straight-Snouted Weevils
(Family Brentidae)

These unusual weevils have elongate, narrow bodies with a snout that protrudes straight forward. The antennae are threadlike or beadlike. Both adults and larvae occur in wood and under bark. Larvae feed on fungi, while adults feed on fungi, sap, or other insects.

The only species found in the Great Lakes region is the oak timberworm, *Arrhenodes minutus* (Drury). It is small to medium sized (7–17 mm) and dark reddish brown with four or more elongate yellowish spots on the elytra. The snout may be narrow and longer than the head (female) or short and broad (male). This species inhabits oak trees and is usually found underneath the bark.

Bark Beetles
(Family Scolytidae)

Most species of bark beetles infest conifer trees, where they burrow under the bark, leaving distinctive engraved patterns. When the new adults emerge they make tiny circular holes in the bark, creating a scattered pattern of "shot holes." They are minute to small (1–9 mm) and brown, reddish brown, or black. The body is more or less elongate and cylindrical and is usually distinctly punctured and grooved (on the elytra). Also, in many species the tip of the abdomen is concave (looks bashed in) and bordered with blunt spines of various sizes. The pronotum is usually large and well developed and may cover the head so it cannot be seen from above. The antennae are short and "elbowed" and end in an abrupt three-segmented club.

The Indiana bark beetle fauna includes 92 species (Deyrup and Atkinson 1987), while 51 species are reported from Ontario (Bright 1976).

The native elm bark beetle, *Hylurgopinus rufipes* (Eichoff), infests elms. It is reddish brown and strongly punctured, the punctures on the elytra arranged in longitudinal rows. The smaller European elm bark beetle, *Scolytus multistriatus* (Marsham), has a large pronotum (almost as long as the elytra), and the underside of the abdomen is concave with a stout spine. This species is involved in the transmission of Dutch elm disease among American elms. The red turpentine beetle, *Dendroctonus valens* LeConte, is light reddish brown to dark brown

Fig. 163. The native elm bark beetle, *Hylurgopinus rufipes* (Eichoff). (From McDaniels 1930.)

Fig. 164. The smaller European elm bark beetle, *Scolytus multistriatus* (Marsham). Courtesy of Michigan State University Extension.)

with tiny little rasplike teeth along the front edge of the elytra. The black turpentine beetle, *D. terebrans* (Olivier), is similar but darker in color (dark reddish brown to black). The southern pine engraver, *Ips grandicollis* (Eichoff), is dark reddish brown to black with the tip of the elytra concave and bordered by 10 blunt teeth. The pine engraver, *I. calligraphus* (Germar), is very similar, but the elytral concavity is bordered by 12 blunt teeth.

For a key to the bark beetles of Canada and the adjacent United States, see Bright 1976.

Snout Beetles and Weevils
(Family Curculionidae)

The Curculionidae is the largest family of beetles. All its members are easily distinguished by the front of the head, which is formed into a downwardly curved snout with tiny little jaws (mandibles) at the tip. This snout may be either short and broad or very long and slender. The antennae are elbowed and clubbed and arise from the base of the snout. Most weevils are associated with plants, feeding on the leaves, stems, wood, flowers, or fruit.

Species identification of this group is often difficult because of the sheer number of closely related species. Examples of a few of the more common and easily recognized species found in the Great Lakes region follow.

The plum curculio, *Conotrachelus nenuphar* (Herbst), is a small weevil associated with fruit trees and hawthorne. It is a short, broad, stubby weevil that is dark brown in color with brownish yellow and whitish hairs forming irregular patches and cross bands. The surface of the elytra are raised and bumpy with tubercles and broken ridges. The rose curculio, *Rhynchites bicolor* (Fabricius), is found in rose flowers. It is red with the snout, legs, and undersides of the body black. The New York weevil, *Ithycerus noveboracensis* (Forster), is shiny black with gray and brown pubescence that forms faint stripes. The scutellum is yellow. It can be found on oak, hickory, beech, and some fruit trees. The strawberry root weevil, *Otiorhynchus ovatus* (L.), is small (5–7 mm), shiny black with reddish legs, and sparsely covered with yellowish pubescence. The pronotum is round and the side margins are ridged. The elytra are coarsely punctate and strongly convex (globular). This species commonly hibernates in houses and can be a significant nuisance problem in the fall. The black vine weevil, *O. sulcatus* (Fabricius), is similar but larger (8–11 mm).

Two weevil species are pests of stored food products and therefore are commonly encountered in granaries, stores, and houses. The rice weevil, *Sitophilus oryzae* (L.), is small and dull brown. The pronotum is deeply punctured and there are four reddish spots on the elytra. The granary weevil, *S. granarius* (L.), is very similar but has scattered elongate punctures on the pronotum and lacks elytral spots.

The corn billbug, *Sphenophorus zeae* Walsh, is medium sized and dark reddish brown or blackish with smooth, raised grayish ridges on the pronotum and elytra. The area between the stripes is yellowish gray and roughened. It is found on corn and grasses.

The rhubarb curculio, *Lixus concavus* Say, is elongate (10–14 mm) and blackish with sparse gray pubescence. The antennae and tarsi are reddish brown. It is found on rhubarb and curly dock. The alfalfa weevil, *Hypera postica* (Gyllenhal), is small and short (5–8 mm). It is brown or grayish brown with a broad, dark brown stripe down the center of the elytra. The acorn weevil, *Curculio sulcatus* (Casey), has a very long, slender snout that is nearly as long as the body. It is found on oak trees and feeds on the acorns.

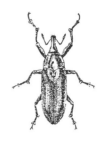

Fig. 165. The rhubarb curculio, *Lixus concavus* Say. (From Felt 1900.)

Identification Resources. There are several general reference works on the beetles of North America, including White 1983, Dillon and Dillon 1972, and Jaques 1951. For keys to aquatic beetles of North America, see Lehmkuhl 1979; for keys to aquatic beetles of spring pools in southern Ontario, see James 1969; for keys to aquatic leaf beetles and weevils of Wisconsin, see Bayer and Brockman 1975. Other identification keys and information for specific families, when available, are mentioned under each respective family.

Twisted-Wing Parasites
(Order Strepsiptera)

Introduction. Most scholars still classify these twisted-winged parasites, also known as stylopids, in the order Strepsiptera, but recently it was proposed by some specialists that these unusual insects should actually be classified as a unique family of parasitic beetles (Stylopidae).

Description. The free-living males are minute, black, and fly-like. The head is very broad (wider than the pronotum) and short, with the compound eyes protruding to the sides. The antennae are fan shaped or comblike. The forewings (elytra) are modified into clublike structures and the hind wings are membranous, fan shaped, and lacking venation. The females are grublike and remain concealed in the body of their host.

Fig. 166. A male twisted-wing parasite of the genus *Stylopus*. (From Bohart 1941.)

Habits. The larvae and the females are parasites of grasshoppers, planthoppers, leafhoppers, treehoppers, spittlebugs, bees, and wasps, usually attacking the top surface of the host's abdomen at the joints, or sutures. The tip of the parasite's abdomen usually protrudes from beneath one of the abdominal segments of the host. Members of the genus *Stylopus* (the largest of the twisted-wing parasite genera) parasitize andrenid bees, *Pseudoxenos* parasitize eumenid and thread-waisted wasps, and *Xenos* parasitize paper wasps (*Polistes* species). Members of the genus *Halictophagus* are parasites of leafhoppers.

Distribution and Faunistics. As you can imagine, because of the unusual parasitic nature of this group they are rarely seen and poorly known in the Great Lakes region. The best way to find stylopids is to collect their hosts and inspect them for the presence of these parasites.

Identification Resources. For keys and further identification information on the stylopids of North America (including the Great Lakes region), see Bohart 1941, Kathirithamby 1989, and Kinzelbach 1991.

Scorpionflies and Hangingflies
(Order Mecoptera)

Introduction. The common name for this insect group might conjure up all sorts of terrifying mental images—scorpions that fly! Well, fortunately not. In fact, these seldom-seen insects are far from dangerous. Their name is derived from the harmless scorpion-like abdomen (genitalia) of the male scorpionflies.

Description. There is no mistaking a scorpionfly for any other insect. The male has a distinct, scorpion-like abdomen, curled forward and swollen at the tip. The abdomen of the female is slender and narrow, without a swelling at the tip. Both the males and the females have a slender, elongate face with chewing mouthparts at the tip. There are two pairs of long, narrow, membranous wings, and these are often spotted or banded. Some species, however, are wingless. The antennae are long and slender, about as long as half the length of the body.

Life Cycle. Scorpionflies have complete metamorphosis, with egg, larva, pupa, and adult stages.

Habits. Adult scorpionflies are either predators or scavengers, feeding on living or dead insects and other invertebrates. A few are known to feed on leaf tissue.

Habitat. Adult scorpionflies frequent damp, shady woods and wetlands with lots of dense vegetation. The larvae are terrestrial and feed as scavengers on many types of dead plant and animal materials.

Ecological and Economic Status. Scorpionflies, even though they are scavengers and predators, have no known impact on the populations of undesirable insects. But, they are not harmful either (despite the males' resemblance to a winged scorpion).

Distribution and Faunistics. Very little is known about the distribution of these insects in the Great Lakes region. It is a small group, with no more than two dozen species occurring in the region. A total of twenty species is known from Michigan (Thornhill and Johnson 1974).

Hangingflies
(Family Bittacidae)

These medium-sized scorpionflies bear a striking resemblance to crane flies, and their name refers to the peculiar habit of hanging from vegetation by the front and middle legs. Only a few species are commonly encountered in the Great Lakes region. The black-tipped hangingfly, *Hylobatticus apicalis* Hagen, which is yellowish brown with dark wing tips, hangs with the wings outstretched. The spotted hangingfly, *Batticus stigmaterus* Say, is similar to the black-tipped hangingfly but hangs with the wings at its side. The bristled hangingfly, *B. strigosus* Hagen, is pale yellowish brown with yellowish wings that have the cross veins margined in gray.

Fig. 167. A hangingfly of the genus *Batticus*. (From Kansas State University Extension 1962.)

Snow Scorpionflies
(Family Boreidae)

These small (about 3 mm) scorpionflies are wingless, with only scalelike (female) or hooklike (male) flaps on the thorax. They are agile and walk or jump about on the ground.

They inhabit mosses, which are also the food for both adults and larvae. They are often active in the late winter and early spring, and they are easily seen as they move about on the snow.

Only two species occur in eastern North America, and *Boreus brumalis* Fitch is the most commonly encountered. *B. nivoriundus* Fitch is the other species. It is dull brown in color, whereas the preceding species is black.

Forcepsflies
(Family Meropeidae)

These unusual scorpionflies are characterized by long forceps-like claspers at the tip of the male's abdomen. The wings are broad and brownish and these insects somewhat resemble a cockroach in appearance. These medium-sized mecopterans are rarely seen, although they occasionally come to lights. There is only one species in North America (and the Great Lakes region), the forcepsfly or earwig scorpionfly, *Merope tuber* Newman.

Fig. 168. A male scorpionfly of the genus *Panorpa*. (From Kansas State University Extension 1962.)

Common Scorpionflies
(Family Panorpidae)

Panorpidae is the largest family of scorpionflies, and species of this family are the ones most frequently encountered by insect collectors. There are many species, including *Panorpa nuptiali* Gerstacher, which is dark reddish brown with wings that have three broad black cross bands, *P. nebulosa* Westwood, which is yellowish brown with spotted wings; and *P. helena* Byers, which is dark yellowish brown, with three black bands on the wings and three smaller spots.

Identification Resources. For identification information on the scorpionflies of North America, see Hine 1901. For identification keys to the scorpionflies of Michigan, see Thornhill and Johnson 1974; for keys to scorpionflies of Illinois, see Webb et al. 1975.

Caddisflies
(Order Trichoptera)

Fig. 169. An adult free-living caddisfly of the genus *Rhyacophila*. (From Ross 1944.)

Introduction. Caddisflies are most familiar to fisherfolk and aquatic entomologists. They are an ancient group of insects and probably shared a common evolutionary branch with the butterflies and moths (Lepidoptera).

Description. Caddisflies are medium sized, more or less elongate, dull colored, mothlike insects with hairy wings and bodies. The wings, four in number, are densely covered with hairs (not flattened scales) and are held rooflike over the body at rest. The hind wings are shorter and broader than the front wings. The antennae are long and threadlike, often longer than the body. The mouthparts are not well developed and consist of segmented palps and a lapping or sponging tongue for liquid foods.

Life Cycle. Caddisflies develop by complete metamorphosis, with egg, aquatic larva, pupa, and adult stages. The eggs are laid in the water, on aquatic plants, in overhanging vegetation, or at the shoreline. As soon as the larvae hatch from the eggs they begin construction of a special protective case made out of silk and sand grains, pebbles, twigs, bits of leaves, and so on. The shape and construction of the case is unique for every species. Only the forward part of the larva's body (head, thorax, and legs) is hardened, or sclerotized. The remainder of the body is soft and flexible. The larvae anchor themselves in the case with hooks at the tip of the abdomen. The cases are enlarged as the larvae grow. When it comes time to pupate, the case is anchored with silk and the opening is sealed up. Pupation then takes place in the larval case. Occasionally a separate silken cocoon is spun. When its development is very nearly complete, the pupa swims to the surface and crawls out of the water so that the final molt into the adult stage may take place. The adults generally live a month or two.

Habits. The adults generally do not feed or fly much, and they usually remain in the general vicinity of the larval habitat. Those that do feed utilize liquid foods, such as nectar and sap. They remain in hiding in cool, damp locations throughout most of the daylight hours and become active starting at dusk. Lights set up near caddisfly habitats will attract many adults.

The aquatic larvae feed on detritus, diatoms, algae, tiny crustaceans or immature aquatic insects, and plant tissue. Some larvae that do not make cases spin silken nets (underwater webs) to strain food out of the flowing water.

Habitat. Adults are generally found in damp woods, wetlands, and other streamside habitats. The larvae are found in all types of moving water (ditches, streams and rivers, and occasionally ponds and lakes).

Ecological and Economic Status. Caddisfly larvae are an important source of food for many fish. Larger fish, like trout, are said to consume them case and all. Large swarms of adults can be a nuisance problem, and the hairy bodies and wings can be irritating to the human respiratory system and skin.

Distribution and Faunistics. Many members of this group have widespread distributions in the Great Lakes region, especially in areas where aquatic habitats are plentiful. Most families have been reasonably well studied and there is pretty good information on the fauna of the region. For example, 184 species of caddisflies are known from Illinois (Ross 1944), 200 species (15 families, 55 genera) are known from Ohio (Huryn and Foote 1983), and 238 species are known from Minnesota (Etnier 1965 and 1968).

Finger-Net Caddisflies
(Family Philopotamidae)

These caddisflies have larvae that form finger-shaped nets for collecting food. The small adults have three ocelli on the head and maxillary palps with the fifth segment two to three times as long as the fourth.

One common species is *Chimarra aterrima* Hagen, which is dark brown with the sides of the abdomen creamy white.

Trumpet-Net Caddisflies
(Family Polycentropodidae)

The larvae of these caddisflies build long tubular nets attached to aquatic vegetation and stones. The small adults have false rings on the fifth (last) segment of the maxillary palp. The middle segment of the thorax has a pair of seta-covered warts.

One common species in northern parts of the region is *Psychomyia flavida* Hagen. It is straw colored with purplish reflections.

Common Net-Spinning Caddisflies
(Family Hydropsychidae)

The larvae of these caddisflies build cup-shaped silk nets between stones to catch food particles. The larvae wait nearby in small silk and pebble retreats. The small- to medium-sized adults have no ocelli. The warts occur on the pronotum only (not the midsection of the thorax).

Many common species marked in mottled patterns of brown, gray, or white occur in the Great Lakes region, including members of the genera *Hydropsyche, Cheumatopsyche,* and *Macronema.*

For identification information on common species in Minnesota, see Denning 1943.

Free-Living Caddisflies
(Family Rhyacophilidae)

Larvae of these caddisflies inhabit cold, fast-flowing waters. The larvae of many species are free-living and do not construct cases until ready to pupate. For this reason they are also known as the "primitive caddisflies." The larger pair of feelerlike appendages (maxillary palpi) of the adult mouthparts have the fourth and fifth segments of nearly equal size. Ocelli are present. Most adults are small (3–10 mm) and mottled brownish.

Purse Casemaker Caddisflies
(Family Hydroptilidae)

The larvae of these caddisflies make thick, purse-shaped nets that are open at both ends. The adults are very small (2–6 mm), and for this reason these caddisflies are also known as the "microcaddisflies." Ocelli may be present or absent. The adults are densely covered with hairs and the wings have a fringe of hair that is longer than the width of the wings. Some of the hairs are clubbed.

One species, *Leucotrichia pictipes* (Banks), is dark brown to blackish with white bands on the antennae and tarsi. There is also a white spot on the base of the forewing.

Giant Casemaker Caddisflies
(Family Phryganeidae)

The larvae of these caddisflies make large cases out of plant materials, and they generally inhabit still or slowly moving waters. The adults are medium to large sized (12–25 mm) and the wings are mottled gray or brown. The maxillary palps of males are four-segmented (females are five-segmented). Ocelli are present. The front tibia has two or more spurs and the middle tibia has four.

One commonly encountered species is *Phryganea cinerea* Walker, which is gray or brown. The front wings have an irregular pattern of dark and light markings along the rear edge of the forewing that form triangular markings when the wings are at rest. The hind wings are uniformly grayish. Another species, *Ptilostomis ocellifera* (Walker), is yellowish brown with darker brown spots.

Fig. 170. An adult male of the brown giant casemaker caddisfly, *Phryganea cinerea* Walker. (From Ross 1944.)

Northern Casemaker Caddisflies
(Family Limnephilidae)

Larvae of these caddisflies make various types of cases constructed of sand, pebbles, bits of snail shells, sticks, bark, or other plant materials. Larvae inhabit both still and flowing waters. The adults range in size from 7 to 23 millimeters. The maxillary palps of males are three-segmented (those of females are five-segmented). Ocelli are present.

One commonly encountered species is *Limnophilus rhombicus* (L.). The adults are large and the forebody is yellowish with brown spots. The head has several brown markings on the front. The forewings are dark brown with oblique, cream-colored stripes.

Longhorned Caddisflies
(Family Leptoceridae)

Larvae of these caddisflies build either cone-shaped cases of sand or squarish cases of small sticks. They inhabit streams, rivers, ponds, and lakes. The small- to medium-sized adults are

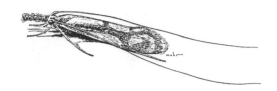

Fig. 171. A longhorned caddisfly of the genus *Triaenodes.* (From Ross 1944.)

slender bodied and have very long antennae, longer than the length of the forewings. Ocelli are absent.

Commonly encountered species include *Nectopsyche exquisita* (Walker). The forebody is tawny and covered with white hairs. The forewings have brownish yellow cross bands and four square spots along the rear edge of the forewing near the tip. Another species, *Oecetis cinerascens* (Hagen), is brown with grayish hairs. The forewings have several dark spots at the forks in the veins. A third species, *Mystacides sepulchralis* (Walker), is distinctively blue black. Both the wings and the thorax have a metallic sheen.

Identification Resources. For keys to families (adults and larvae) of North America, see Lehmkuhl 1979; for keys to the genera of caddisflies in Canada and the adjacent United States (in french), see Schmid 1980. For identification information and keys to the caddisflies of Illinois, see Ross 1944; for a key to the brachycentrid caddisflies of the western Great Lakes region, see Hilsenhoff 1985; for a key to the hydroscaphid caddisflies of Wisconsin, see Schmunde and Hilsenhoff 1986. For an illustrated key to the pupae of six *Hydropsyche* of southern Ontario, see Rutherford 1985.

Butterflies, Moths, and Skippers
(Order Lepidoptera)

Introduction. Many members of the order Lepidoptera, especially the butterflies and certain moths, are the ambassadors of the insect world, because they are immensely popular with the general public. They are so well liked that they are often thought of as being separate from the rest of the insects, and in the minds of some they are treated as if they were not insects at all. But, of course, they are. The highly positive image of butterflies is probably due to their colorful wings, their association with flowers (as pollinators), their provocative habits, and their inability to bite or sting. Most of the other members of the lepidopteran order, the moths and skippers, are generally ignored or underappreciated. Virtually all the pest species of Lepidoptera are moths.

Fig. 172. Members of the order Lepidoptera. *Left,* moth; *center,* skipper; *right,* butterfly. (From USDA 1952.)

Quick Guide to Identification

Butterflies, Skippers, and Moths
(Order Lepidoptera)

Micromoths

Diagnostic Characteristics	Common Name(s)	Scientific Family Name
Head scales rough; second segment labial palps with stout hairs	clothes moths	Tineidae
Much of wing without scales; often wasplike or beelike (small)	clear-winged moths	Sesiidae
Labial palps snoutlike, projecting	snout and meal moths	Pyralidae
Front and hind wings with zigzag lines; palps short	geometer moths	Geometridae
Wings salt-and-pepper, grayish	owlet and cutworm moths	Noctuidae

Macromoths

Diagnostic Characteristics	Common Name(s)	Scientific Family Name
Front and hind wings with zigzag lines; palps short	geometer moths	Geometridae
Vividly striped or spotted	tiger moths	Arctiidae
Front wings salt-and-pepper; hind wings banded	underwing moths	Noctuidae
Wings salt-and-pepper, grayish	owlet and cutworm moths	Noctuidae
Very large; often with eyespots and crescent-shaped spots	silkworm and royal moths	Saturniidae
Wings narrow and elongate, sometimes lacking scales	hawk moths and sphinx moths	Sphingidae

Butterflies and Skippers (Antennae Knobbed)

Diagnostic Characteristics	Common Name(s)	Scientific Family Name
Antennae hooked at tip; stout body	skippers	Hesperiidae
Large; hind wings with "tails"	swallowtail butterflies	Papilionidae
White, yellow, or orange, with black wing tips or margins	whites, sulphurs, and orangetips	Pieridae
Iridescent green, gray, blue, or copper; often with tiny "tails"	hairstreaks, blues, and coppers	Lycaenidae
Shiny, metallic marks on wings	metalmarks	Riodinidae
With snout (elongate labial palps)	snout butterflies	Libytheidae
Angular wings; short forelegs	brush-footed butterflies	Nymphalidae
Brownish with eyespots	satyrs and nymphs	Satyridae
Orange with black wing veins	monarchs	Danaidae

Description. The butterflies, skippers, and moths are soft bodied and small to very large (5 to 200 millimeters in wingspan), with large wings (although a few female moths are wingless). The wings, body, and legs of the Lepidoptera are covered with colorful, flattened scales. The scales occur in just about every color imaginable and are arranged in special color patterns that are unique to each species. They are loosely attached like shingles on a roof. These scales are easily detached (imparting a slippery or "buttery" feeling) and make it difficult for predatory animals to grab and hang on to the wings of these insects. The head bears a coiled mouthpart (proboscis), long antennae, and large compound eyes.

The butterflies, moths, and skippers share many common characteristics, but they are also different in several significant ways. Generally speaking (though there are often exceptions to the following statements), the three groups of lepidopterans can be distinguished on the basis of antennae, coloration, shape, and activity period. For example, the antennae of butterflies are slender and knobbed at the tip, the antennae of skippers are slender and hooked at the tip, and the antennae of moths are more or less broad and often featherlike, but occasionally knobbed or filamentous. Butterflies are usually brightly colored, whereas skippers and moths are often drably colored. The bodies of butterflies are usually slender, whereas the bodies of skippers and moths are usually thicker and more stout. Finally, the adults of butterflies and most skippers are generally active during the daylight hours (busily seeking out flowers), whereas the adults of moths are generally active during the nighttime hours.

Life Cycle. All lepidopterans have a complete metamorphosis, including egg, larva, pupa, and adult stages. The adult lepidopterans often use special sex attractants (pheromones) to bring the sexes together. After mating, the females deposit their eggs either singly or in groups on the appropriate food plant. The larvae, commonly called caterpillars, have chewing mouthparts and consume the leaves or other plant parts. These larvae are elongate and cylindrical, with a series of two to five prolegs on the abdomen. These prolegs are equipped with a series of small hooks (crochets) on the underside. Any "caterpillar" that has more than five pairs of prolegs is actually a sawfly (and a member of the order Hymenoptera). The caterpillars molt five to seven times and then develop into a pupa. The pupa of most butterflies is uncovered and called a chrysalis. The silk-covered pupa of most moths and skippers is referred to as a cocoon. The adult may emerge from the pupa within two weeks or may remain dormant until the following spring, depending on the species. When the new generation of adults appears the cycle starts once again.

Habits. The adults are free-flying, spending most of their time searching for food, avoiding enemies, and seeking a mate. Most are nectar feeders, but they also take other liquids, such as sap, juices of decaying fruits, manure liquids, and puddle water (which is rich in sodium). Some adult lepidopterans have vestigial or poorly developed mouthparts and do not feed at all. They live off food energy obtained as a caterpillar. The wing and body colors of adult lepidopterans are important for protection from enemies. Some color patterns are specially designed to conceal or disguise. Bright reds, oranges, and yellows are meant as a warning, helping predators avoid chemically protected, distasteful species. Many mimics take advantage of these warning colors for their own defense. Large circular eyespots can also be used to startle and distract would be predators. Color patterns may also play a role in temperature regulation and in the mating process, enabling male and female lepidopterans to see each other and to elicit the proper responses so that the mating process will be successful.

The caterpillars are associated with plants and consume the buds, flowers, foliage, fruits, stems, and wood. Some also feed on stored products, such as fabrics, foodstuffs, and beeswax. A very few are predators of other insects. Caterpillars are able to produce silk from a special silk-producing gland in the head that opens through an appendage (spinneret) located on the lower lip of the mouthparts. Silk is used for anchorage; building protective cases; constructing large webs, nests, or other shelters; and wrapping the pupa (forming a cocoon). Caterpillars lack wings as a means of escaping their enemies, so they must rely on other means, especially camouflage, mechanical defenses (like stiff bristles, stinging hairs, and sharp spines), or chemical defenses (usually obtained from the foliage they eat).

Habitat. Butterflies, skippers, and moths occur in all types of habitats, including open areas, grasslands, wetlands, woodlands, and urban and suburban habitats—basically wherever the necessary larval food plants and adult nectar plants are present. At least 17 species of Lepidoptera in the Great Lakes region are aquatic, with larvae that live in the water. Another 17 species are semiaquatic.

Ecological and Economic Status. Adult lepidopterans, especially the butterflies, are important plant pollinators. They serve as food—in both the adult and immature stage—for a host of animals, including birds, mammals, reptiles, amphibians, arachnids, and other insects. Many caterpillars are destructive pests, damaging stored food products and fabrics, defoliating trees and shrubs, and boring into wood, stems, and fruits. Virtually all the caterpillars that cause economic damage are moth larvae. There are only a small number of butterfly and skipper larvae that cause problems, most notably the caterpillars of the cabbage butterfly, alfalfa butterfly, and European skipper.

As an added bonus, many species of butterflies and moths are easily reared in the classroom or at home. Thus it is easy for both youth and adults to view firsthand the miraculous events of insect metamorphosis.

Distribution and Faunistics. As you might guess, because of their great popularity we probably know more about the distributions of butterflies, skippers, and moths than about any other group of insects. Thousands of species of lepidopterans inhabit the Great Lakes region, with the greatest numbers belonging to the many families of moths. Even relatively small geographical areas are likely to have hundreds (possibly over a thousand) species present. For example, in Ohio, 466 species of Lepidoptera were reported from Mohican State Park, 511 species were reported from Stark County, 428 species were reported from Carroll/Tuscarawas Counties, and 472 species were reported from Fulton County, (Rings and Metzler 1989, 1990, and 1991). Similarly, 419 taxa (species and subspecies) of Lepidoptera are known from Richland County, Ohio, and 655 additional species are also thought to occur there (Rings and Metzler 1989). The combined lists of Moore 1930 and Profant 1991 show that 757 species of Lepidoptera occur on the Beaver Islands in northern Lake Michigan.

When looking at just the butterfly and skipper fauna, we see that 149 butterflies and skippers are known to occur in Indiana (Shull 1987) and 144 species of butterflies and skippers are found in Michigan. To date, 78 species of butterflies have been found in Fulton County, Ohio (Rings and Metzler 1992), and at least 97 species of butterflies (no skippers reported) occur in Illinois (Sites and McPherson 1980). In the eastern part of the Great Lakes region, 141 species of butterflies and skippers are known from New York state (Shapiro 1974).

When looking at the moth fauna, we find that 602 species of moths are listed from the Douglas Lake area in northern lower Michigan (Voss 1969, 1981, 1983, and 1992). A total of 244 species of moths were taken by blacklight traps in southwestern Michigan (Rahn 1973). Some individual moth families are very large and contain many species. For example, 395 species of owlet moths (Noctuidae) are known from Minnesota (Knutson 1944) and 570 species are known to occur in Pennsylvania (Tietz 1936). A total of 286 different species of shoot moths (family Tortricidae, subfamily Olethreutinae) were reported from Michigan and Wisconsin (Miller 1987).

It has been estimated that the total number of Lepidoptera to actually occur in Ohio is around 2,500 species. Other states probably have lepidopteran faunas of comparable sizes, although the northern areas are probably not as species rich as the southern areas (like Ohio).

Because there are so many common and familiar butterflies, skippers, and moths occurring in the Great Lakes region, only brief descriptions and comments can be made on some of the more prominent families and species.

Skippers
(Family Hesperiidae)

Skippers are similar to butterflies in having long slender antennae (although with a hooked tip beyond the knobs), but in most of the temperate zone species the drab coloration and stout body shape are more mothlike. These small- to medium-sized lepidopterans fly rapidly and erratically, "skipping" through the air. In the field, you can recognize most skippers by the wing posture—the front wings and hind wings being held at different angles. Another group of skippers hold their wings outstretched and flat at rest.

The silver-spotted skipper, *Epargyreus clarus* (Cramer), has a distinctive orange band on the forewing (above and below) and a large silver band on the underside of the hind wing. Both the southern cloudywing, *Thorybes bathyllus* (Smith), and the northern cloudywing, *T. pylades* (Scudder), occur in the Great Lakes region. They are large, with a wingspan of 30–40 millimeters, and dark brown, and they inhabit open fields and similar areas. The southern cloudywing has a narrow central band with a large transparent spot; the widely distributed northern cloudywing has transparent spots that are small, triangular, and nonaligned.

The hoary edge, *Achalarus lyciades* (Geyer), is dusky brown above with a pale yellowish blotch in the center of the forewing and is especially fond of pine or oak scrubs with sandy soils. The underside of the hind wing has the outer half white. The sootywing, *Pholisora catullus* (Fabricius), is the most common small, black skipper in the region and is found in all types of open habitats, including fields, roadsides, vacant lots, and so on. It is very dark (sooty) with rows of fine white spots near the outer tip of the front wing (and on the hind wing in the female). The least skipper, *Ancyloxypha numitor* (Fabricius), is dark brownish with a large light brown area on the base of the front wing and an orangish area in the center of the hind wing. The hind wing beneath is bright yellowish orange. It inhabits wet, marshy areas. The Hobomok skipper, *Poanes hobomok* (Harris), has a brownish body and orangish wings with irregular black borders. It is commonly encountered in grassy glades near woods or streams.

Mostly a resident of tallgrass prairies, the powesheik skipper, *Oarisma powesheik* (Parker), is most common in extreme western Minnesota and Iowa but is also found in limited numbers in northwestern Indiana and southwestern Michigan. Specimens have also been taken in northwestern Illinois and Essex County, Ontario. It is black above with the leading edge and the veins of the forewing lined in orange; below it has pale brown hind wings with the veins lined in white. The ottoe skipper, *Hesperia ottoe* W. H. Edwards, is another prairie skipper that occurs in southern Wisconsin, northeastern and central Illinois, northwestern Indiana, and southwestern Michigan. Because of its limited habitat it is a species of special concern. It is entirely orangish with a distinct black stigma with grayish hairs in the center of the forewing.

An introduced species, the European skipper, *Thymelicus lineola* (Ochsenheimer), is entirely bright rusty orange with the wings bordered in black. It thrives in all types of open grassy areas and can be extremely abundant at times. The tawny-edged skipper, *Polites themistocles* (Latreille), is brownish with the base of the forewing (including the front edge) orange. An inhabitant of wetlands, the dun skipper, *Euphyes vestris* (Boisduval), is uniformly blackish brown with a black stigma (male) or two small indistinct whitish spots.

Swallowtail Butterflies
(Family Papilionidae)

Most of the species in this family of butterflies are characterized by having long "tails" on the hind wings. A few species lack tails. Also, not all butterflies with tails are swallowtails. The swallowtails are some of our largest native butterflies, reaching wingspans of up to 120 millimeters. Swallowtails can be found in a variety of open and wooded habitats.

Common species include the tiger swallowtail, *Papilio glaucus* L., which is largely yellow with black bands and iridescent blue (females) on the hind wing, and which is found in the southern portion of the region; the Canadian tiger swallowtail, *P. canadensis* Rothschild and Jordan, which is similar to *P. glaucus* but has broad stripes and orange overscaling below and is found in the northern portion of the region; and the black swallowtail, *P. polyxenes* Fabricius, which is black with a yellow band on the hind wings.

Less common species include the pipevine swallowtail, *Battus philenor* (L.), which has a large iridescent green patch and large orange spots on the underside of the hind wings; the zebra swallowtail, *Graphium marcellus* (Cramer), which is pale yellow with broad black stripes, has a long, narrow tail, and is found in southern portions of the region; the Old World swallowtail, *Papilio machaon* L., which is yellow and black with a reddish orange eyespot edged in black on the hind wing, and which is found only north of Lake Superior; and the giant swallowtail, *P. cresphontes* Cramer, which is very large, brown and yellow, and found in the southern portion of the region.

Fig. 173. Tiger swallowtail, *Papilio glaucus* L. (Courtesy of Michigan State University Extension.)

Whites, Sulphurs, and Orangetips
(Family Pieridae)

These small- to medium-sized butterflies are largely white, yellow, or orange, often with black markings at the wing tips or along the margins. The front legs are well developed and the tarsal claws are forked. Most members of this family favor open, sunny habitats. Many species in this family are migratory.

Common species include the mustard white, *Pieris napi* (L.), which is an unmarked white with the tip of the forewing more acutely pointed, not rounded; the West Virginia white, *P. virginiensis* (W. H. Edwards), which is also unmarked white, but with the tip of the forewing broadly rounded; the cabbage butterfly, *P. rapae* (L.), a widespread introduced species that is

Fig. 174. The cabbage butterfly, *Pieris rapae* (L.). (From Felt 1900.)

Fig. 175. The alfalfa butterfly, *Colias eurytheme* Boisduval. (Courtesy of USDA.)

white with black wing tips and two black spots on the front wing; the clouded sulphur, *Colias philodice* Godart, which has yellowish wings with black margins and is widespread throughout the region; the alfalfa butterfly, *C. eurytheme* Boisduval, which is similar to *C. philodice* but is yellowish with orange on the wings and sporadically occurs in northern portions of the region; and the pink-edged sulphur, *C. interior* Scudder, which is yellow with black wing margins and distinctive pink fringe around the wing margins.

One uncommon species is the Olympic marble, *Euchloe olympia* (W. H. Edwards), which is whitish with a faint greenish yellow marbled pattern on the underside of the hind wings, and occurs only in small localized populations. The dogface butterfly, *Colias cesonia* Stoll, is a seasonal migrant into the Great Lakes region. It has a distinctive yellow marking shaped like a dog's head (outlined in black) on the forewing.

Hairstreaks, Coppers, and Blues
(Family Lycaenidae)

These small- to medium-sized butterflies are colored in various shades of blue, gray, or copper, often highlighted with iridescence. Many have short, slender tails at the tip of the hind wing. The eye is notched near the base of the antennae. Several species and subspecies that occur in the Great Lakes region are rare and endangered, including the frosted elfin, Karner blue, northern blue, Edward's hairstreak, early hairstreak, and Henry's elfin. Many members of this family are commonly encountered species.

The harvester, *Feniseca tarquinius* Fabricius, has brownish wings with large pale orange markings and faint white scribble-like markings on the underside of the hind wing. The caterpillars, at least in their early stages, are carnivorous and feed on ants and leafhoppers. The pupa is interesting because its underside resembles a monkey's face.

The coppers lack tails on the hind wings and inhabit open, sunny habitats. The little copper, *Lycaena phlaeas* (L.), has shiny orange front wings with black spots; the hind wings are grayish brown with orange margins. This butterfly, once thought to be a native species, may be a very early introduction from Europe. The bog copper, *L. epixanthe* (Boisduval and LeConte)—which is small, purplish brown above, and grayish with spots below—is a resident of bogs and occurs in localized populations. The bronze copper, *Hyllolycaena hyllus* (Cramer), also inhabits wet areas but is not restricted to bogs. It is purplish brown above with an orange marginal band on the hind wings; the underside of the hind wings is grayish white with orange marginal bands.

The hairstreak butterflies are characterized by a pair of thin, hairlike tails on the hind wings. The coral hairstreak, *Satyrium titus* (Fabricius), which is grayish above with a row of coral red spots on the margin of the hind wing, is a resident of old fields and sparsely wooded hillsides. The brown elfin, *Incisalia augustinus* (W. Kirby), is uniformly chestnut brown and lacks tails, but the tip of the hind wing is lobed. The white M hairstreak, *Parrhasius m-album* (Boisduval and LeConte), has iridescent blue, black-margined wings above, and the underside of the hind wing has a distinct, white, M-shaped marking. It is a woodland resident found primarily in the southern portion of the region. The gray hairstreak, *Strymon melinus* (Huebner), is navy gray above and pale gray below with a banded row of orange spots on the hind wing. It inhabits a wide variety of open habitats. Since it is a subtropical species not adapted to harsh winter temperatures, it must recolonize the northern areas each season.

Various species of blues are among the tiniest of butterflies to be found anywhere in the world. As the name would imply, most species are predominately blue in color. They inhabit a variety of mostly open or forest edge habitats. The eastern tailed blue, *Everes comynatus* (Godart), is bluish brown above; the hind wings have a single tail and the undersides are pale grayish with orange spots along the margin. The spring azure, *Celastrina ladon* (Cramer), which is uniformly blue above, occasionally with a black margin on the forewing, is usually one of the first butterflies to appear in the spring.

Rare species in this butterfly family include the early hairstreak, *Erora laeta* (Edwards), which is black and blue above and jade green or grayish green below, has a narrow band of orange spots on both wings and no tails on the hind wings, and is a bog inhabitant; and the Karner blue, *Lycaeides melissa samuelis* Nabakov, which is blackish with blue wing bases and a continuous black subterminal band around the edge of the wings on the underside, is currently on the federal list of endangered species. It is restricted to local habitats in western New York, northwestern Ohio, southeastern and southwestern Michigan, northwestern Indiana, and central Wisconsin.

Metalmarks
(Family Riodinidae)

Most species in this family have shiny, metallic markings on the undersides of the wings, hence the common name for this family. The front legs on these small- to medium-sized butterflies are reduced and inconspicuous. Most species are highly localized, and at least one species (the swamp metalmark) is rare and possibly threatened.

Metalmarks are primarily tropical butterflies, so the number of species occurring in the Great Lakes region is quite small. The northern metalmark, *Calephelis borealis* (Grote and Robinson), is brownish with two wavy blue bands along the wing margins and occurs in southwestern Pennsylvania, central and northwestern Ohio, and eastern Indiana. The swamp metalmark, *C. muticum* McAlpine is bright rusty red with two blue bands along the margin of the wings and inhabits small areas of wetlands in central Ohio, southern Michigan, and southern Wisconsin.

Snout Butterflies
(Family Libytheidae)

These small butterflies are easily identified by their unusual elongate labial palps that protrude in front of the face in a beaklike or snoutlike fashion. The front legs are small. There is only one species in the United States, and it is an occasional migrant into the Great Lakes region. The American snout butterfly, *Libytheana carinenta* Cramer, has wings that are brownish and orangish; the forewing is squared at the tip and has four large white spots.

Brush-Footed Butterflies
(Family Nymphalidae)

This family of medium- to large-sized butterflies is large. Many of its species are characterized by wings that have irregular, angular edges. The underside of the wings are generally marked distinctly different from the upper surface. The front legs are stunted, held up against the body, and not used for walking. As a result, these and other butterflies with the same condition appear to be four-legged, not six-legged. The front legs of females are modified into sensory structures that are used to identify appropriate larval food plants.

The silvered fritillaries have rounded wing margins and silvery spots ("spangles") on the undersides of the wings. They are the largest of our fritillary species, with the regal fritillary reaching a wingspan of over four inches (100 mm). The most commonly encountered species of fritillary is the great spangled fritillary, *Speyeria cybele* (Fabricius), which is orangish with black markings above and is marked with white spangles and a broad yellowish band on the underside of the hind wing. The Aphrodite fritillary, *S. aphrodite* (Fabricius), is similar to *S. cybele*, but the spangles are slightly larger and there is no yellowish band. The Atlantis fritillary, *S. atlantis* (Edwards), is

Fig. 176. The great spangled fritillary, *Speyeria cybele* (Fabricius). (Courtesy of the Young Entomologists' Society.)

similar to both *S. aphrodite* and *S. cybele*, but it is smaller in size and has a narrow yellowish band on the underside of the hind wing. It is the smallest of our fritillaries and is restricted to northern portions of the region. The regal fritillary, *S. idalia* Drury, the largest of our fritillaries, is disappearing from much of its former range. The forewing is pale orangish with black spots, and the hind wing is dusky brown with orange spots near the margin and white spots near the center of the wing.

Several species of lesser fritillaries are restricted to the northern boreal regions and are frequently associated with bogs. They include the bog fritillary (*Boloria eunomia* Esper), Frigga's fritillary (*B. frigga* Thunberg), and the Freija fritillary (*B. freija* Thunberg). Because of the limited availability of suitable habitats in the Great Lakes region, most members of this genus are of special concern or threatened.

The checkerspots are suitably named for their prominent checkered markings. The most commonly encountered checkerspot butterfly is the Baltimore, *Euphydryas phaeton* (Drury), which has blackish wings marked with orangish spots along the wing margins and many yellowish white spots arranged in rows on the remainder of the wing surfaces.

The anglewings are easily recognized members of the brush-footed butterfly family. Their wings have a distinctively irregular and angular outline. Several species of anglewing butterflies are frequently encountered. The question mark, *Polygonia interrogationis* (Fabricius), is dark orangish and brownish above with black spots on the front wing and a dark hind wing with a white-tipped tail. There is a white shape like a question mark in the lower center of the hind wing on the underside. The comma, or hopmerchant, *P. comma* (Harris), is pale orangish with black spots. The tail of the hind wing is short and the underside of the hind wing is brown with a white comma-shaped marking in the center. The green comma, *P. satyrus* (Edwards), has the underside of the wings irregularly patterned in dark brown with greenish marginal markings on the hind wings. It occurs in the northern portion of the region. The gray comma, *P. progne* (Cramer), is grayish with fine lines of darker gray on the underside of the wings.

The tortoise shell butterflies also have an irregular margin to the wings, but the irregularity is not as pronounced as in the anglewings. Several species of tortoise shell butterflies are common, including the mourning cloak, *Nymphalis antiopa* (L.), which is uniformly dark brown with broad yellowish margin bands. It is one of the first butterflies to leave hibernation in the spring and can be seen as early as March in many parts of the Great Lakes region. Milbert's tortoise shell, *N. milberti* Godart, is dark above with broad orange bands bordered with yellow on the inward side of both wings.

Fig. 177. The green comma, *Polygonia satyrus* (Edwards). (Courtesy of the Young Entomologists' Society.)

Fig. 178. The gray comma, *Polygonia progne* (Cramer). (Courtesy of the Young Entomologists' Society.)

Fig. 179. The mourning cloak, *Nymphalis antiopa* (L.). (Courtesy of the Young Entomologists' Society.)

Fig. 180. The painted lady, *Vanessa cardui* (L.). (Courtesy of Michigan State University Extension.)

Fig. 181. The red admiral, *Vanessa atalanta* (L.). (Courtesy of the Young Entomologists' Society.)

Fig. 182. The viceroy, *Limenitis archippus* Cramer, a mimic of the monarch butterfly. (Courtesy of the Young Entomologists' Society.)

The lady, or thistle, butterflies also have a more or less irregular outline to the wings, but again the irregularity is not as pronounced as in the anglewings. The caterpillars of these butterflies feed on thistles, nettles, and a variety of other plants. Three species of lady butterflies are common in the Great Lakes region, including the American painted lady, *Vanessa virginiensis* (Drury), which is orange, black, and white above, with two large eyespots on the underside of the hind wing, each with blue in it. The painted lady, *V. cardui* (L.), is similar to *V. virginiensis*, but has four small eyespots on the underside of the hind wing, without any blue in them. The red admiral, *V. atalanta* (L.), is blackish above with a bright red band on the forewing and a broad red marginal band on the hind wing.

The buckeye butterflies are primarily subtropical, and one species, the buckeye, *Junonia coenia* Huebner, occasionally ranges into the Great Lakes region during the summer months. It is brownish above with three large eyespots; the dark spot on the forewing sets on a large, irregularly shaped, pale yellowish area.

The admiral and viceroy butterfly group contains at least two members that mimic other distasteful butterflies as a means of protection. The adults have a distinctive flap-and-glide flight. There are several commonly encountered species in this group. The white admiral, *Limenitis arthemis arthemis* Drury—which is black with broad white bands on the wings and red spots located on the hind wing, between a white band and the margin—is the northern subspecies of the white admiral/red-spotted purple butterfly complex. The red-spotted purple, *Limenitis arthemis astyanax* Fabricius, is dark black above with iridescent blue on the hind wing. The underside has prominent reddish spots at the wing bases and margins. It is the southern subspecies and is a mimic of the pipevine swallowtail. Intergrades between the two subspecies are common in many parts of the central Great Lakes region from Minnesota to New York. The viceroy, *Limenitis archippus* Cramer, is an orange and black-veined mimic of the monarch and can be distinguished by the narrow band of black across hind wing and by its generally smaller size.

Nymphs and Satyrs
(Family Satyridae)

These medium-sized, more or less drab brownish or grayish butterflies have large eyespots on the upper surface of the wings near the margins. They frequent shady, wooded habitats and are most often seen erratically flying from one hiding place to another. The front legs are stunted and short.

Commonly encountered species include the eyed brown, *Satyrodes eurydice* (Johansson), which is brownish with four submarginal eyespots that are similar in size and touching one another; the little wood satyr, *Megisto cymela* (Cramer), which is small and has only two eyespots on the forewing; and the common wood nymph, *Ceryconis pegala* Fabricius, which is brownish with a large yellow patch on the forewings that contains two eyespots.

Milkweed Butterflies
(Family Danaidae)

Milkweed butterflies are large butterflies that are generally bright orange, often with black markings. The larvae feed on various species of milkweed, obtaining chemical substances that make themselves and the adult stage distasteful to birds. Milkweed butterflies are primarily tropical and subtropical and must migrate into the Great Lakes region each spring and summer. Only two species occur in the region.

The monarch, *Danaus plexippus* (L.), is orange with black veins. The wing tips are black with white spots. This species has a regular, annual migration to its overwintering sites in Mexico. The monarch butterfly has been selected as the official state butterfly for Illinois. The queen, *D. gilippus* (Cramer), inhabits the southeastern United States but occasionally ranges north into the Great Lakes region during the summer months. It is orangish brown to chestnut brown with scattered white spots on the forewing; only the veins on the hind wing are black.

Fig. 183. The monarch, *Danaus plexippus* (L.), a world-famous migratory butterfly. (Courtesy of the Young Entomologists' Society.)

Clothes Moths
(Family Tineidae)

Although not particularly common these days (probably because of the widespread use of synthetic fibers in clothing), these small moths deserve a brief mention. Clothes moths are reclusive and hide in dark locations most of the time; they never

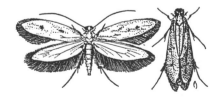

Fig. 184. The case-making clothes moth, *Tinea pellionella* (L.). (From Michelbacher and Furman 1951.)

fly to lights around the house (like the meal moths, or "millers," do). The larvae consume fabrics made from natural fibers, such as wool, furs, and feathers. The webbing clothes moth, *Tineola bisselliela* (Hummel), is shiny yellowish brown with the head rusty reddish and the hind wings paler than the front wings. The case-making clothes moth, *Tinea pellionella* (L.), is similar but often has three dark yellowish brown spots on each forewing.

Clear-Winged Moths
(Family Sesiidae)

These small- to medium-sized moths are readily distinguished by their more or less clear wings. The wing veins and margins are covered with scales, but the areas between the veins (cells) are largely uncovered and transparent. Many species are black-and-yellow or brownish and mimic bees and wasps (in appearance as well as in behavior). The caterpillars are borers in stems, vines, roots, and tree trunks.

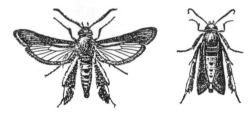

Fig. 185. The squash vine borer, *Melittia curcurbitae* (Harris). *Left,* adult male; *right,* adult female. (From Felt 1900.)

The squash vine borer, *Melittia curcurbitae* (Harris), is black with greenish iridescence and reddish abdomen and legs. The hind wings and a small portion of the base of the front wing are transparent. The hind legs have a long fringe of black hairs. The caterpillars infest vines of melons, squash, pumpkins, and cucumbers. The hornet moth, *Sesia apiformis* (Clerck), is a mimic of the hornets. It is blackish with yellow spots on the thorax and yellow bands on the abdomen. The lilac or ash borer, *Podosesia syringae* (Harris), looks amazingly similar to the paper wasp, having a dark brownish and reddish brown body with a whitish or yellowish streak on the side of the abdomen at the base. The wings are transparent in the basal half and brownish near the tip.

Snout and Meal Moths
(Family Pyralidae)

This very large family contains many economically significant species, including the Indian meal moth, *Plodia interpunctella* Huebner, and the meal moth, *Pyralis farinalis* (L.). The Indian meal moth is small with the forebody and base of the wings silvery tan. The outer half of the wings are brick red. This species commonly occurs in kitchens, infesting all types of foodstuffs, pet foods, seeds, and dried plant materials. The meal moth has the forewings patterned in alternating shades of light and dark brown, separated by irregularly shaped white lines (dark at the base and tip of the wing, light in the middle). It infests all types of stored grains, foodstuffs, and hay.

Fig. 186. The Indian meal moth, *Plodia interpunctella* (Huebner). (From Michelbacher and Furman 1951.)

Geometer Moths
(Family Geometridae)

These small- to medium-sized moths are generally light in color with irregular patterns of zigzag lines and bands on the wings (the patterns generally continuing from the forewing right onto the hind wing without interruption). The shape is also characteristic—a slender body and broadly, sometimes angled, wings. There are many species, and they are often difficult to identify without a good reference. The larvae, as a group, are easily distinguished by the reduced number of prolegs, which cause them to move in an unusual and distinctive looping fashion. First they anchor the front of the body and draw the rear of the body toward it, creating a humped appearance. Then they anchor the rear prolegs and extend the front of the body. This process is repeated over and over, at least until the caterpillar chooses to stop. This form of locomotion has resulted in a variety of interesting common names: loopers, spanworms, inchworms, and measuring worms.

Giant Silkworm and Royal Moths
(Family Saturniidae)

Our largest moths belong to this family, and many species commonly exceed four inches (100 mm) in wingspan. They are colored in various shades of brown, reddish brown, and grayish, often with conspicuous eyespots or crescent-shaped markings.

Fig. 187. The day-flying buck moth, *Hemileuca maia* (Drury). (Courtesy of the Young Entomologists' Society.)

Fig. 188. The luna moth, *Actias luna* (L.). (Courtesy of Michigan State University Extension.)

Fig. 189. The cecropia moth, *Hyalophora cecropia* (L.). (Drawing by S. Andres).

The sexes are easily distinguished by their antennae: those of the males are broadly featherlike, whereas those of the females are narrowly featherlike. The larvae are also large and are commonly encountered as they mature in the late spring and early summer. The mature larvae produce large cocoons of silk and leaf material, which are often attached to twigs and branches. The pupa in the cocoon is the overwintering stage, and the adults emerge the following spring and summer. The adults of most species are seldom seen, because of their nocturnal habits; but occasionally some individuals will be found near porch lights in the coolness of the early morning, unable to fly away. A few species are only locally abundant and are in danger of extirpation.

Several species of royal moths are found in the Great Lakes region. The imperial moth, *Eacles imperialis* (Drury), which is very large and dark yellow with numerous purplish gray specks and irregular wide bands on the wings, is an impressive species. The regal moth, *Citheronia regalis* (Fabricius), not to be outdone, is also very large, with the front wings blackish gray, the veins marked in rusty red, and seven white spots on the outer half of the wing. The hind wings are yellowish brown with lighter yellowish markings. The day-flying buck moth, *Hemileuca maia* (Drury), is large and has a black body and wings, with a broad whitish band on the middle of the wings. The front of the thorax is whitish and the tip of the abdomen is covered with long red hairs. The Io moth, *Automeris io* (Fabricius), is large and mostly yellow, with brownish spots on the front wings and a large, conspicuous eyespot in the middle of the hind wing. The inside edge of the hind wing is bright pink.

Several species of giant silkworm moths are common in the Great Lakes region. The polyphemus moth, *Antheraea polyphemus* (Cramer), is very large and mostly golden brown, with a circular eyespot surrounded by dark brown in the middle of the hind wing. The luna moth, *Actias luna* (L.), is unmistakable: it is very large and uniformly pale green with long tails on the hind wing. The front margin of the forewings and the front of the thorax are purplish red. The introduced Cynthia, or ailanthus, moth, *Samia cynthia* (Drury), is very large and mostly light brownish green with darker wing bases. There is a large, pale, crescent-shaped marking on the center of each wing. The promethea, or spicebush, moth, *Callosamia promethea* (Drury), is also very large and is dark purplish brown or reddish purple, with a dark eyespot near the tip of the forewing. The most commonly encountered species is the cecropia, or robin, moth, *Hyalophora cecropia* (L.). It is very large with the body and a submarginal band on the hind wing reddish and the wings deep brown to brownish red, with a crescent-shaped mark on the center of each wing and an eyespot near the tip of the front wing.

One other species, the Columbian silk moth, *H. columbia* (Smith), occurs in the Great Lakes region and is found in the vicinity of bogs and boreal forests. It is similar to the cecropia moth but lacks the red band on the hind wing.

Hawk Moths and Sphinx Moths
(Family Sphingidae)

Members of this moth family can be recognized by the characteristic shape of the wings, which are long and narrow, with the hind wing significantly smaller than the front wing. The smaller hawk moths (some also known as hummingbird moths) are active during the daylight hours and visit flowers to drink nectar. Some species are very similar in appearance and behavior to hummingbirds and are often mistaken for these birds. Other species, especially the larger sphinx moths, are nocturnal. The larvae have a prominent horn or spine on the tip of the abdomen and are often referred to as hornworms.

Fig. 190. A hawk moth of the genus *Sphinx*. (Drawing by S. Andres.)

Many species of hawk moths and sphinx moths occur in the Great Lakes, and only a few of the more common species can be mentioned here. The tomato hornworm, *Manduca sexta* (L.), and the 5-spotted hawk moth, *M. quinquemaculata* (Haworth), are large and gray with mottled brown markings. The abdomen has five large yellow spots on each side. The twin-spot sphinx, *Smerinthus jamaicensis* (Drury), has distinctive tan and rose hind wings with blue eyespots ringed in black and with a black bar through the center of the blue spot dividing it in two. The big poplar sphinx, *Pachysphinx modesta* (Harris), is large and grayish with a distinctive purplish red area in the center of the hind wing. The common clearwing, or hummingbird, moth, *Hemaris thysbe* (Fabricius), is reddish brown with large areas of the wings transparent and not covered with scales. The tip of the abdomen is dark. This and related species fly during the day and are remarkable mimics of bumble bees. The white-lined sphinx, *Hyles lineata* (Fabricius), is a frequently encountered species that varies considerably in size. The front wing has a

Fig. 192. A clearwing, or hummingbird, moth of the genus *Hemaris*. (From Kansas State University Extension 1962.)

Fig. 191. The tomato hornworm, *Manduca sexta* (L.). *Left,* hornworm caterpillar; *right,* adult moth. (Courtesy of the Young Entomologists' Society.)

prominent tan band from base to tip and white scales along the veins. The hind wing is black on the front and rear margin and pink in the center.

For an illustrated key to the hawkmoths of the eastern United States, see Selman 1975.

Tiger Moths
(Family Arctiidae)

Tiger moths are medium-sized moths that are typically brightly striped or spotted in white, pink, red, orange, or yellow. The caterpillars are equally distinct, with densely hairy bodies that often have long tufts of hairs near the head and tip of the abdomen and dense, brushlike clumps of hairs on the top of the abdomen in the middle of the body.

Fig. 193. The virgin tiger moth, *Apantesis virgo* (L.). (Drawing by S. Andres.)

There are many common species in the region, among them the banded woolly bear, or isabella moth, *Pyrrharctica isabella* (Smith and Abbot). It is buff or light brown with darker spots and stripes. The top of the abdomen has a row of dark spots down the center. The virgin tiger moth, *Apantesis virgo* (L.), has the forebody and forewings darkly colored and marked with white lines and cross bands. The rear of the thorax and the abdomen is bright pink and the hind wing is pinkish with large black spots. The milkweed tussock moth, *Euchaetes egle* (Harris), has grayish white wings and a yellow abdomen with a row of black spots down the center.

One atypical, though very distinct, tiger moth is the Virginia ctenuchid, *Ctenucha virginica* (Esper). It is dark bluish black with the head and thorax orange and the abdomen medium blue. It is a common daytime flier, moving about in a rapid manner.

Owlet, Cutworm, and Underwing Moths
(Family Noctuidae)

Noctuidae is the largest family of Lepidoptera, with hundreds of species occurring in the Great Lakes region. They range in size from small to large. Many of the smaller cutworm moths are salt-and-pepper gray with minute differences in color or smaller markings. Their larvae, known as cutworms, are serious agricultural pests. The underwing moths (*Catocala* species) are larger and also have cryptic salt-and-pepper gray front wings, but the hind wings (which are generally kept concealed beneath the forewings at rest) are brightly patterned in black and either red, orange, yellow, or pink. This latter group, because of their larger size and colorful hind wings, are better known than most other genera within the family Noctuidae.

Fig. 194. An owlet (noctuid) moth of the genus *Mamestra*. (From Felt 1900.)

Many species of underwing moths are known to occur in the Great Lakes region, and only a few can be mentioned here. The white underwing, *Catocala relicta* Walker, has distinctive front wings that are predominantly white with gray markings and black hind wings with a broad white band. The pink underwing, *C. concumbens* Walker, has light gray front wings with white markings and pink hind wings with two broad black bands and a white margin. The darling underwing, *C. cara* (Guenee), has gray brown front wings with black lines and deep pink hind wings with two black bands. The wonderful underwing, *C. mira* Grote, has grayish front wings with a black-bordered white spot near the middle and deep orange yellow hind wings with two irregular black bands.

For identification information on the owlet moths of Ohio, see Rings et al. 1992.

Identification Resources. A number of excellent field guides, identification guides, and keys are available on the Lepidoptera. For butterfly identification consult *A Field Guide to Eastern Butterflies* (Opler and Malikul 1992), *The Audubon Society Field Guide to North American Butterflies* (Pyle 1981), *Butterflies of Indiana* (Shull 1987), *How to Know the Butterflies* (Ehrlich and Ehrlich 1961), *Butterflies of New York* (Shapiro 1974), *A Key to the Butterflies of Illinois* (Sites and McPherson 1980), *The Ontario Butterfly Atlas* (Holmes et al. 1992), and *The Butterflies of the Great Lakes Region* (Douglas, n.d.). For moth identification consult *A Field Guide to the Moths of Eastern North America* (Covell 1984) or *The Moth Book* (Holland 1968).

True Flies
(Order Diptera)

Introduction. The word *fly* is used in the name of many different types of insects, and most of them are not "true flies." The term *true fly* is used to differentiate those insects with a single pair of wings and sponging, sucking, or biting/piercing mouthparts from all other types of insects.

Description. The flies are distinguished from other insects by the single pair of membranous front wings. The hind wings are reduced and modified into knoblike halteres, which are used as organs of balance while in flight. All other two-winged insects (stylopids, male scales, atypical mayflies, and grasshoppers) lack these halteres. True flies are soft bodied and range in size from minute (2 mm) to large (30 mm). Their bodies are often covered with hairs, bristles, or spines. Most have the same general body

Fig. 195. The white underwing moth, *Catocala relicta* Walker. (Drawing by S. Andres.)

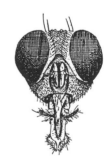

Fig. 196. The head of the house fly, *Musca domestica*, showing the compound eyes, aristate antennae, and sponging mouthparts. (From Knowlton 1951.)

Quick Guide to Identification

True Flies
(Order Diptera)

Antennae Long (Six-Segmented)

Diagnostic Characteristics	Common Name(s)	Scientific Family Name
Mothlike and hairy	moth and drain flies	Psychodidae
Body, wings, and legs with scales	mosquitoes	Culicidae
Scales on margin of wing only	phantom midges	Chaoboridae
Mosquitolike; long-legged; large	crane flies	Tipulidae
Mosquitolike; feathery antenna; small	midges	Chironomidae
Crane fly–like; only one anal vein	phantom crane flies	Ptychopteridae
Small; humpbacked (buffalo-like)	black flies	Simuliidae
Antennae placed low on face	March flies	Bibionidae
Fewer than seven wing veins; long legs	gall midges	Cecidomyidae

Antennae Short (Three-Segmented)

Diagnostic Characteristics	Common Name(s)	Scientific Family Name

Size Minute to Small

Wingless (or scalelike wings)	hippoboscids	Hippoboscidae
Metallic abdomen; long legs	long-legged flies	Dolichopodidae
Metallic sheen; pictured wings	picture-winged flies	Otitidae
Wings banded/spotted; nonmetallic	fruit flies	Tephritidae
Wings faintly spotted; antennae protrude from face	marsh flies	Sciomyzidae
Small; shiny black; shrunken abdomen	black scavenger flies	Sepsidae
Small; dull; break in costal vein	pomace flies	Drosophilidae
Minute; yellowish; break in costa	frit flies	Chloropidae
Small; front of face bulging out	shore flies	Ephydridae
Humpbacked with head placed low on thorax	humpbacked flies	Phoridae

Diagnostic Characteristics	Common Name(s)	Scientific Family Name
Size Small to Medium		
Spurious vein; often beelike	flower flies	Syrphidae
Third antenna segment ringed; biting mouthparts	horse and deer flies	Tabanidae
Abdomen long and tapered; long legs	snipe flies	Rhagionidae
Hairy, pear-shaped body	bee flies	Bombyliidae
Wings faintly spotted; antennae protrude from face	marsh flies	Sciomyzidae
Small; front of face bulging out	shore flies	Ephydridae
Size Medium to Large		
Spurious vein; often beelike	flower flies	Syrphidae
Third antenna segment ringed; biting mouthparts	horse and deer flies	Tabanidae
Third antenna segment elongate; often with yellow spots	soldier flies	Stratiomyidae
Third antenna segment elongate and swollen; large size	mydas flies	Mydidae
"Bearded" face; top of head concave	robber flies	Asilidae
Metallic or with gold hairs on side; antennal arista feathery	blow flies	Calliphoridae
Thorax and abdomen with many bristles; antennal arista bare	parasitic flies	Tachinidae
Gray-and-black "checkerboard" abdomen; antennal arista plumose on base only	flesh flies	Sarcophagidae
Robust, beelike (near livestock)	bot and warble flies	Oestridae
Second anal vein short; dull colors	muscid flies	Muscidae

shape: globular head; stout, humpbacked thorax; and elongate to stout abdomen. The midsection of the thorax is enlarged because of all the wing muscles housed within. The compound eyes are large and often meet in the middle of the head, especially in the males. The antennae are short and inconspicuous, with fewer than six segments (quite often with three segments). The mouthparts are of the sucking variety and may be modified for piercing and sucking, sponging, or lapping. The larvae are legless (often maggotlike) and with or without a distinct head capsule.

Life Cycle. True flies have a complete metamorphosis, with egg, larva, pupa, and adult stages. The life cycle may be as short as two weeks in some species of flies, but for most it requires a full year.

Habits. Adult flies are free-roaming, spending most of their time in the vicinity of their food or larval habitat. Most feed on liquid foods, like blood, sweat, tears, nectar, sap, fruit juices, or manure/carrion liquids. Some are predaceous or parasitic and are found around vertebrate animals. Others are gall makers and leaf miners. Some species of flies mimic bees in appearance and behavior for their own protection.

Habitats. Flies inhabit a wide variety of moist terrestrial and aquatic habitats. Generally they are found associated with other animals as scavengers (on their dung, body fluids, or decaying flesh) or as predators and parasites, or they are found around plants (in fungi, humus, fruit, galls, or leaves). Some, such as mosquitoes, gnats, black flies, and other biting flies, are associated with aquatic habitats, especially in the larval stage.

Ecological and Economic Status. As you might expect with any large group of insects, some species are beneficial while others are pests. Flies and their larvae are beneficial as food for other animals, as pollinators, scavengers, and recyclers, or as predators of undesirable insect pests. Unfortunately, many species are pests, such as leaf miners, gall makers, root borers, bloodsuckers, and internal parasites (endoparasites). Some species are involved in the spread of germs and the transmission of diseases. It is estimated that more than 50 percent of the world's population is ill from fly-borne pathogens and parasites. Likewise, crop losses from fly pests are exceeded by few other insects (Arnett 1985).

Distribution and Faunistics. Most members of this group have widespread distributions in the Great Lakes region. Most of the families containing bloodsucking pests (mosquitoes, black flies, and horse flies) have been reasonably well studied in most areas. No all-inclusive faunal studies on Diptera have been conducted in the region. Most investigations have been carried out on a family-by-family basis. These faunal studies are mentioned under the appropriate families.

Primitive Flies
(Suborder Nematocera)

These flies are characterized by longer antennae, generally with six segments and with a bristlelike arista.

Crane Flies
(Family Tipulidae)

Tipulidae is the largest dipteran family. These medium- to large-sized, long-legged, mosquitolike flies do not bite, and they are not covered with body scales. They are usually brownish, and the wings may be clear or darkened brown or black. The large V-shaped suture (impressed line) on the middle of the thorax is also diagnostic for this group of flies. The adults are usually found in damp, wooded areas near larval habitats (mud, decaying bark, or rotting vegetation).

The most thorough study on crane flies in the Great Lakes region reports 132 species occurring in Wisconsin (Dickenson and Alexander 1932). In southeastern Michigan, at the Edwin S. George Reserve, a total of 198 species of crane flies were found. It was estimated that this accounted for 90 percent of the total number of species that actually occurred in the area (Rogers 1942). Similarly, 153 species of crane flies have been reported from Ohio.

For a key to the crane flies of Wisconsin, see Dickenson and Alexander 1932.

A few recognizable species that occur in the Great Lakes region include the following. The smoky crane fly, *Tipula cunctans* Say, has a thorax with grayish brown stripes and a brown abdomen. The wings are slightly clouded. The three-striped crane fly, *T. trivittata* Say, is brownish. The thorax has a pale stripe down the center and the abdomen has three brown stripes. The wings are tinted brown. The brown crane fly, *Nephrotoma ferruginea* (Fabricius), is brownish yellow with rusty red stripes on the thorax. The abdomen has row of dark spots down the center.

Fig. 197. The brown crane fly, *Nephrotoma ferruginea* (Fabricius). (From Kansas State University Extension 1962.)

Phantom Crane Flies
(Family Ptychopteridae)

These flies resemble crane flies but differ slightly in wing venation. The one species that occurs in the Great Lakes region is so distinctive that no family characteristics are necessary. The phantom crane fly, *Bittacomorpha clavipes* (Fabricius), is grayish or brownish with long legs. The first tarsal segment is greatly enlarged and the legs are striped in black and white.

Since this fly often flies with its legs outstretched, its identity is easily recognized, even from a considerable distance.

Moth Flies
(Family Psychodidae)

These small, mothlike flies are distinctive and easily recognized. The body and wings are hairy and held rooflike over the body. Since the larvae are aquatic or semiaquatic, the adults are usually found in damp, shady places near water. The drain fly, *Telmatoscopus albipunctatus* (Williston), commonly breeds in drains, garbage disposals, toilets, and sewers and is frequently found in buildings.

March Flies
(Family Bibionidae)

Although commonly referred to as march flies, it would be quite exceptional to have these flies active in the Great Lakes region during the month of March, since they are generally dependent on flowers. These dark-colored, hairy flies have short antennae that arise low on the face, and they have tibia spurs on the legs. Only one common species is likely to be encountered in the Great Lakes region. The white-legged march fly, *Bibio albipennis* Say, is shiny black with dense grayish hairs. The veins and wing spots are yellowish brown, and the legs are long and white.

Gall Midges
(Family Cecidomyidae)

These minute flies have long legs and antennae. The wings have greatly reduced venation with fewer than seven longitudinal veins. Most of the members of this family are gall makers (with a characteristically shaped gall for each species), but some are stem borers and leaf feeders. A few are predaceous. The pinecone willow gall (actually a greatly enlarged terminal bud) is caused by *Rhabdophaga strobiloides* (Osten Sacken). The maple blister gall (circular yellow leaf blisters with a red margin) are caused by *Cecidomyia ocellaris* Osten Sacken.

Phantom Midges
(Family Chaoboridae)

These flies closely resemble mosquitoes but have the scales confined to the margins of the wings and have short, nonbiting mouthparts. One of the species likely to be encountered is *Chaoboris flavicans* (Meigen), which is pale with dark longitudinal markings on the middle of the thorax. The wings are transparent.

For a key to the phantom midge larvae of northern Michigan, see Roth 1967.

Mosquitoes
(Family Culicidae)

Mosquitoes are familiar to anyone who has been outdoors during the spring and summer months. These small, delicate flies have long legs and scale-covered bodies and wing veins. The mouthparts of the female are elongate and adapted for blood sucking (and nectar sipping). The mouthparts of the male are shorter and are used exclusively for feeding on nectar. The males also have feathery (plumose) antennae. Mosquito larvae are aquatic and inhabit many types of still water, where they feed on algae, protozoans, and detritus. Female mosquitoes are major vectors of several diseases, including yellow fever, encephalitis, human malaria, bird malaria, and heartworm in dogs.

Many faunal studies of mosquitoes in the Great Lakes region have been conducted over the years, including those for Illinois (Ross 1947), Indiana (Hart 1968), Michigan (Cassani and Bland 1978, Irwin 1941, Matheson 1924, and Sabrosky 1944), Minnesota (Barr 1957 and Owen 1937), New York (Felt 1904), Ohio (Evans 1909, Masters 1949, Restifo 1982, and Venard and Mead 1953), Ontario (Benedict 1962, Judd 1950b, Steward and McWade 1961, and Wood et al. 1979), and Wisconsin (Amin and Hageman 1974 and Dickenson 1944). At least 64 species of mosquitoes in 10 genera occur in the Great Lakes region.

There are many significant species of the Great Lakes region, and only a few can be discussed here. The pitcher plant mosquito, *Wyeomyia smithii* (Coquillet), is brightly covered with silvery thoracic scales. The larvae develop inside the pitcher plant and are immune to the digestive juices of this unique bog-inhabiting plant. The gallinipper, *Psorophora ciliata* (Fabricius), is large and dark bodied; its head and thorax have patches of white and yellow, and the abdomen has patches of iridescent blue. The floodwater mosquito, *Aedes sticticus* (Meigen), has a pale yellowish head, two yellow brown lines down the center of the thorax, and a white patch on the first segment and narrow white cross bands on the rest of the abdomen. The vexans mosquito, *A. vexans* (Meigen), has a brown forebody and a black abdomen with white rings. There are white rings at the tip of the tarsi. The northern house mosquito, *Culex pipiens pipiens* L., is brownish with the abdomen dark with pale basal bands connected to white lateral patches. It commonly overwinters inside buildings and adults may be seen at just about any time of the year.

Fig. 198. The mosquito. *Top,* aquatic larva; *bottom,* aerial adult. (Courtesy of Michigan State University Extension.)

For identification keys to the mosquitoes of Ontario, see Steward and McWade 1961 and Wood et al. 1979. For a key to mosquitoes of Ohio, see Restifo 1982. For a key to container-breeding species of mosquitoes in Michigan, see Wilmot et al. 1992. For a key to first instar aedine larvae of Minnesota, see Price 1960.

Black Flies
(Family Simuliidae)

Fig. 199. The striped black fly *Simulium vittatum* Zetterstedt. (From Knowlton and Rowe 1934.)

These minute biting flies are typically associated with the north country but are actually common wherever larval habitat is available. They are stocky, humpbacked, and short legged, with short antennae. The wing venation is simple, with the veins near the front of the wing thick and distinct, the others thin and indistinct. Females are blood feeders. The larvae inhabit streams and cling to stones, sticks, and other objects by means of tiny hooks at the tip of the abdomen.

Because of their biting ability and annoying behavior, many faunal studies of black flies in the Great Lakes region have been conducted over the years, including those for Michigan (Merritt et al. 1978), Minnesota (Nicholson and Mickel 1950), Ontario (Davies et al. 1961 and Wood et al. 1963), Pennsylvania (Adler et al. 1982 and Frost 1949), and the northeastern United States (Cupp and Gordon 1983). In total, at least 61 species of black flies are known from northeastern North America (the Great Lakes region plus New England).

There are many common species, and only a few of the significant ones can be mentioned here. *Simulium vittatum* Zetterstedt is dark gray with five dark longitudinal thoracic stripes and the basal half of the tarsi yellow. *S. venustum* Say is shiny black with thorax and legs yellowish or yellowish and white.

For identification keys to black fly larvae of Ontario, see Wood et al. 1963; for adults, see Davies et al. 1961. For keys to the larvae of Michigan black flies, see Merritt et al. 1978.

Midges
(Family Chironomidae)

These small, mosquitolike flies have feathery antennae (males) and elongate (often hairy) front tarsi. The proboscis is very short and they do not bite. They are generally common and abundant (dancing about in huge swarms), occurring most frequently in moist environments. They are frequently attracted to lights, where they can be a considerable nuisance problem. The reddish larvae live in oxygen-poor aquatic environments. The red coloration of these so-called bloodworms is due to the presence of hemoglobin. Midge larva are one of the few insects known to have hemoglobin in their body.

There are two commonly encountered species in the Great Lakes region. The common midge, *Chironomus attenuatus* Walker, is small (3.5–4 mm) and pale green to light brown, with thoracic markings and abdominal bands of darker brown. The plumose midge, *C. plumosus* (L.), is larger (6–7.5 mm) and pale brown, with thoracic markings and abdominal bands of darker brown.

For identification keys to the genera of larval midges from Canada and the adjacent United States, see Oliver and Roussel 1983. For a key to the larvae of *Polypedilum* midges of northeastern North America, see Boesel 1985; for a key to larvae of the *Cricotopus* midges, see Boesel 1983.

Short-Horned Flies
(Suborder Brachycera)

The flies in this suborder have short, three-segmented antennae. A few families have the terminal segment of the antennae ringed, with the appearance of indistinct segmentation.

Horse and Deer Flies
(Family Tabanidae)

This infamous group of medium- to large-sized biting flies is well known to most people. Their habit of circling the head (always seemingly out of reach) and unexpectedly biting in the middle of the back or on the leg is an all too frequent summertime experience. Color varies from black to yellowish brown, often with stripes on the abdomen. The eyes are large and often iridescent green (hence the common name "greenheads") or purple. The males are not bloodsucking and feed primarily on nectar. The predaceous larvae live in semiaquatic (muddy soil) or aquatic habitats.

It is often mentioned that certain insects have affected human history, either directly or indirectly. Horse flies are more or less responsible for the selection of July 4th as Independence Day. It is said that the Declaration of Independence was signed on July 4, 1776—instead of at a later date that would permit further discussion—because the horse flies in Philadelphia were biting so fiercely at the time that the delegates decided to adjourn just to get away from them (Arnett 1985).

The horse fly and deer fly fauna of the region is reasonably well studied, including that of Illinois (Pechuman et al. 1983), Indiana (Burton 1975 and Myers and Sanders 1975), Michigan (Hays 1956 and Strickler and Walker 1993), Minnesota (Pechuman 1931 and Philip 1931), New York (Pechuman 1957 and 1972), Ohio (Drees 1982), Ontario (Judd 1958 and Pechuman et al. 1961), Pennsylvania (Frost and Pechuman 1958), and Wisconsin (Amin and Hageman 1974, Morris and DeFoliart 1971, and Roberts and Dicke 1958). The number of species reported from the region is close to 100, with 86 species for Illinois (Pechuman et al. 1983), 87 species for Ontario (Pechuman et al. 1961), 90 species for Ohio (Drees 1982), and 96 species for Michigan (Hays 1956).

Many species are common, and only a few of the more widely distributed tabanids in the Great Lakes region can be included here.

Deer flies are characterized by their smaller size (10–12 mm), blackish or brownish coloration, and distinctively banded wings. Significant species of deer flies in the Great Lakes region include the black deer fly, *Chrysops niger* Macquart, which has a black body with some white pubescence and wings banded in black. The striped deer fly, *C. vittatus* Weidemann, which is yellowish with four black stripes on the abdomen and with the wings banded in black. The deer fly *C. callidus* Osten Sacken is black with yellow triangular markings down the center of the abdomen, large yellow spots on the sides of the abdomen at the base, and brown banding on the wings. The carbon-colored deer fly, *C. carbonarius* Walker, is black with yellow gray and green gray markings and with black banding on the wings.

Fig. 200. A deer fly of the genus *Chrysops.* (Courtesy of Michigan State University Extension.)

Horse flies are characterized by their larger size (greater than 13 mm), brownish or blackish bodies, and unbanded wings. Common species of horse flies include the greenheaded horse fly, *Hybomitra lasiophthalma* (Macquart), which has a black thorax with gray stripes and a brown abdominal tip. The eyes are green with purple bands and the wings are lightly spotted with brown. This species loves to bite wet skin and is a real problem at beaches. The American horse fly, *Tabanus americanus* Forester, is reddish brown with clear wings. The black horse fly, *T. atratus* Fabricius, is uniformly black with the wings dark brown or blackish. The striped horse fly, *T. lineola* Fabricius, is brown with a reddish tint. The thorax and abdomen are striped with yellow or yellowish red, and the abdomen has a distinct white stripe down the center. The compound eyes are purple with three green stripes.

Fig. 201. A horse fly of the genus *Tabanus.* (Courtesy of Michigan State University Extension.)

For identification keys to the horse flies and deer flies of Illinois, see Pechuman et al. 1983; for Indiana, see Burton 1975; for Michigan, see Hays 1956; for Ohio, see Drees 1982; for Ontario, see Pechuman et al. 1961; for Pennsylvania, see Frost and Pechuman 1958; for Wisconsin, see Amin and Hageman 1974.

Snipe Flies
(Family Rhagionidae)

These uncommon flies are small to medium sized and generally long legged. The abdomen is also elongate and tapers toward the tip. Adults of most species of snipe flies are predaceous, but a few bite humans. The only significant species to occur in the Great Lakes region is the snipe fly, *Rhagio mystacea* (Macquart). It is small (6–8 mm) and has a thorax that is yellow with four dark stripes and an abdomen that is dark with light stripes. The eyes are large and reddish brown.

Soldier Flies
(Family Stratiomyidae)

These small- to medium-sized flies are usually darkly colored, but some are marked with yellow, green, or blue markings and stripes on the abdomen. The third segment of the antennae is usually elongate and prominent. The most commonly encountered species are wasplike in appearance, but others have a dorsoventrally flattened abdomen. Soldier flies are most commonly seen on flowers. The soldier fly fauna of the Great Lakes region is small; for example, only thirteen species were found at Cedar Point in Sandusky, Ohio, when studied by Fulton (1911).

The black soldier fly, *Hermetia illuscens* (L.), is medium sized and black with two large yellowish spots on the base of the abdomen. The wings are smoky. Meigen's soldier fly, *Stratiomys meigenii* Weidemann, is black with gray pubescence on the side of the thorax, curved yellow stripes on the sides of the abdomen, and a yellow line in the center of the abdomen at the tip.

Mydas Flies
(Family Mydidae)

These large, distinctive flies have the third antennal segment elongate and swollen. The hind legs are longer and thicker than in most other types of flies. Most species are black and wasplike and may have orange or yellow markings on the abdomen. Only one species, the clavate mydas fly, *Mydas clavatus* (Drury), is likely to be encountered in the southern portions of the Great Lakes region. It is black with yellow spots on the second abdominal segment and red and yellow markings on the hind legs.

Fig. 202. A robber fly of
the genus *Promachus*
showing elongate body
and "bearded" face.
(From USDA 1925.)

Robber Flies
(Family Asilidae)

These medium-sized flies are variable in their body shape and coloration, but they all have a densely hairy, "bearded" face with the top of the head concave. Most species have relatively long, spiny legs for grabbing their prey. Some species are stout and hairy and resemble bumble bees. At least a dozen Ohio species are reported to be mimics of various Hymenoptera (both bees and wasps) (Bromley 1950). One species, *Bombomima thoracica* Fabricius, is a very good mimic of the female bumble bee and prompts many beekeepers to complain about "the bumble bee that catches my honey bees." Other species have slender or tapered abdomens.

All robber flies are predaceous and actively attack other insects, such as flies, bees, grasshoppers, and beetles. The bumble bee robber fly *B. thoracica* has also been known to capture 17-year cicadas when they are available. Robber flies prefer to sit on sunlit perches, often flying off suddenly to catch nearby prey. When attacking beetles they must attack while the beetles are on the wing. When beetles have their flight wings folded and their elytra in place there is no way the robber fly can insert its mouthpart into the beetle to feed. But when the beetle flies, its wings are held upward and outward exposing the soft surface of the thorax and abdomen. By the time the beetle is aware it is being attacked, it is too late. Other species of robber flies are known as "beewolves" or "beekillers" because they prey on honey bees. In Ohio 20 different species of robber flies were observed with honey bee prey (Bromley 1950).

Nearly a hundred robber fly species occur in the Great Lakes region. For example, 79 species (72 documented and 7 probable) are reported from Michigan and 94 species are reported from Ohio.

Some of the commonly encountered species of robber flies include *Leptogaster flavipes* (Loew), which is medium sized, wasplike, and dull reddish brown; *Laphria flavicollis* Say, which is medium sized and beelike and has a black body with long yellow hairs; *Efferia aestuans* (L.), which is large, has a grayish thorax and darker abdomen with yellowish spots near the tip, and is especially fond of catching muscid flies; and *Asilus sericeus* Say, which is large and has a brown body, brownish wings, and pale brown legs.

For an identification key to the robber flies of Michigan, see Baker and Fischer 1975.

Bee Flies
(Family Bombyliidae)

These small- to medium-sized flies are characterized by stout, hairy bodies and slender, patterned wings. These flies are often seen hovering above the ground, suddenly darting a few feet away and hovering again. The adults visit flowers and feed on nectar and pollen. The larvae are parasites of grasshoppers, beetles (especially larval tiger beetles), bees, and wasps.

A few prominent species include the black-tailed bee fly, *Bombylius major* L., which has a blackish body covered with dense, paler hairs, wings with a dark brown front half, and a proboscis that is long and protruding; the grasshopper bee fly, *Systoechus vulgaris* Loew, which is beelike with a short proboscis; the analis bee fly, *Anthrax analis* Say, which is blackish, with the tip and sides of the abdomen covered with pale hairs and the base of wings dark brown, and which is a parasite of larval tiger beetles; and the tiger bee fly, *A. tigrinus* (De Geer), which is reddish brown with some black hairs at the base of the abdomen and with white hairs on the upper half and is a parasite of carpenter bees.

Long-Legged Flies
(Family Dolichopodidae)

These small flies are recognized by their slender and usually metallic green abdomen and by their long legs. Although relatively common, they are seldom seen or collected, because of their small size. Males of some species are decorated with tufts of scales on their legs and antennae and/or palps, and these attributes are used in the mating display. Dolichopodidae inhabit shaded areas near running water but are also common in wet meadows and swamps. Most of the species in the family occurring in the Great Lakes region belong to the genus *Dolichopodus*.

Humpbacked Flies
(Family Phoridae)

These small flies have a distinctive body shape—a large, humpbacked thorax and a small head attached low on the thorax. Most species are yellowish, brownish, or blackish. The wing

veins are few in number and only those near the front edge of
the wing at the base are thickened and distinct. The hind femora
are flattened. Humpbacked flies are found around decaying vegetation and other organic debris, including drains and sewers.

Flower Flies
(Family Syrphidae)

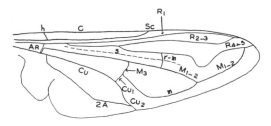

Fig. 203. The right wing of a flower fly. Note the spurious vein running lengthwise
near the middle of the wing (labeled "s"). (From Telford 1939.)

Fig. 204. A beelike flower
fly of the genus *Syrphus*.
(From Metcalf 1913.)

Flower flies are medium sized and generally black with yellow
spots or bands on the abdomen. Some are smooth with few
hairs, appearing somewhat wasplike, while others are densely
hairy and beelike. Some species even make a buzzing sound
when in flight. Despite the variation in size, color, or shape
there is one constant diagnostic characteristic: a spurious vein.
This more or less indistinct vein runs lengthwise in the middle
of the wing and is not connected to any other vein at either end
of its length. Many flower flies hover in the air above flowers,
darting about from time to time. They feed on pollen and nectar
and they are important pollinators.

More than a hundred species of flower flies inhabit the
Great Lakes region. For example, 113 species are reported from
Ohio (Metcalf 1913) and 135 species are reported from Min-
nesota (Telford 1939).

A few species are widespread and commonly encountered.
The drone fly, *Eristalis tenax* (L.), is brownish and yellow,
resembling a large honey bee. The American flower fly, *Meta-
syrphus americanus* (Weidemann), is metallic green, with three
yellow abdominal cross bands. The oblique flower fly, *Allo-
grapta obliqua* Macquart, has a shining green thorax with a yellow stripe on each side. The abdomen is dark brown with two
yellow orange cross bands.

For an identification key to the flower flies of Ohio, see
Metcalf 1913; for a key to the Minnesota species, see Telford
1939.

Picture-Winged Flies
(Family Otitidae)

These small- to medium-sized flies are recognized by their bright, metallic bodies and spotted ("pictured") wings. The anal cell (near the hind angle at the base of the wing) has a fingerlike projection pointing toward the rear edge of the wing. Adults are generally found in moist habitats, such as wet meadows and other wetlands.

One of the most common species is *Delphinia picta* (Fabricius), which is reddish brown with a yellow scutellum and stout bristles on the thorax. The wings are "pictured" in a reddish brown and opaque white pattern.

Fruit Flies
(Family Tephritidae)

This family contains many small- to medium-sized flies that are agricultural (orchard) pests. The wings are banded or spotted, and the anal cell (near the hind angle at the base of the wing) has a fingerlike projection, as in the picture-winged flies. The fruit flies have a subcostal vein that is short and bends forward at a right angle before ending just short of the leading edge of the wing (costal vein). The larvae infest many types of fruits, including apples, crab apples, cherries, blueberries, and citrus fruits (in the south).

Two species of significance are the cherry fruit fly, *Rhagoletis cingulata* (Loew), which has a thorax that has an irregular stripe on the sides and wings with four cross bands, bands three and four united at the front edge of the wing with a small spot at the tip of the wing; and the apple maggot fly, *R. pomonella* (Walsh), which has a dark brown to black thorax that is paler on the underside and wings with four cross bands, bands one and two united at the rear edge of the wing and bands two through four united along the front edge of the wing.

Fig. 205. The apple maggot fly, *Rhagoletis pomonella* (Walsh). (Courtesy of Michigan State University Extension.)

Marsh Flies
(Family Sciomyzidae)

Marsh flies are small- to medium-sized flies that are typically brownish, yellowish, or grayish, often with dark spots on some of the cross veins. The antennae project forward in a distinctive fashion. The adults are commonly found around marshes, swamps, wet meadows, ponds, bogs, or damp woodlands.

One of the more common species is the marsh fly, *Poecilographa decora* (Loew). It has a yellowish body and indistinct brown spots on the wings. The body only has a few bristles, but those on the head and thorax arise from conspicuous brown spots.

Black Scavenger Flies
(Family Sepsidae)

These small, shiny, black flies have a characteristic body shape. The head is round and the abdomen is elongate, narrow, and constricted at the base. Palps are lacking. These flies are generally found around manure, carrion, and rotting vegetation.

One of the most common species is *Sepsis punctum* (Fabricius), which is purplish black with a dull black spot on the front edge of the wing near the tip. The abdomen has several prominent bristles.

Fig. 206. The black scavenger fly, *Sepsis punctum* (Fabricius). (From Howard 1911.)

Pomace or Vinegar Flies
(Family Drosophilidae)

These small, yellowish flies are commonly attracted to overripe fruits and vegetables in the house (and outdoors as well). The key diagnostic characteristic is the break in the front edge of the wing (costal vein), about one third of the way from the base to the tip.

Fig. 207. The common fruit fly, *Drosophila melanogaster* Meigen. *Left,* larva, or maggot; *center,* pupal case, or puparium; *right,* adult fly with enlargement of antenna to show detail. (From Howard 1911.)

The common, or laboratory, fruit fly, *Drosophila melanogaster* Meigen, is common throughout the region. They are yellowish brown with dark cross bands on the abdomen and have red eyes. The arista on the last antennal segment is featherlike. They can be easily reared on any overripe fruit and complete their entire life cycle in as little as two weeks.

A little over a dozen species occur in the Great Lakes region; for example, Band 1993 reports 14 species from Michigan.

Shore Flies
(Family Ephydridae)

These small flies are generally dark in color and the front of the face generally bulges outward. The front wing margin (costal vein) has two breaks in it, one near the base and another about one-third of the distance between the base and the tip of the wing. These flies can be exceedingly numerous around fresh and salt water, where they feed on algae and other organic materials that accumulate on the shoreline. Some species are leaf miners in aquatic plants, and two species were found in hydroponic greenhouses in Illinois (Steinly et al. 1987). Because they frequently occur in such large numbers they are an important food for fish and waterfowl.

Fewer than two dozen species occur in the Great Lakes region. For example, only 16 species are found in Illinois (Steinly et al. 1987).

One interesting species found in the Great Lakes region is the mantid shore fly, *Ochthera mantis* (De Geer). It has the front legs greatly enlarged and spined (like the praying mantid). It feeds on other insects, especially flies (both adults and larvae).

Frit Flies and Grass Flies
(Family Chloropidae)

These flies are some of the smallest inhabiting the Great Lakes region, ranging in size from 1 to 4 millimeters in length. Most species are blackish or grayish, with yellow markings and stripes. The front edge of the wing (costal vein) has a break in it, about one third of the way between the base and the tip. The area around the simple eyes (ocelli) is often raised and shiny and extends down the face toward the antennae. Adults are generally found in grasses and low, dense vegetation.

The eye gnat, *Liohippelats pusio* Loew, is attracted to the secretions associated with the eyes of livestock and humans. These gnats are small (1.5 mm) and blackish with the legs and the tip of the abdomen pale. The frit fly, *Oscinella frit* (L.), is associated with grasses and cereal crops (larvae live inside the plants' stems). Adult flies are minute (1–2 mm) and black with yellow halteres. The legs are also frequently marked with yellow.

House Flies
(Family Muscidae)

Members of this group are often abundant in and around dwellings. Most are medium sized and grayish or brownish in

Fig. 208. The house fly, *Musca domestica* L. *Left,* larva, or maggot; *center,* pupal case, or puparium; *right,* adult fly with enlargement of antenna to show detail. (From Howard 1911.)

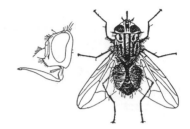

Fig. 209. The stable fly, *Stomoxys calcitrans* (L.). *Left,* larva, or maggot; *center,* pupal case, or puparium; *right,* adult fly with enlargement of head to show biting mouthparts. (From Howard 1911.)

color (never metallic). They are often difficult to distinguish from other families of closely related flies. One approach is to consider all the other possible identifications first; if none of these possibilities seems to fit, the subject is probably a muscid fly. Members of this family are best distinguished on the basis of their wing venation and the presence of certain thoracic bristles. On the wings, the two short veins (anal veins) near the hind angle of the wing never reach the wing margin, and on the large thoracic plate above and between the middle and hind pair of legs there is more than one stout bristle.

Many muscid flies are important agricultural and household pests, contaminating food, spreading diseases, and in some cases biting.

Some of the more notable species are the house fly, *Musca domestica* L., which has a blackish thorax with four grayish stripes, and which has brown markings on the abdomen; the little house fly, *Fannia canicularis* (L.), which is similar to the house fly but is noticeably smaller; the face fly, *Musca autumnalis* De Geer, which is very similar to the house fly but is darker gray with the sides of the abdomen either black (in the female) or orange (in the male); and the stable fly, *Stomoxys calcitrans* (L.), which is gray with a greenish yellow sheen, and which has a biting proboscis that protrudes forward from the face.

Blow Flies
(Family Calliphoridae)

Most of the flies in this family can be readily distinguished by their metallic coloration (green, blue, or black). A few species are not noticeably metallic, so you need to know that the hairlike arista of the last antennal segment is featherlike. Most species are associated with decomposing materials (dung, carrion, and rotting plant material), but a few species are parasites of other animals.

Some notable species occurring in the Great Lakes region are the black blow fly, *Phormia regina* (Meigen), which has a shiny blackish green body; the greenbottle fly, *Phaenicia sericata* (Meigen), which is metallic yellowish green or bluish green, sometimes with a coppery tint; the blue blow fly, *Calliphora vicina* Robineau-Desvoidy, which has a blue gray thorax and a metallic blue abdomen, red eyes, and orange on the sides of the face; and the cluster fly, *Pollenia rudis* (Fabricius), which has a dark, blackish thorax covered with a dense patch of golden hairs on the sides beneath the wing base, and a black and silver abdomen.

Bot and Warble Flies
(Family Oestridae)

These flies are medium to large sized, stout bodied, and covered with dense hairs (but no bristles). They are swift in flight and difficult to catch. Some resemble bumble bees, carpenter bees, or honey bees. The larvae are internal parasites of warm-blooded animals, including rabbits, rodents, livestock, domestic pets, and occasionally humans. The larvae may be internal parasites (bots) or may cause boillike swellings in the skin (warbles).

The distribution of bot and warble flies is determined by the distribution and availability of suitable hosts. Many of the common species are associated with domestic animals and livestock and consequently are cosmopolitan in distribution. A few examples of those that occur in the Great Lakes region include the horse bot fly, *Gasterophilus intestinalis* (De Geer); the rabbit bot fly, *Cuterebra cuniculi* (Clark); the sheep bot fly, *Oestrus ovis* L.; the common cattle grub, *Hypoderma lineatus* (de Villers); and the northern cattle grub, *H. bovis* (L.).

Flesh Flies
(Family Sarcophagidae)

Flesh flies are mostly medium sized and black or gray in color (never metallic), often with yellow hairs. The arista of the last

Fig. 210. The greenbottle fly, *Phaenicia sericata* (Meigen). (From Howard 1911.)

Fig. 211. The cluster fly, *Pollenia rudis* (Fabricius). (From Kansas State University Extension 1962.)

Fig. 212. The sheep bot fly, *Oestrus ovis* L. (From Kansas State University Extension 1962.)

Fig. 213. The red-tailed flesh fly, *Sarcophaga haemorrhoidalis* (Fallen). (From Kansas State University Extension 1962.)

antennal segment is featherlike on the basal half only. Most larvae are associated with animal flesh (dead and alive), but some are parasites of other insects (grasshoppers and beetles). The adults are generally found near plants, feeding on nectar, sap, honeydew, or overripe fruits.

Some of the flesh flies commonly found in the Great Lakes region include the red-bellied flesh fly, *Senotainia rubriventris* (Macquart), which has a gray thorax with indistinct stripes and a red abdomen (except at the base); the red-tailed flesh fly, *Sarcophaga haemorrhoidalis* (Fallen), which has a gray thorax with three to five distinct black stripes and a red abdominal tip; and the rapax flesh fly, *Helicobia rapax* (Walker), which has a gray thorax with three to five distinct stripes and a black and silvery gray checkerboard pattern on the abdomen.

Parasitic Flies
(Family Tachinidae)

These medium- to large-sized flies can be recognized by the combination of an enlarged, bulbous swelling on the rear of the thorax (postscutellum)—best seen from the side—the bare arista of the last antennal segment, and the stout bristles on the thorax and abdomen. Tachinidae is the second largest family of flies, and there are many undescribed species waiting to be named (so it could actually be the largest dipteran family). The larvae of all species are internal parasites of other insects, especially caterpillars, beetle larvae, sawflies, and other immature insects. The adult flies feed on nectar, sap, honeydew, and overripe fruit.

A few of the more common species known to occur in the Great Lakes region include the red-tailed tachina fly, *Winthemia rufopicta* (Bigot), which is gray and black with the sides and tip of the abdomen reddish brown; the tachina fly *Archytas apicifer* (Walker), which has a gray thorax with five indistinct stripes and a shiny metallic blue abdomen; the tachina fly *A. analis* (Fabricius), which has a powdery yellow thorax and shiny black abdomen; and the parasitic fly *Compsilura concinnata* (Meigen), which has a whitish thorax with four black stripes and a whitish head with a black stripe and orange palps.

Fig. 214. The tachina fly *Archytas apicifer* (Walker). (Courtesy of Michigan State University Extension.)

Louse Flies
(Family Hippoboscidae)

These highly unusual flies are unique among the Diptera in being flattened and occasionally wingless. Even when wings are present, they are often broken off. The tarsal claws are well developed with a series of small teeth. Louse flies are specialized external parasites of birds and mammals. Each species is usually restricted to a specific host animal.

A little more than two dozen species are currently known to occur in the Great Lakes region. Fifteen species of hippoboscids have been reported from Wisconsin and Michigan (MacArthur 1948). One species, the pigeon fly, *Pseudolynchia canariensis* (Macquart), was reported as a problem at a school in Indiana (Sanders and Petersen 1974). The sheep ked, or louse fly, *Melophagus ovinus* (L.), which has a cosmopolitan distribution, also occurs in the region and infests domestic sheep.

For an identification key to the species of louse flies from the western portion of the Great Lakes region, see MacArthur 1948.

Identification Resources. For a key to the Diptera with aquatic and semiaquatic larvae, see Lehmkuhl 1979. Other sources of identification keys and information for specific families, when available, are mentioned under each respective family.

Fig. 215. The sheep ked (louse fly), *Melophagus ovinus* (L.), an unusual, wingless, ectoparasitic fly that infests sheep. (Courtesy of Michigan State University Extension.)

Fleas
(Order Siphonaptera)

Introduction. Fleas are highly specialized ectoparasites whose exact evolutionary origins and relationships to other insects are unknown. No winged forms, past or present, have ever been found. Even so, they are still presumed to have had winged ancestors, closely related to the Diptera.

Description. Fleas are minute to small (1–10 mm), uniformly colored brown or blackish, and highly compressed from side to side. The head and thorax are small, but the abdomen is large and oval shaped (though compressed). The body is covered with short spines and bristles, all of which point toward the rear of the flea. These projections enable the flea to easily navigate through the dense hairs or feathers of the host, while at the same time preventing the host from easily dislodging or removing them. Wings are absent. The compound eyes and short, three-segmented antennae are inconspicuous. The mouthparts are of the piercing and sucking variety.

The tiny larvae are whitish, with chewing mouthparts. The body is elongate, legless, and wormlike, with two curved spines near the tip of the abdomen.

Life Cycle. Fleas have complete metamorphosis, with egg, larva, pupa, and adult stages. The complete life cycle may take as little as two weeks or as much as several months, depending on temperatures, humidity, and food supply.

Habits. Adult fleas feed on the blood of their hosts, and they have no alternate food supply. They often roam about on their hosts, sometimes moving between hosts, and occasionally leaving the hosts. Unlike lice, they will not perish if separated from the host. They are able to go several weeks, and sometimes months, between blood meals with no apparent harm.

Larval fleas are scavengers, feeding on organic debris associated with host animals (blood stains, urine, shed hair and feathers, bits of food, etc.).

Habitat. Adult fleas are known to infest a wide variety of animals, including bats, bears, birds, cats, chipmunks, dogs, flying squirrels, humans, mice, moles, opposums, pigs, rabbits, raccoons, rats, shrews, squirrels, voles, weasels, and woodchucks. Larval fleas are found in the nest of hosts or in other nearby areas.

Ecological and Economic Status. As bloodsucking ectoparasites of mammals and birds, fleas can cause extreme irritation to their hosts. The so-called sand fleas reported by homeowners are actually nothing more than cat or dog fleas that become abundant in the house and yard when the normal host pet is absent for a period of time and the fleas emerge after its departure. These hungry fleas, in need of a blood meal, will readily feed on the legs and bodies of any human being available, in particular women and small children. This phenomenon is common at the end of a vacation period, especially in the summer. It is wise upon returning home to let the pet enter first (Arnett 1985). Fleas are also involved in the transmission of disease. The larva are scavengers, feeding on various types of organic debris.

Distribution and Faunistics. Fleas are dependent on their host animals for their survival, so the distribution of fleas in the Great Lakes region is highly influenced by the distribution of suitable host animals. Investigations on the flea fauna, especially in nonurban areas, are few in number, and there is much we still do not know about the distribution of fleas in the region. In total, at least 46 species of fleas are currently known from within the Great Lakes region.

Fig. 216. The cat flea, *Ctenocephalides felis* Stiles and Collins. (From Kansas State University Extension 1962.)

Common Fleas
(Family Pulicidae)

The cat flea, *Ctenocephalides felis* Stiles and Collins, is widely distributed and feeds on dogs as well as cats. It is light brown with 8 spines on the front of the head just above the mouthparts (genal comb) and 16 spines on the rear edge of the pronotum (pronotal comb). The dog flea, *C. canis* (Curtis), is also common. It is similar to the cat flea, but the first spine in the genal comb is much shorter than the second. The Oriental rat flea, *Xenop-*

sylla cheopis (Rothschild), lacks both a genal comb on the front of the head and a pronotal comb on the rear of the pronotum; as its name implies, it is associated with rats.

Rodent Fleas
(Family Hystrichopsyllidae)

These fleas infest rodents and other closely related mammals. The mammal flea *Ctenophthalmus pseudagryrtes* Baker is common and has been reported from rats, mice, squirrels, voles, chipmunks, and shrews in the Great Lakes region. It has a genal comb with three spines.

Fig. 217. The human flea, *Pulex irritans* L., another member of the common flea family. (Courtesy of Michigan State University Extension.)

Bird and Rodent Fleas
(Family Ceratopsyllidae)

The northern rat flea, *Nosopsyllus fasciatus* (Bosc), is associated with large populations of rats. It lacks a genal comb but has a pronotal comb of 18–20 spines. The gray squirrel flea, *Orchopeas howardii* (Baker), is common on gray squirrels in the region. It also lacks a genal comb and has a pronotal comb, and the middle and hind thoracic segments have two rows of bristles.

Identification Resources. For identification information on fleas of the Great Lakes region, see Ewing 1929.

Ants, Bees, Sawflies, and Wasps
(Order Hymenoptera)

Introduction. The members of this insect order are familiar to most people, from the bees and wasps that sting to the ants that raid our kitchens and picnics. We all learn at an early age how to recognize those species that sting, by instruction from our parents and periodic reinforcement with lessons from the insects themselves. Many other insects—most notably flies, moths, and some beetles—mimic stinging hymenopterans, relying on other animals' avoidance of stings to protect them also. The complex social habits of these insects have intrigued humans for a very long time, and it is hard to resist comparing their colonies to our own towns and cities. Lastly, their roles as plant pollinators and predators and parasites of plant pests and their production of useful products make this group one of the most important groups of insects.

Description. Hymenopterans are generally soft bodied and

Quick Guide to Identification

Ants, Bees, Sawflies, and Wasps
(Order Hymenoptera)

Broad waists

Diagnostic Characteristics	Common Name(s)	Scientific Family Name
Large; hornetlike; clubbed antenna	cimbicid sawflies	Cimbicidae
Stout-bodied; 13-segmented serrate or pectinate antennae	conifer sawflies	Diprionidae
Antennae nine-segmented, threadlike	common sawflies	Tenthredinidae
Elongate, cylindrical, with "horntail"	horntails	Siricidae

Narrow Waists (Minute Parasitic Wasps)

Diagnostic Characteristics	Common Name(s)	Scientific Family Name
Wings very narrow, hairy; antennae long	fairyfly wasps	Mymaridae
Only hind wing very narrow; antennae short	trichogrammatid wasps	Trichogrammatidae
Hind femora greatly enlarged and toothed on underside	chalcid wasps	Chalcididae
Spur on tibia large and curved; five-segmented tarsi	pteromalid wasps	Pteromalidae

Narrow waists (Larger Parasitic Wasps)

Diagnostic Characteristics	Common Name(s)	Scientific Family Name
Second recurrent vein absent in front wing	braconid wasps	Braconidae
Second recurrent vein present; often with long ovipositor	ichneumon wasps	Ichneumonidae

Narrow waists (Nonparasitic Wasps and Ants)

Diagnostic Characteristics	Common Name(s)	Scientific Family Name
Pronotum with donut-shaped swelling	gall wasps	Cynipidae
Abdomen very small, stalk attached at top of thorax	ensign wasps	Evaniidae

Diagnostic Characteristics	Common Name(s)	Scientific Family Name
Abdomen long and slender; head attached to thorax by distinct neck	gasteruptiid wasps	Gasteruptiidae
Abdomen and antennae very long	pelecinid wasps	Pelecinidae
Metallic and coarsely punctured	cuckoo wasps	Chrysididae
Antlike; very hairy, with red/orange	velvet ants	Mutillidae
Node on first abdominal segment; "elbowed" antennae	ants	Formicidae
Long legs; transverse suture on side of thorax above middle leg	spider wasps	Pompilidae
Wings fold longitudinally	true wasps	Vespidae
Pronotum collarlike; thin abdomen	digger, sand-loving, and thread-waisted wasps	Sphecidae

Hairy Bodies (Bees and Others)

Diagnostic Characteristics	Common Name(s)	Scientific Family Name
Antlike; very hairy, with red/orange	velvet ants	Mutillidae
Lapping tongue short; yellow on face	yellow-faced and plasterer bees	Colletidae
Two sutures beneath antennae	andrenid bees	Andrenidae
Metallic; one suture under antennae	sweat bees	Halictidae
Pollen basket on underside of abdomen	leafcutter and mason bees	Megachilidae
Distinct tibial spurs on hind legs	digger and carpenter bees	Anthophoridae
Robust; hairy, black-and-yellow or brownish with hairy eyes	bumble bees and honey bees	Apidae

elongate and vary in size from minute (less than 0.5 mm) to large (more than 30 mm). Most are brown or black and many are marked with red, orange, or yellow warning colors. Metallic green and blue colors are not uncommon. The first abdominal segment is usually stalklike, creating an hourglass constriction between the thorax and abdomen. Only the more primitive sawflies and horntails have a broad waist. Virtually all species have four membranous wings with simple venation. The wings are often used in unison, being hooked together with tiny barbs. The females generally have a long ovipositor, and in some cases the ovipositor is modified into a stinger. The antennae are "elbowed," with one or more long basal segments and many small beadlike segments beyond the bend.

Life Cycle. All ants, bees, sawflies, and wasps have a complete metamorphosis, passing through egg, larva, pupal, and adult stages. The larvae are either caterpillarlike (such as the sawflies) or grublike (such as the ants, bees, and wasps). The caterpillarlike larvae of sawflies can be distinguished from the true caterpillars of butterflies and moths by the paired, leglike, cylindrical outgrowths of the abdomen (prolegs). Sawflies have six or more pairs of prolegs that lack tiny little hooks (crochets). True caterpillars have five or fewer pairs and the prolegs have one or more rows of crochets on the bottom surface. Parthenogenesis is quite common among the hymenopterans, and many species control the ratio of males to females by regulating fertilization: fertilized eggs develop into males, whereas unfertilized eggs develop into females.

Habits. Many species of Hymenoptera are social and construct nests in soil, wood, and plant stems or build special structures out of mud or paper. Some of the cuckoo bees lay their eggs in the nests of other bees and let these foster parents raise their young. Many hymenopterans are predators and parasites. Some are plant feeders, consuming foliage or gathering nectar and pollen. A few, such as ants, are scavengers or omnivores, consuming all types of plant and animal materials.

Habitats. Hymenopterans are found in all types of woodlands and open areas. Many species are associated with plants and will be found on foliage or flowers. Others are ground dwellers, building nests in the soil. The parasitic species is found searching for their hosts—for example, ichneumon wasps are often found around woodpiles and logs where there are plenty of wood borers.

Ecological and Economic Status. This order is certainly one of the most important insect orders, from both an ecological and an economic standpoint. Ants, which are probably the most numerous creatures on the earth, are tillers of the soil and important scavengers and predators. Wasps are predators and parasites of many of the most important plant pests. Bees are probably the most significant and reliable pollinators of plants, and most flowering plants would not be able to survive without them. They also produce honey and beeswax. A few hymenopterans are pests: sawflies defoliate hardwood and coniferous trees, and horntails and wood wasps are wood borers. Some hymenopterans sting, causing pain and even death (the latter in cases of severe allergic reactions and/or multiple stings).

Distribution and Faunistics. Most members of this group have widespread distributions in the Great Lakes region. Most families have been reasonably well studied and we have pretty good information on the fauna of the region. For example, 166 species of bees (Apoidea) are known from northern Wisconsin (Graenicker 1911). Among the wasp fauna, 43 species of spider wasps (Pompilidae) and 80 species of scoliid wasps (velvet ants, tiphiid wasps, scoliid wasps, and thread-waisted wasps) are found in northwestern Pennsylvania along Lake Erie (Kurczewski and Kurczewski 1963). Other examples are given in the discussions on common hymenopterous families.

Cimbicid Sawflies
(Family Cimbicidae)

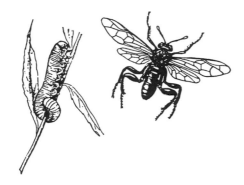

Fig. 218. The elm sawfly, *Cimbex americanus* Leach. *Left,* caterpillarlike sawfly (larva); *right,* wasplike adult. (Courtesy of Michigan State University Extension.)

These large sawflies resemble hornets (but lack the narrow waist) and have clubbed antennae. The plump, caterpillarlike larvae feed on the foliage of hardwood trees. One of the most common species is the elm sawfly, *Cimbex americanus* Leach. The body of the adult is dark blue and marked with yellow spots on the side of the abdomen (male) or with a large white spot near the prothorax (female). The wings are smoky colored.

Conifer Sawflies
(Family Diprionidae)

The larval stage of these sawflies are seen more often than the adult stage, and the larvae can be found feeding on the needles of spruce, pine, and balsam fir. The adults are small to medium and stout bodied with long sawlike or featherlike antennae.

The adult red-headed pine sawfly, *Neodiprion lecontei* (Fitch), is small with a reddish brown forebody and a black abdomen (female) or with an all black body (male). The caterpillarlike larvae have distinctive reddish orange heads. The European pine sawfly, *N. sertifer* (Geoffrey), is an introduced species and is common throughout the Great Lakes region. The larvae, which are dusky green with shiny black heads, feed on the needles of pine in the spring. The adults are medium sized and yellowish brown to brown with dark wing venation.

For an identification key to the sawfly larvae infesting spruce and fir in Ontario, see Lindquist and Miller 1972.

Common Sawflies
(Family Tenthredinidae)

Fig. 219. An adult common sawfly of the genus *Tenthredella*. (From Ruggles 1918.)

Tenthredinidae is the largest sawfly family and its members are indeed quite common. They are characterized by having nine-segmented, threadlike antennae and two spurs on the front tibia. Some species are gall makers or leaf miners in the larval stage, but most are leaf feeders. The adults frequent a variety of plant foliage, including their host plants.

The elm leaf miner, *Fenusa ulmi* Sundevall, and the birch leaf miner, *F. pusilla* (Lepletier), are commonly associated with their respective hosts. The pear-slug sawfly, *Caliroa cerasi* (L.), and the rose-slug sawfly, *Endolomyia aethiops* (Fabricius), are very atypical sawfly larvae. They are sluglike and slimy, and they skeletonize the leaves of pear and cherry trees and roses.

Horntails
(Family Siricidae)

Fig. 220. The pigeon tremex, or horntail wasp, *Tremex columba* (L.). (Courtesy of Michigan State University Extension.)

These large to very large hymenopterans are unmistakable, with their elongate, cylindrical body and stout spine on the last abdominal segment. In addition, the females have a short, stout ovipositor. The eggs are laid in dead and dying trees, and the larvae are wood borers.

The pigeon tremex, or horntail wasp, *Tremex columba* (L.), is very large and dark colored. The forebody is reddish brown and black and the abdomen is black with yellow cross bands. The wings are smoky yellow. The blue horntail, *Sirex cyaneus* Fabricius, is large and metallic blue black. The legs are reddish and the wings are clouded or smoky yellowish near the tip.

Braconid Wasps
(Family Braconidae)

Fig. 221. A parasitic braconid wasp of the genus *Apanteles*. (Courtesy of Michigan State University Extension.)

These mostly minute to medium sized wasps (2–12 mm) are parasites of other insects. The wing venation is greatly reduced, and there are generally fewer than seven closed cells in the forewing. Most are dark colored, but some have spots of white or red. Members of the genus *Apanteles* are important parasites of caterpillar pests. For example, the pupae of *Apanteles congre-*

gaturs (Say) parasitize hornworms (Lepidoptera: Sphingidae) and form the often seen white cocoons on the outside of these caterpillars, while *A. glomeratus* (L.) parasitizes caterpillars of the cabbage butterfly (Lepidoptera: Pieridae).

Ichneumon Wasps
(Family Ichneumonidae)

This very large family of parasitic wasps is well represented in the Great Lakes region. They are similar to braconid wasps but generally have seven closed cells in the forewing and on average are larger (5 to 40 mm). The antennae are long (16 or more segments) and extend for more than one-half the length of the body. The abdomen is long and slender and is often compressed. The females of many species have long, nonretractile ovipositors that are as long as, or longer than, the body. Most female ichneumon wasps will attempt to sting when handled, but those with very long ovipositors usually do not succeed. The species with shorter ovipositors are usually more successful at stinging with their ovipositor (although they have no venom). Most species parasitize caterpillars, but a few attack beetles and even other hymenopterans. This family of Hymenoptera is the largest in the region, and identification at the species level is often difficult. Only a few examples of common species are given here.

Fig. 222. An ovipositing female of the giant ichneumon, *Megarhyssa macrurus macrurus* (L.). (From McDaniels 1930.)

The giant ichneumon, *Megarhyssa macrurus macrurus* (L.), is distinctive in both size and coloration. The body exceeds one inch (25 mm) in length and is marked with brown, yellow, and some black. The ovipositor of the female is exceptionally long (much longer than the body). The white-footed ichneumon, *Cryptus albitarsis* (Cresson), is small and black with a red abdomen. The face, sides of the pronotum, and tibia of the first two pairs of legs are white. The black ichneumon, *Therion morio* (Fabricius), is large and black, with the antennae, hind tibia, and tarsi yellowish. Ichneumonids of the genus *Ophion* are medium sized, uniformly orangish tan, and laterally compressed. They are common at lights.

Fig. 223. An ichneumon wasp of the genus *Ophion*. (From Ruggles 1918.)

Pteromalid Wasps
(Family Pteromalidae)

These small (2–4 mm) wasps are metallic black, green, blue, or bronze with five-segmented tarsi and a large, curved spine (tibial spur) at the "knee joint" of the front legs. The wing venation is greatly reduced. Three species known to inhabit the Great Lakes region include *Dibrachys cavus* (Walker), which is minute with a shiny green forebody and a black abdomen;

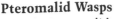

Pteromalus puparium (L.), which is minute to small with a green body, gold abdomen (males), bronzed legs, and yellow, brown, and black antennae; and *Perilampus hyalinus* Say, which is small and bluish with greenish golden legs and yellowish tarsi. This last species is commonly found on goldenrod flowers in the late summer.

Chalcidid Wasps
(Family Chalcididae)

These small parasitic wasps have the hind femora greatly enlarged and toothed on the underside. The wings are held outstretched at rest, not folded longitudinally. The most likely species to be encountered is *Brachyomera ovata* (Say), which is small (3–7 mm) and black and yellow with the tip of the hind femora yellow or whitish.

Fairyfly Wasps
(Family Mymaridae)

These are among the smallest parasitic wasps in our region, with most species shorter than 1 millimeter. They are blackish and the antennae are long and clubbed. The venation of the wings is reduced to a single thick vein at the leading edge of the wings. The hind wing is very long and narrow. The tiny hairs on the wings are arranged in a random pattern over the surface of the wing. The fairyfly wasp *Patasson conotracheli* (Gerrault) is an egg parasite of the plum curculio and other weevils and presumably can be found throughout the range of its host. The fairyfly wasp *Caraphractus cinctus* Walker is an egg parasite of the predaceous diving beetles (*Dytiscus* species). The female wasp has peculiar paddle-shaped wings that help it swim through the water as it hunts for the eggs of diving beetles.

Fig. 224. The minute egg parasite, *Trichogramma minutum* Riley. (Courtesy of Michigan State University Extension.)

Trichogrammatid Wasps
(Family Trichogrammatidae)

These minute wasps are egg parasites of other insects. They have stocky, dark-colored bodies and short antennae. The wing venation is greatly reduced to a single thick vein at the leading edge of the front wings. The hind wing is short and narrow. The tiny hairs on the wings are arranged in definite rows from the middle of the wing out toward the tip. The most commonly encountered species is the minute egg parasite, *Trichogramma minutum* Riley, which is used in the biological control of many pest insects.

Gall Wasps
(Family Cynipidae)

These wasps are responsible for the formation of many plant galls, most of them on oaks. The adult wasps are small and shiny black with long antennae (11 to 16 segments). The shield-shaped plate on the thorax (scutellum) has an unusual donut-shaped swelling on it. The abdomen is large, oval, and compressed laterally. The wing venation is greatly reduced, but at least several distinct veins are discernible.

Each gall has a characteristic shape and most of the wasps are named after the type of gall they produce. A gall is a plant abnormality caused by the feeding or egg-laying activity of certain insects and mites (or other causal agents, such as plant pathogens). For example, the spongy oak apple gall is found on oak leaves and is caused by the gall wasp *Amphibolips confluenta* (Harris). The jumping oak gall is caused by the gall wasp *Neuroterus saltatorius* (H. Edwards) and is found on the underside of white oak leaves. This tiny, rounded, brown gall is easily broken off the leaf and when dislodged does a pretty good imitation of the Mexican jumping bean. When the wasp pupates the jumping ceases.

Ensign Wasps
(Family Evaniidae)

These small parasitic wasps are black with a large squarish thorax and a small, elongate abdomen attached to the upper hind part of the thorax. The wing venation is reduced, but there are many veins and even a few closed cells. The antennae are quite long. The most commonly encountered species is the cockroach parasite, *Evania appendigaster* (L.), which is the largest species in the family (5 mm). It is most abundant where any of its hosts, the various species of cockroaches, are plentiful.

Gasteruptiid Wasps
(Family Gasteruptiidae)

These peculiar wasps vaguely resemble some of the long-abdomen ichneumons. The gasteruptiid wasps have a long, slender abdomen (attached to the upper hind part of the thorax) and a head attached to the thorax by a slender neck. The antennae are generally quite short. Most species are black and long bodied (13–20 mm), and the ovipositor is usually as long as the body. Only the genus *Gasteruption* occurs in North America, and the most widespread species is *G. assectator* (L.). It is black and the wing veins are dark. The wings are folded lengthwise (like the social wasps) and they are generally found on umbelliferous flowers in the late summer.

Pelecinid Wasps
(Family Pelecinidae)

These long, skinny wasps are instantly recognizable, even when seen flying through the air in the distance. Females have a very long, slender abdomen and reach a total length of two inches (50 mm); males are shorter (25 mm or more) with a club-shaped abdomen. The only species in the Great Lakes region is *Pelecinus polyturator* (Drury). It is uniformly shiny black, although the males have white tarsi. The larvae are parasites of scarab beetles and some wood borers.

Fig. 225. A female *Pelecinus polyturator* Drury. (From Ruggles 1918.)

Cuckoo Wasps
(Family Chrysididae)

Cuckoo wasps are generally small (6–10 mm) and brilliant metallic green or blue. The surface of the body is coarsely and densely pitted. These wasps are harmless and when disturbed curl up into a ball. They are called "cuckoo" wasps because of their parasitic nature. Like the European cuckoo bird, they raid the nests of other related animals and substitute their offspring for those of the host. The cuckoo wasps are external parasites of other wasps and bees (mud daubers, solitary bees, etc.).

One of the most commonly encountered species is the blue cuckoo wasp, *Chrysis coerulans* Fabricius. This medium-sized wasp is metallic blue or blue green. The tip of the abdomen has four toothlike projections.

Velvet Ants
(Family Mutillidae)

The small- to medium-sized, wingless, hairy females of these wasps resemble ants in general appearance. They are often brightly patterned with warning colors of black and orange (or black and red or yellow) and have a well-developed sting. The males are also very hairy, but they are winged and larger in size. The males, of course, do not have a sting, but if you hold one between your fingers they will attempt to mimic stinging by moving the abdomen in a threatening manner and making a squeaking noise. This behavior is perfectly harmless but still unnerving. Females, however, can give a very painful sting. The

Fig. 226. Velvet ants of the genus *Dasymutilla. Left,* adult male; *right,* wingless adult female. (From Ruggles 1918.)

females are generally found crawling across the ground in areas with open, sandy soils. The males can be seen flying close to the ground in meandering back-and-forth or circular patterns in the same sandy areas where females are present.

The common eastern velvet ant, *Dasymutilla occidentalis* (L.), is medium sized and black with red orange hairs on the thorax and the midsection of the abdomen. The winged male is black with red hairs on the pronotum.

Ants
(Family Formicidae)

Ants are the most abundant form of life in the Great Lakes region, so it is no wonder that everyone is familiar with them. The wingless worker ants seem to be everywhere, both outdoors in barren fields, grasslands, woodlands, wetlands, forests, and urban and suburban yards, and indoors in kitchens, bathrooms, and basements. Ants are characterized by having the first and sometimes the second abdominal segments small and narrowed with swollen, bumplike nodes. The head and mandibles are usually large and prominent, but the compound eyes are generally small. Ocelli are also present on the head. The antennae are "elbowed." The wingless workers (sterile females) and winged reproductives (males and females) are essentially alike, except for the reproductive organs, wings, and differing sizes. The wings, four in number, are membranous with reduced venation. The hind wings are distinctly smaller than the front wings. Owing to their modified ovipositor, some female ants are able to sting for defense; most other ants have to settle for biting with their well-developed mandibles. Ants range in size from minute (2 mm) to large (25 mm).

Ants are social insects living in small to large colonies generally located in soil or wood. The colonies are composed of individuals with specialized functions (castes). Members of the worker caste construct and maintain the nest, care for the young (eggs, larvae, and pupae), and defend the colony against intruders. Some colonies have specialized workers for defense, the soldiers. The males of the reproductive caste are short-lived, surviving only long enough to mate with a queen or a female supplemental reproductive. The queen is only mated once and from that point on is able to produce eggs continually throughout the remainder of her lifetime (generally up to two or three years). When an ant colony becomes overcrowded, large numbers of reproductive males and females are produced. These winged reproductives leave the nest in a large swarm and individual males and females pair up and leave the area to search for a suitable site to establish a nest of their own.

Ants consume a wide range of foods, including nectar, sap, and honeydew from aphids, scale insects, and treehoppers (which they frequently tend and protect). They also utilize foliage, seeds, fungi, and dead or live insects and other small animals.

There have only been a few studies on the ant fauna of the Great Lakes region. A total of 41 species of ants were found in Ashtabula County, in northeastern Ohio (Headley 1943) and 87 species were recorded from the Edwin S. George Reserve in southeastern Michigan (Talbot 1975).

Fig. 227. The pavement ant, *Tetramorium caespitum* L. (From Eckert and Mallis 1941.)

Ants can be subdivided into groups based on the number of nodes that they have between the thorax and abdomen. All the following ants have two nodes between the thorax and the main part of the abdomen. The little black ant, *Monomorium minimum* (Buckley), is a minute black ant that is common throughout houses in the region. The closely related pharaoh ant, *M. pharaonis* (L.), is also minute, but is a lighter yellowish or light brown color. It is also a serious household pest. The pavement ant, *Tetramorium caespitum* (L.), is a small black ant that also occurs in houses, as well as outdoors, and that lives under slabs of concrete and asphalt. They are easily distinguished by the many longitudinal grooves all over the body. The lined acrobat ant, *Crematogaster lineolata* (Say), has a distinctly heart-shaped abdomen that is attached to the petiole on the top front surface

Fig. 228. The lined acrobat ant, *Crematogaster lineolata* (Say). (From Eckert and Mallis 1941.)

Fig. 229. The black carpenter ant, *Camponotus pennsylvanicus* (De Geer). *Left,* adult female reproductive; *right,* sterile female worker. (From Back 1946 and Ruggles 1918.)

(not on the middle or lower surface as in most other ants). This ant lives under stones, under bark, or in old carpenter ant nests.

All the following ants have a single abdominal node. The black carpenter ant, *Camponotus pennsylvanicus* (De Geer), is a large, uniformly black ant that nests in wood, including structural timbers. The red carpenter ant, *C. ferruginneus* (Fabricius), is similar, but the middle of the body (the thorax and the front part of the abdomen) is reddish. The cornfield ant, *Lasius alienus* (Foerster), is a medium-sized, brownish ant that nests in the soil and tends root aphids. Their nests are often seen around the roots of corn plants. The small black and reddish Allegheny mound ant, *Formica exsectoides* Forel, builds large mounds in fields and open woodlands and is often reported by wary landowners as fire ants (which are of subtropical origin and do not occur in the Great Lakes region).

True Wasps
(Family Vespidae)

Most of the members of this family, especially the hornets and yellowjackets, are medium to large sized (8–20 mm) and boldly patterned in a black-and-yellow/red/white warning coloration. Their potent venom causes painful stings, and most animals avoid these insects whenever possible. Many other nonstinging insects from different orders (e.g., bugs, beetles, and moths) have developed similar patterns of warning coloration and enjoy a remarkable degree of protection as a result. Members of this wasp family can be distinguished by the unique way they fold their wings. The wings are folded longitudinally and usually rest over the back.

The antelope potter wasp, *Ancistrocerus antilope* (Panzer), nests in the pithy stems of sumac and feeds on the larvae of the silver spotted skipper. It is medium sized and black with narrow yellow bands on the abdomen. The legs are yellow. The potter wasp, *Eumenes fraternus* Say, is also medium sized and black, but the thorax and abdomen have yellow markings. The wings are smoky brown with a violet tinge. They make a pot-shaped nest of clay.

The German yellowjacket, *Vespula germanica* (Fabricius), is a medium-sized introduced species that is well established in many parts of the Great Lakes region. It usually builds its nest in wall voids and attics, so it is commonly encountered around buildings and is one of the most common uninvited guests at summertime picnics. It is boldly marked in black and yellow and is separated from other species by the shape of the black markings on the abdomen. The abdomen is marked with black diamond shapes (at the front) and triangles (beyond the middle)

Fig. 230. The common yellowjacket, *Vespula maculifrons* (L.). (Courtesy of Michigan State University Extension.)

Fig. 231. The bald-faced
hornet, *Dolichovespula
maculata* (L.). (Courtesy
of Michigan State
University Extension.)

down the center with black spots on each side. The common yellowjacket, *V. maculifrons* (Buysson), is often less common than its imported cousin. The black marking on the first abdominal segment is diamond shaped, although transversely stretched and narrowed. The remainder of the black markings have the triangles and spots coalesced into one broad transverse band on each abdominal segment.

The bald-faced hornet, *Dolichovespula maculata* (L.), is easily distinguished by its large size (greater than 14 mm) and black-and-white coloration. This species builds a large aerial nest of grayish paper (that is bigger than a basketball by summer's end). The aerial yellowjacket, *D. arenaria* (Fabricius), builds a smaller aerial paper nest (between the size of a softball and the size of a volleyball) that can be located in trees and shrubs or attached to buildings. This species is marked in black and yellow, with the black markings on the abdomen less pronounced and narrow. The paper wasp, *Polistes fuscatus* (Fabricius), is commonly seen around buildings and on goldenrod flowers in the late summer. These wasps are medium sized, slender bodied, and long legged. The body is dark brown or blackish and there are two large, reddish spots on the side of the abdomen near the thorax. Their small, umbrella-shaped nests (with no significant outer wrapper of paper) are usually attached to the eaves of buildings or other similar overhangs.

For an identification key to the social wasps of Indiana, see MacDonald and Deyrup 1989.

Spider Wasps
(Family Pompilidae)

These small- to medium-sized solitary wasps are often seen running across the ground in search of spider prey. Their bodies are usually black or blue and their wings are dark with bluish reflections. These wasps are long legged, with the hind femora elongated and extending beyond the tip of the abdomen. There is an impressed line or groove (transverse suture) on the side of the thorax above the middle leg. The wings are not folded longitudinally at rest.

The tornado wasp, *Episyron biguttatus* (Fabricius), is medium sized, black, and marked with some buff color. Parts of the thorax and front of the abdomen are covered with scales. The pubescence has a bluish tint. These wasps hunt for orb weaver spiders but can often be found on many types of umbelliferous flowers. The American spider wasp, *Anoplius americanus* (Beauvois), is medium sized and dull black with the base of the abdomen brick red. The wings are smoky-colored. The two-spotted spider wasp, *Ceropales bipunctata* Say, is medium sized and black, marked with creamy white on the head and thorax. The claws of the hind tarsi are bent at right angles.

For identification keys to the subfamilies and tribes of spider wasps in Michigan, see Dreisbach 1948a.

Digger, Sand-Loving, and Thread-Waisted Wasps
(Family Sphecidae)

These small- to medium-sized wasps are generally blackish or reddish brown, but a few are metallic blue or green. The pronotum is usually short and collarlike. Some members of the group have a very long, slender petiole. These solitary wasps may be found around flowers and they nest in the ground, in cavities, or in nests made of mud.

The blue mud dauber, *Chlorion aerarium* Patton, is a distinctively colored, shiny, purplish blue or bronze wasp. It nests in sandy soil. Another blue species is the blue mud wasp, *Chalybion californicum* (Saussure). It is entirely metallic blue, violet blue, or blue green with an elongate threadlike petiole (which is as long as the rest of the abdomen). The wings are bluish. The black-and-yellow mud dauber wasp, *Sceliphron caementarium* (Drury), is a medium-sized wasp characterized by a very long abdominal petiole (which is longer than the rest of the abdomen). The body is dull black and covered with grayish hairs. The base of the antennae, front and middle of the pronotum, front of the bulbous part of the abdomen, and parts of the legs are yellow.

Fig. 232. The black-and-yellow mud dauber wasp, *Sceliphron caementarium* (Drury). (Courtesy of Michigan State University Extension.)

The great golden digger wasp, *Sphex ichneumoneus* (L.), is medium sized and black with the base of the abdomen and legs red. The body, especially the abdomen, is covered with golden hairs. The wings are smoky yellow. The cicada killer wasp, *Sphecius speciosus* (Drury), is a very large wasp that is blackish with reddish orange legs and large yellow spots on the abdomen. It nests in sandy soil and provisions its nest with cicadas. Though not aggressive, its large size alarms many people. The common sand wasp, *Bembix americana spinolae* Lepeletier, is medium sized, stout bodied, and yellowish with black markings on the top of the head, thorax, and abdomen. The wings are smoky brown. It nests in sandy or gravelly soils.

Fig. 233. The cicada killer wasp, *Sphecius speciosus* (Drury). (Courtesy of the Young Entomologists' Society.)

For identification keys to the Michigan species of Philanthine wasps, see Dreisbach 1947; for a key to the sand wasps (tribe Bembicini) of Michigan, see Dreisbach 1951; for a key to sand wasps (tribes Stizini and Bembicini) of Indiana, see Chandler 1966; for a key to Michigan thread-waisted wasps, see Dreisbach 1944.

Bees
(Superfamily Apoidea)

The eight, closely related families of bees in this superfamily are the most highly evolved of the Hymenoptera. Bees differ from wasps in having hairs that are branched or featherlike (plumose). Wasps are generally naked or have simple bristlelike hairs. Also, in bees the first segment of the hind tarsi are enlarged and flattened and serve as pollen baskets. (One exception are the leafcutter bees, which use the dense hairs on the underside of the abdomen as pollen baskets.) Wasps, which are primarily predators, have no pollen baskets. While bees have chewing mandibles present, the rest of the mouthparts are modified into a tongue for feeding on nectar. Bees feed exclusively on plant materials, primarily nectar and pollen, and store quantities of these materials in their nests for the larvae to eat. Incidentally, bees may be sexed simply by examining the antennae: male bees have 13-segmented antennae and females have 12-segmented antennae. Of course, the females also have the ovipositor modified into a stinger.

Yellow-Faced and Plasterer Bees
(Family Colletidae)

These bees have a short lapping tongue that is bilobed or squared off at the tip. They are either slender, wasplike, and black with yellow or white markings on the face (yellow-faced bees) or robust, hairy, and brownish (plasterer bees). These bees nest either in the ground or in the crevices of bricks or stones.

The black yellow-faced bee, *Hylaeus modestus* Say, is small and black with yellow markings on the front (male) or sides (female) of the face. The rear margin of the abdominal segments are usually brownish. The common plasterer bee, *Colletes compactus* Cresson, is black with some yellow on the head and with white markings on each of the abdominal segments. The body is only sparsely hairy and the wings are transparent.

Fig. 234. A bee of the genus *Andrena*. (From Ruggles 1918.)

Andrenid Bees
(Family Andrenidae)

Andrenid, or mining, bees resemble honey bees because most are yellowish, brownish, or reddish brown and about the same size. The most reliable diagnostic feature for recognizing andrenid bees is the pair of impressed lines (sutures) under each antenna. Since these bees are often quite hairy, these sutures are not as easily seen as one might imagine. These bees generally nest in the soil.

The most commonly encountered members of this family in the Great Lakes region are members of the large genus *Andrena*. One species, *Andrena vicina* Smith, occurs in the eastern part of the region. It is dark in color with yellow markings on the lower part of the face (clypeus) only. They are common at flowers of hawthorne, apple, wild cherry, and related plants. They nest in the soil and at times can be common in lawns.

Sweat Bees
(Family Halictidae)

Although only a few species of this family seem to be attracted to perspiration, all members of the group are still referred to as sweat bees. They are small (4–10 mm) and blackish or brownish, often with metallic reflections. They are similar to some andrenid bees but have a single suture at the base of each antenna.

The metallic sweat bee, *Augochloropsis metallica metallica* (Fabricius), is a brilliant metallic green and inhabits the eastern portion of the region. The splendid sweat bee, *Agapostemon splendens* (Lepeletier), is brilliant blue green with the abdomen tinged in bluish or purplish (female) or in black with a yellow base (male). The top of the thorax (scutum) is coarsely punctured. The virescent green sweat bee, *Agapostemon virescens* (Fabricius), is similar to other brightly colored sweat bees. It has a brilliant green forebody, but the side of the thorax is covered with white hairs and the abdomen is black (female) or striped with yellow bands (male). The legs are striped in black and yellow. The sweat bee *Halictus ligatus* Say is black with yellowish white pubescence (with the hairs on the abdomen arranged into cross bands).

For an identification key to the green Halictine bees of Michigan, see Dreisbach 1945.

Leafcutter and Mason Bees
(Family Megachilidae)

Most of the members of this family have the underside of the abdomen covered with a thick patch of hairs that serves as a pollen basket. These medium- to large-sized solitary bees have a lapping tongue that is long and slender. They are important pollinators of many types of plants.

The mason bees, *Osmia* species, are generally robust and metallic blue or green. They have a membranous appendage between the claws. They nest in a variety of situations, including under stones, in the soil, in twigs, and in plant galls. One widespread species is *Osmia lignaria* Say, which is shiny dark blue green with long white hairs. The compound eyes, antennae, and legs are black. The leafcutter bees, *Megachile* species, are so named because they cut more or less circular holes out of leaves and use these disks in the construction of special brood chambers in their nests. The leafcutter bee *Megachile latimanus* Say is black with white bands on the abdomen. The body is covered with dense yellowish brown pubescence; the pollen basket is pale reddish. The leafcutter bee *M. brevis* Say is also black and the sixth abdominal segment is slightly concave. The abdomen is covered with sparse black pubescence and the pollen basket is pale. Several species of leafcutter bees have been introduced into the United States and can be found in the Great Lakes region, including the alfalfa leafcutter, *M. centuncularis* (L.), and the pale leafcutter, *M. concinna* Smith.

Digger and Carpenter Bees
(Family Anthophoridae)

Some of these medium- to large-sized bees resemble bumble bees because they are robust and hairy. The maxillary palps are long and there are three submarginal cells on the front wings.

The eastern digger bee, *Ptilothrix bombiformis* (Cresson), is medium to large sized and black, without pale markings (although some of the hairs on the thorax and abdomen are pale). The common carpenter bee, *Xylocopa virginica* (L.), is black with some white or yellow on the face. The forebody is hairy, but the abdomen is essentially bald and shiny black or blue black. These bees construct tunnellike nests in softwood lumber and sound deadwood of trees.

Bumble Bees and Honey Bees
(Family Apidae)

These hairy bees range in size from medium sized (12–15 mm), in the case of the honey bees, to large (15–27 mm), in the case of the bumble bees. The bumble bees are generally black with large areas of yellow or sometimes orange. There are many common species, including *Bombus americanorum* (Fabricius), on which the head, the rear two-thirds of the pronotum, and the back half of the abdomen are black and the front of the thorax and abdomen are yellow; *B. bimaculatus* Cresson, on which the thorax and base of the abdomen are yellow, the remainder of the abdomen is black, and the yellow on the abdomen is mushroom shaped; *B. griseocollis* De Geer, on which the base of the abdomen is grayish, the rest of the abdomen and the head are black, and the wings are brownish; *B. pennsylvanicus* De Geer, on which the rear half of the thorax is black and the front two-thirds of the abdomen is yellow with a black tip; and *B. ternarius* Say, on which the body is yellow with a black band across the thorax and tip of the abdomen and an orange cross band on the middle of the abdomen.

Although wild colonies exist in nature, the cosmopolitan honey bee, *Apis mellifera* L., is one of the only two domesticated insects (the silk moth is the other). Honey bees vary in color from light to very dark brown according to the geographic origin of each race, although they are generally golden brown with the abdomen banded in lighter and darker shades. They are unique among the bees in having hairy compound eyes. In recognition of the highly beneficial nature of honey bees, the state of Wisconsin has designated the honey bee as its official state insect.

For an identification key to the bumble bees of Michigan, see Milliron 1939; for Indiana, see Chandler 1950; for Illinois, see Frison 1919.

Identification Resources. Further information on the identification of the sawflies, ants, wasps, and bees of the Great Lakes region is available in Washburn 1918. Other sources of identification information on specific families of Hymenoptera, when available, are mentioned under each respective family.

Fig. 235. The honey bee, *Apis mellifera* L. (From USDA 1976.)

Glossary

Abdomen The hind section of the three main body parts of the insect

Acute Pointed; forming an angle of less than 90 degrees

Annulated Having ringlike segments or subdivisions

Antenna (pl., antennae) A pair of segmented appendages located on the head, usually sensory in function

Anterior front; in front of

Apical At the end, tip, or opposite to the base; for example, the apical segment of the antennae is the segment farthest from the head

Arista A large bristle located on the top side of the last antennal segment

Basal At the base, near the point of attachment; opposite to apical

Beak Protruding structure of an insect having piercing/sucking mouthparts

Bilobed tarsus A second-to-last tarsal segment divided into two lobes

Brace vein A slanting cross vein just behind the inner end of the stigma in some dragonflies

Caste A form or type of an adult social insect that carries out special tasks in its colony

Cell A space in the membrane of the wing that is surrounded by veins either partly (open cell) or completely (closed cell)

Cercus (pl., cerci) A filamentous, segmented appendage located near the tip of the abdomen

Chrysalis (pl. chrysalides) The pupa of a butterfly

Clubbed Expanded or enlarged terminal antennal segments

Clypeus The front part of an insect's head to which the labrum is attached

Cocoon A silken case inside which a pupa is formed

Collophore A tubelike organ located on the underside of springtails

Common species A species that is abundant and frequent in its habitat

Compound eye The major insect eye, composed of many individual facets, or lenses, called ommatidia

Corium The elongated, usually thickened, basal part of the front wing in some Heteroptera

Cornicle One of a pair of tubular structures extending from the posterior part of an aphid's abdomen

Costal vein The thickened anterior vein of the wing

Coxa (pl. coxae) The basal segment of the insect leg, located closest to the body

Crepuscular Active before sunrise and at twilight

Crochets Hooked spines at the tip of the prolegs of caterpillars (lepidopterous larvae)

Cross band A more or less broad stripe or marking across the body

Cross vein A vein connecting adjacent longitudinal veins

Cuneus A more or less triangular apical piece of the corium, set off from the rest of the corium by a suture; part of the hemelytron in Heteroptera

Diurnal Active during the daytime

Dorsal Top or uppermost; pertaining to the back or upper side

Dunes A mound or ridge of sand, accumulated by action of water and wind

Ectoparasite A parasite that lives on (and feeds on) the skin, fur, or feathers of a host

Elytron (pl., elytra) A thickened, armorlike forewing, found in beetles (Coleoptera) and earwigs (Dermaptera)

Femur (pl., femora) The third leg segment, located just above the tibia

Front The portion of the face between the antennae, eyes, and ocelli; the *frons*

Furcula The forked, springlike apparatus of Collembola

Gall An abnormal growth (tumor) of plant tissues caused by the feeding or egg laying of an insect or by the presence of a fungus, bacteria, or other plant

Genal comb A row of heavy spines on the lower side of a flea's head

Genitalia The sexual organs and associated structures; the external sexual organs

Glacial outwash Material carried by glacial meltwater and deposited as broad plains and deltas

Gregarious Living in groups

Halter (pl., halteres) A small knoblike balancing organ derived from the original hind wings of primitive Diptera

Head The anterior body region; the section that bears the compound eyes, ocelli, and mouthparts

Hemelytron (pl., hemelytra) An insect wing characterized by leathery basal section and a membranous wing tip, typical of insects in the order Hemiptera

Honeydew A sugary liquid discharged from the anus of certain homopterous insects

Host The plant or animal in (or on) which a parasite lives

Humerus (pl. humeri) The shoulder; the lateral angle at the front of the forewing

Imago (pl. imagoes) The adult reproductive stage of an insect

Labial palp One of a pair of feelerlike or antennalike structures arising from the labium, or lower "lip"

Labium The lower lip of an insect's mouthparts

Labrum The upper lip, lying just below the clypeus on the front of the head

Lamella (pl. lamellae) A thin leaflike or platelike process

Lamellate A type of antennae characterized by terminal antennal clubs composed of leaflike plates

Larva (pl. larvae) The immature stage, between egg and pupa, of an insect that undergoes complete metamorphosis

Lateral On or pertaining to the side (the left or right side)

Leaf miner An insect that feeds on the leaf cells located between the upper and lower surfaces of the leaf

Local species A species found in a restricted habitat type and generally not found elsewhere

Longitudinal Lengthwise of the body or of an appendage (leg, wing, etc.)

Mandibles The pincerlike jaws of an insect with chewing mouthparts

Maxilla (pl. maxillae) One of the paired mouthpart structures immediately behind the mandibles

Maxillary palp A small feelerlike structure arising from the maxilla

Membranous Wings made of a thin film of tissue, usually transparent

Mesosternum The lower plate of the middle, or second, segment of the thorax

Mesothorax The middle, or second, segment of the thorax

Metamorphosis Change in form during growth and development

Metathorax The rear, or third, segment of the thorax

Mimicry The ability of an animal to imitate or mimic another animal or plant species in form and color and sometimes in behavior

Moniliform Beadlike, with rounded segments

Moraine The accumulation of rocky material at the leading edge of an ice sheet, often left behind in a series of ridges or piles

Naiad An aquatic nymph

Nocturnal Active at night

Nodus A strong cross vein near the middle of the costal border of the wing; a stigma

Notum (pl. nota) The dorsal or top surface of the thorax

Nymph The immature wingless stage, between egg and adult, of an insect that undergoes gradual or incomplete metamorphosis

Ocellus (pl., ocelli) A simple eye of an insect or other arthropod

Ovipositor The egg-laying apparatus (the external genitalia of a female insect)

Palp A segmented, fingerlike extension on the mouthparts, located on either the labium or the maxilla

Parasite An animal that lives in or on the body of another animal and does not kill its host or consume a large portion of its tissues

Pectinate Comblike; with branches or processes like the teeth of a comb; used in reference to antennae or tarsal claws

Petiole The stalk or stem by which the abdomen is attached to the thorax in Hymenoptera

Pheromones Substances that are given off by an animal and cause a specific behavioral response from individuals of the same species; includes trail-making substances, sex attractants, and alarm substances

Plumose Featherlike

Posterior Hind or rear

Predator An animal that attacks and feeds on other, usually smaller, animals

Proboscis The extended, coiled mouthparts of Lepidoptera and some other insects

Proleg One of the fleshy abdominal legs of caterpillars and sawflies

Pronotal comb A row of heavy spines on the top rear of a flea's head

Pronotum The dorsal body plate of the thorax, frequently enlarged and prolonged in many insects

Prothorax The anterior, or first, segment of the thorax; never bearing wings

Pubescent Covered with short, fine hairs

Punctate Outside surface covered with minute impressed points appearing as pinpricks

Raptorial Fitted for grasping prey

Rare Extremely uncommon

Scarce Infrequently seen or found; more common than rare

Sclerotized Hardened or armorlike

Scutellum A segment of the pronotum, appearing as a more or less triangular segment behind the pronotum (Heteroptera, Homoptera, and Coleoptera)

Segment A subdivision of the body or of an appendage, between joints (sutures)

Serrate Sawlike

Setaceous Bristlelike

Spine A sharp, pointed, thornlike process of the exoskeleton

Spinneret A small tubelike organ that produces silk

Spiracle An external opening of the tracheal respiratory system; a breathing pore

Spittle The watery material produced by the nymphs of certain homopterous insects (spittlebugs or froghoppers)

Spurious vein A thickening of the insect wing that resembles a vein

Sternum A segment on the underside of the body (located between the legs)

Stigma A thickening of the wing membrane along the front margin of the wing, near the apex

Striate Marked with fine, parallel, depressed lines or furrows

Stylet A needlelike structure; one of the piercing structures in sucking mouthparts

Subcostal vein The second major wing vein, usually located immediately behind the costal vein (leading edge of the wing)

Subimago A winged developmental stage immediately preceeding the adult in mayflies (Ephemeroptera)

Subspecies A subdivision of a species, usually a geographic race that varies in size, color, or other characteristics; ordinarily not sharply differentiated within species, because subspecies are capable of interbreeding and intergrade with one another

Suture An external linelike groove in the exoskeleton or a narrow membranous area between segments

Tarsal formula The number of tarsal segments on the front, middle, and hind tarsi, respectively; for example, 5-5-4

Tarsus (pl., tarsi) The leg segment beyond the tibia, composed of three to five subsegments

Tegmen (pl. tegmina) The thickened, leathery forewing of an orthopteran or homopteran insect

Glossary

Terminal At the end or tip; the last of a series

Terrestrial Living on the land

Thorax The body region behind the head and in front of the abdomen, which bears the legs and wings

Tibia The fourth segment of the leg, located between the femur and tarsus; usually long and slender

Transverse Across the body or appendage, at right angles to the longitudinal axis

Trochanter The second segment of the leg, after the coxa

Truncate Cut off, squared at the end

Tubercles Small pimplelike bumps found on the outside of an insect's body

Tympanum (pl. tympana) A vibrating membrane; an auditory membrane, or eardrum

Uncommon species A species from which individuals are seen fairly regularly but never seem plentiful

Vein A thickened line in the wings

Ventral Lower or underneath; pertaining to the underside of the body

Vertex The top of the head

Vestigial Small, poorly developed, nonfunctional

Appendixes

Entomological Organizations in the Great Lakes Region

Over the years entomologists of the region have banded together for mutual support and networking. The first entomological societies in the region were the Entomological Society of Ontario and the American Entomological Society of Philadelphia (both founded in 1863). Other societies that formed in the region include the Detroit Entomological Society, the Entomological Society of Chicago, the Michigan Entomological Society, the Minnesota Butterfly and Moth Society, the New York Entomological Society, the North Central Branch of the Entomological Society of America, the Ohio Coleopterists, the Ohio Lepidopterists, the Pennsylvania Entomological Society, the Toronto Entomological Association, and the Wisconsin Entomological Society.

Many of the original entomological organizations are still active in the Great Lakes region, and additional groups have been recently established. The following organizations provide valuable service to insect enthusiasts of diverse backgrounds by sponsoring meetings and by publishing up-to-date information on insects of the region.

AMERICAN ARACHNOLOGICAL SOCIETY
Department of Entomology, American Museum of Natural History, Central Park West at 79th Street, New York, NY 10024. Publications: *Journal of Arachnology* and *American Arachnology*. Activities: annual meeting.

AMERICAN ENTOMOLOGICAL SOCIETY
1900 Benjamin Franklin Pkwy, Philadelphia, PA 19103. Phone: 215-561-3978. Publications: *Transactions of the American Entomological Society, Memoirs of the American Entomological Society,* and *Entomological News.* Activities: publication discounts to members, library facilities, and regular meetings.

THE COLEOPTERISTS SOCIETY
Field Museum of Natural History, Roosevelt Road at Lakeshore Drive, Chicago, IL 60605. Publication: *Coleopterists Bulletin.* Activities: annual meeting, occasional field trips, and youth incentive grants.

ENTOMOLOGICAL SOCIETY OF CANADA
393 Winston Avenue, Ottawa, Ontario K2A 1Y8, Canada. Publications: *Canadian Entomologist, Bulletin of the Entomological Society of Canada,* and *Memoirs of the Entomological Society of Canada.* Activities: annual meeting and publication discounts.

ILLINOIS MOSQUITO CONTROL ASSOCIATION
P.O. Box 1030, Harvey, IL 60426. Phone: 312-333-4120. Publications: *Proceedings of the Illinois Mosquito Control Association* and a biannual newsletter. Activities: annual meeting and photo salon.

INSECT MIGRATION ASSOCIATION
University of Toronto, Scarborough Campus, Scarborough, Ontario M1C 1A4, Canada. Activities: Monarch butterfly tagging and research.

INTERNATIONAL SOCIETY OF HYMENOPTERISTS
J. Wooley, Department of Entomology, Texas A&M University, College Station, TX 77843.

ISOPTERA SOCIETY
c/o Michael Haverly, United States Department of Agriculture Forest Service, P.O. Box 245, Berkeley, CA 94701.

LEPIDOPTERISTS SOCIETY
3838 Fernleigh Avenue, Troy, MI 48083. Publications: *Lepidopterists News* and *Journal of the Lepidopterists Society*. Activities: meetings and field trips.

MICHIGAN ENTOMOLOGICAL SOCIETY
Department of Entomology, Michigan State University, East Lansing, MI 48824. Publications: *Great Lakes Entomologist* and *Newsletter of the Michigan Entomological Society*. Activities: annual meeting.

MICHIGAN MOSQUITO CONTROL ASSOCIATION
812 Livingston Street, Bay City, MI 48708. Activities: annual meeting.

MINNESOTA BUTTERFLY AND MOTH SOCIETY
c/o Gary Pechan, Box 98, Savage, MN 55378.

NEW YORK ENTOMOLOGICAL SOCIETY
Department of Entomology, American Museum of Natural History, Central Park West at 79th Street, New York, NY 10024. Publication: *Journal of the New York Entomological Society*.

OHIO LEPIDOPTERISTS
1241 Kildale Square North, Columbus, OH 43229. Publication: newsletter. Activities: meetings, field trips, and book discounts.

OHIO MOSQUITO CONTROL ASSOCIATION
3306 Warner Avenue, Toledo, OH 43615. Phone: 419-726-7891. Publication: *Proceedings of the Ohio Mosquito Control Association*. Activities: annual meeting.

THE ORTHOPTERISTS SOCIETY
S. K. Gangwere, Department of Biological Sciences, Wayne State University, Detroit, MI 48202. Publications: *Proceedings of the Orthopterists Society* and *Metalepta* (newsletter). Activities: annual meeting.

WISCONSIN ENTOMOLOGICAL SOCIETY
Department of Entomology, 237 Russell Labs, 1630 Linden Drive, Madison, WI 53706. Publication: *Newsletter of the Wisconsin Entomological Society*. Activities: meetings, photo salon, and field trips.

THE XERCES SOCIETY
10 Southwest Ash Street, Portland, OR 97204. Phone: 503-222-2788. Publications: *Wings* (newsletter) and *ATALA* (journal). Activities: annual meeting and field trips.

YOUNG ENTOMOLOGISTS' SOCIETY, INC.
1915 Peggy Place, Lansing, MI 48910. Phone: 517-887-0499. Publications: *Young Entomologists' Society Quarterly*, *Insect World*, *Flea Market*, and special publications series. Activities: competitions, contributing articles, meetings, field trips, and educational materials for insect enthusiasts.

Periodicals Featuring Information on Great Lakes Insects

American Midland Naturalist: c/o Theodore Hesburgh Library, University of Notre Dame, Notre Dame, IN 46556.

Bulletin of the Milwaukee Public Museum: 800 West Wells Street, Milwaukee, WI 53233.

Great Lakes Entomologist: Michigan Entomological Society, c/o Department of Entomology, Michigan State University, East Lansing, MI 48824.

Illinois Natural History Survey Bulletin: c/o Illinois Natural History Survey, 607 East Peabody, Champaign, IL 61820.

Journal of the New York Entomological Society: c/o Department of Entomology, American Museum of Natural History, Central Park West at 79th Street, New York, NY 10024.

Newsletter of the Ohio Lepidopterists: Ohio Lepidopterists, 1241 Kildale Square North, Columbus, OH 43229.

New York State Museum Bulletin: c/o University of the State of New York, State Education Department, Albany, NY 12230.

Ohio Journal of Science: c/o Department of Biological Sciences, Bowling Green State University, Bowling Green, OH 43403.

Proceedings of the Entomological Society of Ontario: c/o Department of Environmental Biology, University of Guelph, Gelph, Ontario, Canada.

Proceedings of the Illinois State Academy of Science: Illinois State Museum, Springfield, IL 62706.

Proceedings of the Indiana Academy of Science: c/o John Wright Memorial Library, 140 North Senate Street, Indianapolis, IN 46204.

Proceedings of the Pennsylvania Academy of Sciences: c/o Department of Biology, Lafayette College, Easton, PA 18042.

Institutional Insect Collections in the Great Lakes and Adjacent Regions

Many public and private institutions (museums and universities) in the Great Lakes region house insect collections. These collections are a tremendous resource for understanding and interpreting the distribution, biology, and ecology of insects of the Great Lakes region.

Museums with Interpretive Displays

Academy of Natural Science of Philadelphia: Nineteenth and Parkway, Philadelphia, PA 19103.

Cranbrook Institute of Science and Nature Center: 500 Lone Pine Road, P.O. Box 801, Bloomfield Hills, MI 48013. Phone: 313-645-3225.

Field Museum of Natural History: Lakeshore Drive and Roosevelt Road, Chicago, IL 60605. Phone: 312-922-9410.

Illinois State Museum: Corner of Spring and Edwards, Springfield, IL 62706.

The Insectarium: 8046 Frankfort Avenue, Philadelphia, PA 19136. Phone: 215-338-3000.

Milwaukee City Public Museum: 800 West Wells Street, Milwaukee, WI 53233.

University of Michigan Museum of Zoology: University of Michigan, Ann Arbor, MI 48104.

Museums and Universities with Insect Collections

Academy of Natural Science of Philadelphia: Nineteenth and Parkway, Philadelphia, PA 19103.

American Museum of Natural History: Department of Entomology, Central Park West at 79th Street, New York, NY 10024.

Bell Museum of Natural History: University of Minnesota, Minneapolis, MN 55455.

Buffalo Museum of Science: Humbolt Parkway, Buffalo, NY 14211.

Canadian National Collection of Insects: Biosystematics Research Centre, Research Branch, Agriculture Canada, Ottawa, Ontario K1A 0C6, Canada.

Carnegie Museum: 900 Forbes Avenue, Pittsburgh, PA 15213.

Chicago Academy of Sciences: Museum of Natural History, Lincoln Park, 2001 North Clark Street, Chicago, IL 60614.

Cincinnati Museum of Natural History: 1720 Gilbert Street, Cincinnati, OH 45202.

Cleveland Museum of Natural History: Wade Oval, University Circle, Cleveland, OH 44106.

Cornell University: Department of Entomology, Cornell University, Ithaca, NY 14850.

Cranbrook Institute of Science and Nature Center: 500 Lone Pine Road, P.O. Box 801, Bloomfield Hills, MI 48013. Phone: 313-645-3225.

Dayton Museum of Natural History: 2629 Ridge Avenue, Dayton, OH 45414.

Field Museum of Natural History: Lakeshore Drive at Roosevelt Road, Chicago, IL 60605. Phone: 312-922-9410.

Grand Rapids Public Museum: Grand Rapids, MI 49502.

Illinois Natural History Survey: 607 East Peabody Drive, Champaign, IL 61820.

Indiana State Museum: Department of Natural Resources, 202 North Alabama Street, Indianapolis, IN 46204.

Indiana University: Department of Biology, Indiana University, Bloomington, IN 47401.

Kalamazoo Public Museum: 315 Rose Street, Kalamazoo, MI 49001.

McMaster University: Biology Department, McMaster University, 1280 Main Street West, Hamilton, Ontario L8S 4K1, Canada.

Miami University: Department of Zoology, Miami University, Oxford, OH 45056.

Michigan State University: Department of Entomology, Michigan State University, East Lansing, MI 48824.

Milwaukee City Public Museum: 800 West Wells Street, Milwaukee, WI 53233.

National Museum of Natural Sciences: Invertebrate Zoology Division, National Museum of Natural Sciences, Ottawa, Ontario K1A 0M8, Canada

New York State Museum: Biological Survey, 3132 Cultural Education Center, Albany, NY 12230.

Ohio Historical Society: 1982 Velma Avenue, Columbus, OH 43211.

Ohio State University: Department of Entomology, 1736 Neil Avenue, Ohio State University, Columbus, OH.

Pennsylvania State Department of Agriculture: Bureau of Plant Industry, 2301 North Cameron Street, Harrisburg, PA 17120.

Pennsylvania State University: Frost Entomological Museum, Pennsylvania State University, University Park, PA 16802.

Purdue University: Department of Entomology, Purdue University, West Lafayette, IN 47907.

Royal Ontario Museum: Department of Entomology, Royal Ontario Museum, 100 Queen's Park Crescent, Toronto, Ontario M5S 2C6, Canada.

St. Cloud University: Department of Biology, St. Cloud University, St. Cloud, MN 56301.

Science Museum of Minnesota: 30 East Tenth Street, St. Paul, MN 55101.

Southern Illinois University: Department of Zoology, Southern Illinois University, Carbondale, IL 62901.

State University of New York: Department of Environmental Science and Forestry, State University of New York, Syracuse, NY 13210.

United States National Museum: Department of Entomology, Smithsonian Institution, Washington, DC 20560.

University of Guelph: Department of Environmental Biology, University of Guelph, Guelph, Ontario N1G 2W1, Canada.

University of Michigan: Museum of Zoology, University of Michigan, Ann Arbor, MI 48104.

University of Minnesota: Department of Entomology, Fisheries and Wildlife, University of Minnesota, St. Paul, MN 55101.

University of Western Ontario: Department of Zoology, University of Western Ontario, London, Ontario N6A 5B7, Canada.

University of Wisconsin: Department of Entomology, University of Wisconsin, Madison, WI 53706.
Western Illinois University: Department of Biological Sciences, Western Illinois University, Macomb, IL 61455.

Insect Zoos and Butterfly Houses in the Great Lakes Region

Insect zoos and butterfly houses also provide the public with a better understanding of the insect world by allowing them to interact with living insects and other arthropods. For more information on the educational value of butterfly houses and insect zoos see "The Case Study for Live Public Butterfly Habitats in the United States" (Cotham 1992).

Binder Park Zoo: 7400 Division Street, Battle Creek, MI 49017.
Brookfield Zoo: 3300 Golf Road, Brookfield, IL 60513. Phone 708-485-0263.
Cincinnati Zoo Insectarium: 3400 Vine Street, Cincinnati, OH 45220. Phone 513-281-4700.
Cleveland Metroparks Zoo Rain Forest Exhibit: Education Department, 3900 Brookside Park Drive, Cleveland, OH 44109. Phone 216-661-6500.
Columbus Zoo: 9990 Riverside Drive, P.O. Box 400, Powell, OH 43065. Phone: 614-645-3400.
Field Museum of Natural History: Lakeshore Drive and Roosevelt Road, Chicago, IL 60605. Phone 312-922-9410.
Illinois State Museum: Corner of Spring and Edwards, Springfield, IL 62706.
The Insectarium: 8046 Franfort Avenue, Philadelphia, PA 19136. Phone 215-338-3000.
John Ball Zoological Garden: 201 Market Street, Grand Rapids, MI 49503.
Mackinac Island Butterfly House: Sawyers Greenhouse, 1308 McGaulpin, Mackinac Island, MI 49757. Phone 906-847-3972.
Musser's Butterfly Farm: Rural Route no. 7, 13200 Fulton Road, Sidney, OH 45365.
Philadelphia Zoo: 3400 West Gerard Avenue, Philadelphia, PA. Phone 215-387-6400.
Saginaw Children's Zoo: 1435 South Washington, Saginaw, MI 48601.
Toledo Botanical Gardens: 5403 Elmer Drive, Toledo, OH 43615. Phone 419-536-8365.

Appendix E

Regulations on Collecting Insects on Public Lands in the Great Lakes Region

CANADA

National Parks: Permit Required. Contact Environment Canada, Canadian Parks Service, Ottawa, ON K1A 0H3, Canada.

UNITED STATES

National Parks: Permit required. Contact the National Park Service, Office of Public Affairs, Department of the Interior, Washington, DC 20240, or the park superintendent of the park where any proposed study would take place.

National Forests: No permit required; collectors are urged to contact local forest supervisors for anything more than casual collecting. Contact the Forest Service Information Office, United States Department of Agriculture, P.O. Box 2417, Washington, DC 20013.

National Wildlife Refuges: Permission required. Contact the Assistant Director–Public Affairs, United States Fish and Wildlife Service, Department of the Interior, Washington, DC 20240, or the refuge manager of the wildlife refuge where any proposed study would take place.

Illinois State Parks and State Forests: Permit required. Contact the Department of Conservation, Division of Land and Historic Sites, 605 State Office Building, Springfield, IL 62706.

Indiana State Parks: Permission required. Contact the Indiana Department of Natural Resources, Division of State Parks, 616 State Office Building, Indianapolis, IN 46204.

Indiana State Forests: Permission required. Contact the Indiana Department of Natural Resources, State Forester, 613 State Office Building, Indianapolis, IN 46204.

Michigan State Parks and Forests: No restrictions at this time. Courtesy call to unit manager suggested.

Minnesota State Parks: Permission required. Discuss plans with park manager, then submit application to the Director of Parks and Recreation, Minnesota Department of Natural Resources, Box 39, Centennial Building, St. Paul, MN 55101.

Minnesota State Forests: No restrictions at this time.

New York State Parks: Permission required. Contact the Office of Parks and Recreation, Executive Department, Empire Plaza, Albany, NY 12233.

New York State Forests and State Forest Recreation Areas: Permission required. Contact the Department of Environmental Conservation, Division of Lands and Forests, 50 Wolf Road, Albany, NY 12233.

Ohio State Parks: Permit required. Contact the Division of Parks and Recreation, Ohio

Department of Natural Resources, Fountain Square Building C, Columbus, OH 43224.

Ohio State Forests: Permit required. Contact the Division of Forestry, Ohio Department of Natural Resources, Fountain Square Building C, Columbus, OH 43224.

Pennsylvania State Parks: Permission required. Contact the Environmental Management Section, Bureau of State Parks, Department of Environmental Protection, Harrisburg, PA 17120.

Pennsylvania State Forests: No restrictions at this time.

Wisconsin State Parks and Forests: Permission required. Contact the Wisconsin Department of Natural Resources, 101 South Webster Street, P.O. Box 7921, Madison, WI 53707.

Endangered, Threatened, and Special Concern Insects of the Great Lakes Region

Insects in the Great Lakes region that are currently listed as state or federal endangered species include the following.

Mayflies: aneopore flat-headed mayfly (*Aneoporus simplex*), a flat-headed mayfly (*Raptoheptagenia cruentata*), pecatonia mayfly (*Acanthametropus pecatonica*), a mayfly (*Epeorus namatus*), a mayfly (*Pseudiron centralis*), sand-filtering mayfly (*Homoeoneuria ammophila*), homoplectran mayfly (*Homoplectra doringa*), and sand-dwelling mayfly (*Pseudiron centralis*).

Dragonflies: extra-striped snaketail dragonfly (*Ophiogomphus anomalus*) and pygmy snaketail dragonfly (*O. howei*).

Leafhoppers: leafhopper (*Paraphlepsius lupalus*)

Beetles: Hungerford's crawling water beetle (*Brychius hungerfordi*), American burying beetle (*Necrophorus americanus*), Kramer's cave ground beetle (*Pseudanophthalmus krameri*), Ohio cave ground beetle (*Pseudanophthalmus ohioensis*), Knobel's riffle beetle (*Stenelmis knobeli*), cobblestone tiger beetle (*Cicindela marginipennis*), and serpentine rove beetle (*Lissobiops serpentinus*).

Butterflies and Skippers: frosted elfin butterfly (*Lycaeides irus*), Karner blue butterfly (*Lycaeides melissa samuelis*), northern blue (*Lycaeides idas nabokovi*), silvery blue (*Glaucopsyche lygdamus couperi*), veined white (*Artogeia napi oleracea*), Mitchell's satyr butterfly (*Neonympha mitchellii mitchellii*), Appalachian brown (*Satyrodes appalachia*), persius dusky wing (*Erynnis persius persius*), purplish copper (*Epidemia helloides*), regal fritillary (*Speyeria idalia*), swamp metalmark (*Calephelis muticum*), powesheik skipper (*Oarisma powesheik*), arogos skipper (*Atrytone arogos*), and two-spotted skipper (*Euphyes bimacula*).

Moths: three-staff underwing moth (*Catocala amestris*), phlox moth (*Schinia indiana*), leadplant moth (*Schinia lucens*), unexpected cycnia (*Cycnia inopinatus*), graceful underwing (*Catocala gracilis*), pointed sallow (*Epiglaea apiata*), prominent moth (*Hyperaeschra tortuosa*), Spartiniphaga inops, Hypocoena enervata, silphium borer moth (*Papaipema silphii*), eryngium stem borer (*Papaipema eryngium*), Papaipema beeriana, pinkpatched looper moth (*Eosphoropteryx thyatyroides*), Lithophane semiusta, Trichoclea artesta, Tricholita notata, Ufeus plicatus, Ufeus satyricus, and Erythroceia hebardi.

Spongillaflies: a spongilla fly (*Climacia* sp.).

Caddisflies: a caddisfly (*Setodes oligius*), northern case-making caddisfly (*Goera stylata*) and Ross' northern case-making caddisfly (*Pycnopsyche rossi*).

Scorpionflies: earwig scorpionfly (*Merope tuber*).

Insects in the Great Lakes region that are currently considered threatened include the following.

Dragonflies: grayback dragonfly (*Tachopteryx thoreyi*) and elfin skimmer (*Nannothemis bella*).
Grasshoppers: Lake Huron locust (*Trimerotropis huroniana*).
Spittlebugs: great plains spittlebug (*Lepyronia gibbosa*).
Leafhoppers: red-veined prairie leafhopper (*Aflexia rubranura*).
Butterflies and Skippers: pipevine swallowtail (*Battus philenor*), silver-bordered fritillary (*Boloria selene*), frosted elfin (*Incisalia irus*), olympia marblewing (*Euchloe olympia*), mottled dusky wing (*Erynnis martialis*), Duke's skipper (*Euphyes dukesi*), dusted skipper (*Atrytonopsis hianna*), cobweb skipper (*Hesperia metea*), and ottoe skipper (*Hesperia ottoe*).
Moths: a lytrosis moth (*Lytrosis permagnaria*).
Spongillaflies: a spongilla fly (*Sisyra* sp.).
Caddisflies: sponge-feeding caddisfly (*Ceraclea* sp.).

Species of special concern or status unknown include the following.

Spittlebugs: angular spittlebug (*Lepyronia angulifera*) and red-legged spittlebug (*Prosapa ignipectus*).
Dragonflies: false spiketail dragonfly (*Cordulegaster erronea*), white-lined clubtail dragonfly (*Gomphus lineatifrons*), warpaint emerald dragonfly (*Somatochlora incurvata*), amnicola snaketail dragonfly (*Stylurus amnicola*), notable snaketail (*Stylurus notatus*), plagiated snaketail (*Stylurus plagiatus*), and Canadian bog skimmer (*Williamsonia fletcheri*).
Grasshoppers: secretive locust (*Appalachia arcana*), Davis's shield-bearing grasshopper (*Atlanticus davisi*), bog cone-head grasshopper (*Melanoplus flavidus*), tamarack tree cricket (*Oecanthus laricis*), pinetree cricket (*Oecanthus pini*), red-faced meadow katydid (*Orchelimum concinnum*), delicate meadow katydid (*Orchelimum delicatum*), barrens locust (*Orphulella pelidna*), Hoosier locust (*Paroxya hooseri*), Atlantic coast locust (*Psiniidia fenestralis fenestralis*), and pine katydid (*Scudderia fasciata*).
Beetles: beach tiger beetle (*Cicindela hirticollis*), a tiger beetle (*C. ancocisconensis*), a tiger beetle (*C. cursitans*), a tiger beetle (*C. cuprascens*), and a tiger beetle (*C. macra*), six-banded long-horned beetle (*Dryobius sexnotatus*), Cantrall's bog beetle (*Liodessus cantralli*), black lordithon rove beetle (*Lordithon niger*), and Douglas stenelmis riffle beetle (*Stenelmis douglasensis*).
Butterflies and Skippers: Edward's hairstreak (*Satyrium edwardsii*), early hairstreak (*Erora laeta*), Freya's fritillary (*Boloria freija*), frigga fritillary (*Boloria frigga*), gorgone checkerspot (*Chlosyne gorgone carlota*), grizzled skipper (*Pyrgus centaurei wyandot*), Henry's elfin (*Incisalia henrici*), hoary comma (*Polygonia gracilis*), large marble butterfly (*Euchloe ausonides*), Macoun's arctic (*Oeneis macounii*), Olympia marblewing (*Euchloe olympia*), red-disked alpine (*Erebia discoidalis*), regal fritillary (*Speyeria idalia*), swamp metalmark (*Calephilis muticum*), tawny crescentspot (*Phyciodes batesii*), and wild indigo dusky wing (*Erynnis baptisiae*).
Moths: Aweme borer (*Papaipema aweme*), barrens buckmoth (*Hemileuca maia*), black arctiid moth (*Holomelina nigricans*), boreal branchionyncha moth (*Branchionyncha borealis*), Clemen's sphinx (*Sphinx luscitiosa*), Columbian silk moth (*Hyalophora columbia*), corylus dagger moth (*Acronicta falcula*), Culver's root

borer moth (*Papaipema sciata*), Doll's merolonche moth (*Merolonche dolli*), Doris tiger moth (*Grammia doris*), fadus sphinx (*Aellopos fadus*), gold moth (*Basilodes pepita*), Magdalen underwing (*Catocala illecta*), maritime sunflower borer moth (*Papaipema maritima*), Newman's brochade moth (*Meropleon ambifusca*), northern pine sphinx (*Lapara bombycoides*), oithona tiger moth (*Grammia oithona*), one-eyed sphinx (*Smerinthus cerisyi*), pine imperial moth (*Eacles imperialis pini*), pure lichen moth (*Crambidia pura*), quiet underwing moth (*Catocala dulciola*), Riley's lappet moth (*Heteropacha rileyana*), Robinson's underwing (*Catocala robinsoni*), slender clearwing moth (*Hemaris gracilis*), small heterocampa moth (*Heterocampa subrotata*), spartina moth (*Spartiniphaga inops*), Sprague's pygarctia moth (*Pygarctia spraguei*), tersa sphinx (*Xylophanes tersa*), three-striped oncocnemia moth (*Oncocnemis piffardi*), wild cherry sphinx (*Sphinx drupiferarum*), and yellow-banded day sphinx (*Proserpinus flavofasciata*).

Bibliography

Adams, C. C. 1909a. Annotations on certain Isle Royale invertebrates. Pp. 249–77 in An ecological survey of Isle Royale, Lake Superior, *Ann. Rept. Mich. Geol. Surv.* (1908), Lansing.

———. 1909b. The Coleoptera of Isle Royale, Lake Superior, and their relation to the North American centers of dispersal. Pp. 157–203 in An ecological survey of Isle Royale, Lake Superior, *Ann. Rept. Mich. Geol. Surv.* (1908), Lansing.

Adler, P. H., et al. 1982. Seasonal emergence patterns of black flies (Diptera: Simuliidae) in Northwestern Pennsylvania. *Great Lakes Entomol.* 15 (4): 253–60.

Alexander, Richard D. 1957. The taxonomy of the field crickets of the eastern United States (Orthoptera: Gryllidae: *Acheta*). *Annals Entomol. Soc. Amer.* 50:584–602.

Alexander, Richard D., and Thomas E. Moore. 1962. The evolutionary relationships of 17-year and 13-year cicadas, and three new species (Homoptera, Cicadidae, *Magicicada*). *Misc. Publ. Mus. Zool. Univ. Mich.* 121:1–59.

Alexander, Richard D., et al. 1972. The singing insects of Michigan. *Great Lakes Entomol.* 5 (2): 33–69.

Alrutz, R. W. 1992. Additional records of dragonflies (Odonata) from Ohio. *Ohio J. Sci.* 92 (4): 119–20.

Amin, Omar M. 1976. Lice, mites, and ticks of southeastern Wisconsin mammals. *Great Lakes Entomol.* 9 (4): 195–98.

Amin, O. M., and A. G. Hageman. 1974. Mosquitoes and tabanids in southeast Wisconsin. *Mosquito News* 34:170–77.

Anderson, Robert S. 1982. Resource partitioning in the carrion beetle (Coleoptera: Silphidae) fauna of southern Ontario: Ecological and evolutionary considerations. *Can. J. Zool.* 60:1314–25.

Anderson, Robert S., and Stewart B. Peck. 1985. The insects and arachnids of Canada. Part 13. The carrion beetles of Canada and Alaska (Coleoptera: Silphidae and Agyrtidae). *Agric. Can. Publ.* No. 13.

Andrews, A. W. 1916. Results of the Mershon expedition to the Charity Islands, Lake Huron: Coleoptera. Pp. 65–108 in Misc. papers on the zoology of Michigan. *Mich. Geol. and Biol. Surv. Publ.* No. 20.

———. 1923. The Coleoptera of the Shiras expedition to Whitefish Point, Chippewa Co., Michigan. *Pap. Mich. Acad. Sci., Arts, Ltrs.* 1:293–390.

Arnett, Ross H. 1985. *American Insects.* New York: Van Nostrand Reinhold.

Baker, Norman T., and Roland L. Fischer. 1975. A taxonomic and ecologic study of the Asilidae of Michigan. *Great Lakes Entomol.* 8 (2): 31–91.

Band, Henrietta Trent. 1993. Drosophilidae (Diptera) collected in spring in Michigan. *Great Lakes Entomol.* 26 (3): 237–40.

Banks, N. 1903. A revision of the Nearctic Chrysopidae. *Trans. Entomol. Soc. Amer.* 29:137–62.

———. 1906. A revision of the Nearctic Coniopterygidae. *Proc. Entomol. Soc. Wash.* 8:77–86.

———. 1927. A revision of the Nearctic Myrmeleontidae. *Bull. Mus. Comp. Zool.* 68:1–84.

Barnes, Burton V., and Warren H. Wagner, Jr. 1981. *Michigan trees: A Guide to the Trees of Michigan and the Great Lakes Region.* Ann Arbor: Univ. Michigan Press.

Barr, A. R. 1957. The mosquitoes of Minnesota (Diptera: Culicidae: Culicinae). *Univ. Minn. Agric. Exp. Sta. Tech. Bull.* 228:1–154.

Bayer, Lutz J., and H. Jane Brockman. 1975. Curculionidae and Chrysomelidae found in aquatic habitats in Wisconsin. *Great Lakes Entomol.* 8 (4): 219–26.

Beatty, G. H., and A. F. Beatty. 1968. Checklist and bibliography of Pennsylvania Odonata. *Proc. Penn. Acad. Sci.* 43:120–29.

Bednarik, Andrew F., and W. P. McCafferty. 1977. A checklist of stoneflies, or Plecoptera, of Indiana. *Great Lakes Entomol.* 10 (4): 223–26.

Benedict, W. C. 1962. Mosquitoes in and about Windsor, Ontario. *Proc. Entomol. Soc. Ont.* 93:82–84.

Bennet, D. V., and E. F. Cook. 1981. The semiaquatic Hemiptera of Minnesota (Hemiptera: Heteroptera). *Univ. Minn. Agric. Exp. Sta. Tech. Bull.* 332:1–59.

Bergman, Edward A., and William L. Hilsenhoff. 1978. *Baetis* (Ephemeroptera: Baetidae) of Wisconsin. *Great Lakes Entomol.* 11 (3): 125–36.

Bernard, Ernest C. 1975a. A new genus, six new species, and records of Protura from Michigan. *Great Lakes Entomol.* 8 (4): 157–82.

———. 1975b. New species and additional records of Protura from Michigan. *Great Lakes Entomol.* 8 (4): 187–96.

Bickley, W. E., and E. G. MacLeod. 1956. A synopsis of the Nearctic Chrysopidae with a key to the genera (Neuroptera). *Proc. Entomol. Soc. Wash.* 58:177–202.

Bland, Roger G. 1972. New orthopteroid records in Michigan derived from sampling a small field. *Great Lakes Entomol.* 5 (3): 109–10.

———. 1973. Chronological and geographical distribution of Orthopteroid populations in an abandoned Michigan field. *Environ. Entomol.* 2 (5): 737–42.

———. 1978. *How to know the insects.* Dubuque, Ia.: William C. Brown Co.

———. 1989. An annotated list of the Orthoptera of Beaver Island, Lake Michigan. *Great Lakes Entomol.* 22 (1): 39–43.

Blatchley, W. S. 1891. Catalogue of butterflies known to occur in Indiana. *Ann. Rept. Ind. Dept. Geol. Nat. Hist.* 17:365–408.

———. 1893. The Blattidae of Indiana. *Proc. Ind. Acad. Sci.,* 1892, 153–65.

———. 1903. The Orthoptera of Indiana. An illustrated descriptive catalogue of the species known to occur in the state, with bibliography, synonymy and descriptions of new species. *Ann. Rept. Ind. Dept. Geol. and Nat. Res.* 27: 125–471 and 669–77.

———. 1910. *An illustrated and descriptive catalog of the Coleoptera or beetles of Indiana (exclusive of the Rhyncophora).* Indianapolis: Nature Publ. Co.

———. 1920. *The Orthoptera of northeastern America with especial reference to the fauna of Indiana and Florida.* Indianapolis: Nature Publ. Co.

———. 1926. *Heteroptera or true bugs of eastern North America with special reference to the faunas of Indiana and Florida.* Indianapolis: Nature Publ. Co.

Blickenstaff, C. C. 1946. A list of the Thysanoptera known to occur in Indiana. *Amer. Midl. Nat.* 36 (3): 668–70.

Boesel, M. W. 1983. A brief review of the genus *Cricotopus* in Ohio, with a key to adults of the northeastern United States (Diptera: Chironomidae). *Ohio J. Sci.* 83 (3): 74–90.

———. 1985. A brief review of the genus *Polypedilum* in Ohio, with keys to known

stages of species occurring in the northeastern United States (Diptera: Chironomidae). *Ohio J. Sci.* 85 (5): 245–62.

Bohart, Richard M. 1941. A revision of the Strepsiptera with special reference to the
species of North America. *Univ. Calif. Publ. Entomol.* 7 (6): 91–160.

Borror, Donald J., and Richard E. White. 1970. *A field guide to the insects of America
north of Mexico.* Boston: Houghton-Mifflin.

Borror, Donald J., D. W. DeLong, and C. Triplehorn. 1976. *An introduction to the study
of insects.* New York: Holt, Rinehart, and Winston.

Bradley, J. C., and B. S. Galil. 1977. The taxonomic arrangement of the Phasmatodea
with keys to the subfamilies and tribes. *Proc. Entomol. Soc. Wash.* 79:176–
208.

Braun, E. Lucy. 1989. *The woody plants of Ohio.* Columbus: Ohio State Univ. Press.

Bright, Donald E., Jr. 1976. The insects and arachnids of Canada. Part 2. The bark
beetles of Canada and Alaska (Coleoptera: Scolytidae). *Agric. Can. Publ.* No. 1576.

Brimley, J. F. 1930. Coleoptera found in the Rainy River district, Ontario. *Can. Field
Nat.* 44:135–40.

Bromley, S. W. 1950. Ohio robber flies V. *Ohio J. Sci.* 50:229–34.

Brown, W. J. 1930. Coleoptera of the north shore of the Gulf of St. Lawrence. *Can.
Entomol.* 62:231–37 and 239–46.

———. 1932. Additional records of Coleoptera of the north shore of the Gulf of St.
Lawrence. *Can. Entomol.* 64:198–209.

Burgess, A. F. 1906. Preliminary report on the mosquitoes of Ohio. *Ohio Nat.* 6 (3):
438–40.

Burks, B. D. 1953. The mayflies, or Ephemeroptera, of Illinois. *Bull. Ill. Nat. Hist. Surv.*
26 (1): 1–216.

Burton, John J. S. 1975. The deer flies of Indiana (Diptera: Tabanidae: *Chrysops*). *Great
Lakes Entomol.* 8 (1): 1–29.

Caldwell, John S. 1938. The jumping plant-lice of Ohio (Homoptera: Chermidae). *Bull.
34, Ohio Biol. Surv.* 6 (5).

Cantrall, Irving J. 1943. The ecology of the Orthoptera and Dermaptera of the George
Reserve, Michigan. *Misc. Publ. Mus. Zool. Univ. Mich.* 54:1–182.

———. 1965. *Phyllophaga* at light traps on the E. S. George Reserve, Michigan. *Pap.
Mich. Acad. Sci., Arts, Ltrs.* 50:95–103.

———. 1968. An annotated list of the Dermaptera, Dictyoptera, Phasmatoptera, and
Orthoptera of Michigan. *Mich. Entomol.* 1 (9): 299–346.

———. 1972. *Forficula auricularia* (Dermaptera: Forficulidae) in Michigan. *Great
Lakes Entomol.* 5 (3): 107–8.

Carpenter, F. M. 1931. Revision of Nearctic Mecoptera. *Bull. Mus. Comp. Zool.*
72:205–77.

———. 1940. A revision of the Nearctic Hemerobiidae, Berothidae, Sisyridae, Polystoechotidae, and Dilaridae (Neuroptera). *Proc. Amer. Acad. Arts and Sci.*
74:193–280.

Cassani, John R., and Roger G. Bland. 1978. New distribution records for mosquitoes in
Michigan (Diptera: Culicidae). *Great Lakes Entomol.* 11 (1): 51–52.

Chamberlain, T. R., et al. 1943. Species, distribution, flight, and the host preferences of
the June beetles in Wisconsin. *J. Econ. Entomol.* 36:674–80.

Chandler, L. 1950. The Bombidae of Indiana. *Proc. Ind. Acad. Sci.* 60:167–77.

———. 1956. The order Protura and Diplura in Indiana. *Proc. Ind. Acad. Sci.* 66:112

———. 1966. Wasps of the tribes Stizini and Bembecini in Indiana. *Proc. Ind. Acad. Sci.*
75:141–47.

Christiansen, K. A., and P. F. Bellinger. 1981. The Collembola of North America north of the Rio Grande. Grinnell, Ia.: Grinnell College.

Cleland, C. E. 1975. *A brief history of Michigan Indians.* Lansing, Mich.: Mich. Dept. State, Hist. Div.

Cockerell, T. D. A. 1916. Bees from the northern Lower Peninsula of Michigan. *Occ. Pap. Mus. Zool. Univ. Mich.* 23:1–10.

Cotham, W. Mark. 1992. The case for live public butterfly habitats in the United States. Special Publication No. 7. Young Entomologists' Society. Lansing, Mich.

Cooper, R. H. 1938. A breeding record for the red-barred sulphur (*Callidryas philea* Linn.) from Indiana. *Entomol. News* 49 (6): 261.

Corbet, Philip S. 1967. The Odonata of Ontario. *Proc. North Central Branch Entomol. Soc. Amer.* 22:116–17.

Covell, Charles V., Jr. 1970. An annotated check list of the Geometridae (Lepidoptera) of Wisconsin. *Trans. Wisc. Acad. Sci., Arts, Ltrs.* 58:167–83.

———. 1984. *A field guide to the moths of eastern North America.* Boston: Houghton-Mifflin.

Cupp, E. W., and A. E. Gordon, eds. 1983. Notes on the systematics, distribution, and bionomics of black flies (Diptera: Simuliidae) in the northeastern United States. *Search: Agric.* 25:1–75.

Dahl, Robert A., and Daniel L. Mahr. 1991. Light trap records of *Phyllophaga* (Coleoptera: Scarabaeidae) in Wisconsin, 1984–1987. *Great Lakes Entomol.* 25 (1): 1–8.

Davies, D. M., et al. 1961. The black flies (Diptera: Simuliidae) of Ontario. Part I. Adult identification and distribution with descriptions of six new species. *Proc. Entomol. Soc. Ont.* 92:70–154.

Dawson, R. W., and W. Horn. 1928. Notes on the tiger beetles of Minnesota. *Univ. Minn. Agric. Exp. Sta. Tech. Bull.* No. 56: 1–13.

Deay, H. O., and G. E. Gould. 1936. The Hemiptera of Indiana. Family Gerridae. *Amer. Midl. Nat.* 17:753–69.

DeCoursey, R. M. 1971. Keys to families and subfamilies of nymphs of North American Hemiptera-Heteroptera. *Proc. Entomol. Soc. Wash.* 73:413–28.

DeLong, D. M. 1948. The leafhoppers, or Cicadellidae, of Illinois. *Bull. Ill. Nat. Hist. Surv.* 24 (2): 97–376.

DeLong, D. M., and J. G. Sanders. 1930. A revision of the North American *Empoasca* known to occur north of Mexico with special study of the internal male genital structures. *U.S.D.A. Tech. Bull.*

Denning, Donald G. 1943. Hydropsychidae of Minnesota (Trichoptera). *Entomol. Amer.* 23:101–71.

Deonier, D. L., and J. T. Regensberg. 1978. New records of Ohio shore flies (Diptera: Ephydridae). *Ohio J. Sci.* 78 (3): 154–55.

Deyrup, Mark. 1981. Annotated list of Indiana Scolytidae. *Great Lakes Entomol.* 14 (1): 1–10.

Deyrup, Mark, and Thomas H. Atkinson. 1987. New distribution records of Scolytidae from Indiana and Florida. *Great Lakes Entomol.* 20 (3): 67–68.

Dickenson, W. E. 1944. The mosquitoes of Wisconsin. *Bull. Publ. Mus. Milwaukee* 8:269–365.

Dickenson, W. E., and C. P. Alexander. 1932. The crane-flies of Wisconsin. *Bull. Publ. Mus. Milwaukee* 8:139–266.

Dietz, H. F., and H. Morrison. 1916. The Coccidae or scale insects of Indiana. *Ind. State Entomol. Rpt.* No. 8 (1914–1915): 195–321.

Dillon, E. S., and L. S. Dillon. 1972. *A manual of common beetles of eastern North America.* 2 pts. New York: Dover Publications.

Dodge, H. R. 1938. The bark beetles of Minnesota. *Univ. Minn. Agric. Exp. Sta. Tech. Bull.* 132:1–60.

Dorr, John A., Jr., and Donald F. Eschman. 1970. *Geology of Michigan.* Ann Arbor: Univ. Michigan Press.

Douglas, M. M. n.d. The butterflies and skippers of the Great Lakes region. Manuscript.

Douglass, John F. 1983. New northernmost records of the spicebush swallowtail in Lower Michigan. *Jack-Pine Warbler* 61:15–16.

Downie, N. M. 1956. Records of Indiana Coleoptera I. *Proc. Ind. Acad. Sci.* 66:115–24.

———. 1958. Records of Indiana Coleoptera II. *Proc. Ind. Acad. Sci.* 68:155–58.

Downie, N. M., and C. E. White. 1966. Records of Indiana Coleoptera III. *Proc. Ind. Acad. Sci.* 76:308–16.

Drake, Carl J. 1917. A survey of North American species of *Merragata* (Hebridae). *Ohio J. Sci.* 17 (3): 101–5.

Drake, C. J., and H. C. Chapman. 1953. Preliminary report on the Pleidae (Hemiptera) of the Americas. *Proc. Biol. Soc. Wash.* 66:53–59.

———. 1958. New neotropical Hebridae including a catalogue of the American species (Hemiptera). *J. Wash. Acad. Sci.* 48:317–26.

Drake, C. J., and F. C. Hottes. 1950. Saldidae of the Americas (Hemiptera). *Great Basin Nat.* 10:51–61.

Drake, C. J., and D. R. Lauck. 1959. Descriptions, synonymy and check-list of American Hydrometridae (Hemiptera: Heteroptera). *Great Basin Nat.* 19:43–52.

Drees, B. M. 1982. A checklist of horse flies and deer flies (Diptera: Tabanidae) of Cedar Bog, Champlain Co., Ohio. *Ohio J. Sci.* 82 (4): 170–76.

Dreisbach, R. R. 1942. The genus *Vespula* (Hymenoptera: Vespidae) in Michigan, with keys and distribution. *Pap. Mich. Acad. Sci., Arts, Ltrs.* 28:323–29.

———. 1944. The thread-waisted wasps (Hymenoptera-Sphecinae) of Michigan, with keys and distribution. *Pap. Mich. Acad. Sci., Arts, Ltrs.* 29:265–75.

———. 1945. The green halictine bees of the genera *Agapostemon, Augochlora, Augochlorella,* and *Augochloropsis* (Hymenoptera: Apoidea) in Michigan, with keys and distribution. *Pap. Mich. Acad. Sci., Arts, Ltrs.* 30:221–27.

———. 1947. The genera *Aphilanthops, Oclocletes,* and *Philanthus* (Hymenoptera: Philanthidae) in Michigan, with keys and distribution. *Pap. Mich. Acad. Sci., Arts, Ltrs.* 31:141–45.

———. 1948a. Keys to the subfamilies and tribes of Pompilidae (Hymenoptera: Vespoidea) in North America and to the genera in Michigan. *Pap. Mich. Acad. Sci., Arts, Ltrs.* 32:239–47.

———. 1948b. The genera *Ceropales, Calicurgus,* and *Allocyphonyx* (Hymenoptera: Pompilidae) in Michigan, with keys and distribution. *Pap. Mich. Acad. Sci., Arts, Ltrs.* 32:249–50.

———. 1951. The family Bembicidae (Hymenoptera: Sphecoidea) in Michigan, with keys to genera and species of the state and distribution records. *Pap. Mich. Acad. Sci., Arts, Ltrs.* 35:101–7.

Dunn, Gary A. 1982. Ground beetles (Coleoptera: Carabidae) collected by pitfall trapping in Michigan small-grain fields. *Great Lakes Entomol.* 15 (1): 37–38.

———. 1983. Collections of hibernation ground beetles (Coleoptera: Carabidae). *Great Lakes Entomol.* 16 (1): 34.

————. 1984. Ground beetles collected in a Michigan asparagus field. *Young Entomol. Soc. Quart.* 1 (3): 5–9.

————. 1994. *A beginner's guide to observing and collecting insects.* Lansing, Mich.: Young Entomologists' Society.

————. n.d. Entomological literature for the Great Lakes region, 1850–1994. *Great Lakes Entomol.*, forthcoming.

Dury, Charles. 1879a. Catalogue of Lepidoptera in the vicinity of Cincinnati. *Cincinnati Soc. Nat. Hist.* 1.

————. 1879b. List of Coleoptera observed in the vicinity of Cincinnati. *Cincinnati Soc. Nat. Hist.* 1.

————. 1882. Coleoptera of the vicinity of Cincinnati. *Cincinnati Soc. Nat. Hist.* 5:218–20.

————. 1884. Notes on Coleoptera, with additions to the list of the collection of Cincinnati. *Cincinnati Soc. Nat. Hist.* 7:91–92.

————. 1902. A revised list of Coleoptera observed near Cincinnati, Ohio. *Cincinnati Soc. Nat. Hist.* 20:107–96.

————. 1904. Notes on Coleoptera. *Entomol. News* 15 (2): 52–53.

————. 1906. Additions to the list of Cincinnati Coleoptera. *Cincinnati Soc. Nat. Hist.* 20:257–60.

————. 1909. New species and additions to the list of Cincinnati Coleoptera. *Cincinnati Soc. Nat. Hist.* 21:64–67.

————. 1912. New Coleoptera from Cincinnati, Ohio. *Cincinnati Soc. Nat. Hist.* 21:64–67.

————. 1916. Two new beetles from Cincinnati, Ohio. *Cincinnati Soc. Nat. Hist.* 22:14–15.

Eertmoed, Gary, and Elizabeth Eertmoed. 1983. A collection of Psocoptera from Voyageurs National Park, Minnesota. *Great Lakes Entomol.* 16 (4): 123–26.

Ehrlich, P. R., and A. E. Ehrlich. 1961. *How to know the butterflies.* Dubuque, Ia.: William C. Brown Co.

Etnier, D. A. 1965. An annotated list of the Trichoptera of Minnesota, with description of a new species. *Entomol. News* 76 (3): 141–52.

————. 1968. Range extension of Trichoptera in Minnesota, with descriptions of two new species. *Entomol. News* 79 (4): 188–92.

Evans, J. D. 1895. The insect fauna of the Sudbury District, Ontario. *Can. Entomol.* 27 (6): 141–43.

Evans, W. E., Jr. 1909. Mosquitoes of Ohio. *J. Columbus Hort. Soc.* 24:117–30.

Everly, R. T. 1927. A check list of the Carabidae of Columbus, Ohio, and vicinity. *Ohio J. Science* 27 (2): 155–56.

————. 1939. Spiders and insects found associated with sweet corn, with notes on the food and habits of some species. Pt. 4. Hymenoptera. *Ohio J. Sci.* 39 (1): 48–51.

Ewing, Henry E. 1929. *A manual of external parasites.* Springfield, Ill.: Chas. Thomas Publ.

————. 1940. The Protura of North America. *Annals Entomol. Soc. Amer.* 33:495–551.

Favinger, John J. 1984. Anecdotal history of Entomology in Indiana. *Proc. Ind. Acad. Sci.* 94:307–11.

Fay, H. T. 1862. On winter collecting. *Proc. Entomol. Soc. Philadelphia.* 1:194–98.

Felt, E. P. 1904. Mosquitoes or Culicidae of New York State. *New York State Mus. Bull.* 79:239–400.

Ferkinhoff, W. D., and R. W. Gunderson. 1983. A key to the whirligig beetles of Minnesota and adjacent states and Canadian provinces (Coleoptera: Gyrinidae). *Scientific Pub. Sci. Mus. Minn.* 5 (3): 1–53.

Fernekes, V. 1906. List of Lepidoptera occurring in Milwaukee County. *Bull. Wisc. Nat. Hist. Soc.* 4:39–58.

Flowers, R. Wills, and William L. Hilsenhoff. 1975. Heptageniidae (Ephemeroptera) of Wisconsin. *Great Lakes Entomol.* 8 (4): 201–18.

Fluke, C. L., Jr. 1921. Syrphidae of Wisconsin. *Trans. Wisc. Acad. Sci. Arts, Ltrs.* 20:215–53.

Forbes, S. A. 1916. A general survey of the Maybeetles (*Phyllophaga*) of Illinois. *Ill. Agric. Exp. Sta. Bull.* 186:215–57.

Forbes, W. T. M. 1948. Lepidoptera of New York and neighboring states. Part II. Geometridae, Sphingidae, Notodontidae, Lymantriidae. *Cornell Univ. Agric. Exp. Sta. Mem.,* 274:1–263.

———. 1954. Lepidoptera of New York and neighboring states. Part III. Noctuidae. *Cornell Univ. Agric. Exp. Sta. Mem.,* No. 329.

———. 1960. Lepidoptera of New York and neighboring states. Part IV. Agaristidae through Nymphalidae, including butterflies. *Cornell Univ. Agric. Exp. Sta. Mem.,* 371:1–424.

Fracker, S. B. 1918. The Alydinae of the United States. *Annals Entomol. Soc. Amer.* 11:255–88.

Franklin, H. J. 1912. The Bombidae of the New World. *Trans. Amer. Entomol. Soc.* 38:177–486.

Frison, T. H. 1919. Keys to the separation of the Bremidae, or bumblebees, of Illinois, and other notes. *Trans. Ill. State Acad. Sci.* 12:157–66.

———. 1929. The fall and winter stoneflies, or Plecoptera, of Illinois. *Bull. Ill. Nat. Hist. Surv.* 18 (2): 345–409.

———. 1935. The stoneflies, or Plecoptera, of Illinois. *Bull. Ill. Nat. Hist. Surv.* 20 (4): 281–471.

Frost, S. W. 1949. The Simuliidae of Pennsylvania (Dipt.). *Entomol. News* 60 (5): 129–31.

Frost, S. W., and L. L. Pechuman. 1958. The Tabanidae of Pennsylvania. *Trans. Amer. Entomol. Soc.* 84:169–215.

Fulton, Bently B. 1911. The Stratiomyidae of Cedar Point, Sandusky. *Ohio Nat.* 11:299–301.

Furth, David G. 1974. The stink bugs of Ohio (Hemiptera: Pentatomidae). *Bull. Ohio Biol. Surv.* (N.S.) 5 (1): 1–60.

Gall, Wayne K. 1992. Further eastern range extension and host records for *Leptoglossus occidentalis* (Heteroptera: Coreidae): Well documented dispersal of a household nuisance. *Great Lakes Entomol.* 25 (3): 159–72.

Garman, Philip. 1917. The Zygoptera, or damselflies, of Illinois. *Bull. Ill. State Lab. Nat. Hist.* 12:411–587.

Gaufin, Arden R. 1956. An annotated list of the stoneflies of Ohio (Plecoptera). *Ohio J. Sci.* 56 (6): 321–24.

Gosling, D. C. L. 1973. An annotated list of the Cerambycidae of Michigan (Coleoptera). Part I. Introduction and subfamilies Paradrinae, Prioninae, Spondylinae, Aseminae, and Cerambycinae. *Great Lakes Entomol.* 6 (3): 65–84.

———. 1977. An annotated list of the Cerambycidae of Michigan (Coleoptera). Part II. The subfamilies Lepturinae and Lamiinae. *Great Lakes Entomol.* 10 (1): 1–37.

———. 1980. An annotated checklist of the checkered beetles (Coleoptera: Cleridae) of Michigan. *Great Lakes Entomol.* 13 (2): 65–76.

———. 1983. New state records of Cerambycidae (Coleoptera). *Great Lakes Entomol.* 16 (4): 187.

———. 1984. Flower records for anthophilous Cerambycidae in a southwestern Michigan woodland (Coleoptera). *Great Lakes Entomol.* 17 (2): 79–82.

Graenicher, S. 1911. Bees of northern Wisconsin. *Bull. Publ. Mus. Milwaukee* 1 (3): 221–49.

Graves, Robert C. 1963. The Cicindelidae of Michigan (Coleoptera). *Amer. Midl. Nat.* 69:492–507.

———. 1965. The distribution of tiger beetles in Ontario (Coleoptera: Cicindelidae). *Proc. Entomol. Soc. Ont.* 95:63–70.

Graves, Robert C., and David W. Brzoska. 1991. The tiger beetles of Ohio (Coleoptera: Cicindelidae). *Bull. Ohio Biol. Surv.* (N.S.) 8 (4): 1–42.

Gurney, A.B. 1951. Praying mantids of the United States, native and introduced. *Rept. Smithsonian Inst.* 1952:305–25.

Hamilton, John. 1895. Catalogue of Coleoptera of southwestern Pennsylvania, with notes and descriptions. *Trans. Amer. Entomol. Soc.* 22:317–81.

Hamilton, K. G. Andrew. 1982. The insects and arachnids of Canada. Part 10. The spittlebugs of Canada (Homoptera: Homoptera). *Agric. Can. Publ.* No. 1740.

Hanna, Murray. 1970. An annotated list of the spittlebugs of Michigan (Homoptera: Cercopidae). *Mich. Entomol.* 3 (1): 2–16.

Hanna, M., and T. M. Moore. 1966. The spittlebugs of Michigan (Homoptera: Cercopidae). *Pap. Mich. Acad. Sci., Arts, Ltrs.* 51:39–73.

Harden, P. H., and C. E. Micket. 1952. The stoneflies of Minnesota (Plecoptera). *Univ. Minn. Agric. Exp. Sta. Tech. Bull.* 201:1–82.

Harmston, F. C., and Knowlton. 1946. New and little-known Dolichopodidae from Indiana (Diptera). *Amer. Midl. Nat.* 63:671–74.

Harrington, W. H. 1884a. Additions to Canadian lists of Coleoptera. *Can. Entomol.* 16:44–47.

———. 1884b. List of Ottawa Coleoptera. *Trans. Ottawa Field Nat. Club* 5:67–85.

———. 1890. On lists of Coleoptera published by the Geological Survey of Canada, 1842–1888. *Can. Entomol.* 22:135–40, 153–54.

Hart, Charles A. 1919. The Pentatomoidea of Illinois with keys to the nearctic species. *Bull. Ill. Nat. Hist. Surv.* 13:157–223.

Hart, James W. 1968. A checklist of the mosquitoes of Indiana with a record of the occurrence of *Aedes infirmatus* D & K. *Proc. Ind. Acad. Sci.* 78:257–59.

———. 1970. A checklist of Indiana Collembola. *Proc. Ind. Acad. Sci.* 80:249–52.

———. 1971. New records of Indiana Collembola. *Proc. Ind. Acad. Sci.* 81:246

———. 1972. New records of Indiana Collembola. *Proc. Ind. Acad. Sci.* 82:231.

Hatch, M. H. 1924. A list of Coleoptera from Charlevoix Co., Michigan. *Pap. Mich. Acad. Sci., Arts, Ltrs.* 4:543–86.

———. 1927. Studies on the carrion beetles of Minnesota, including new species. *Univ. Minn. Agric. Exp. Sta. Tech. Bull.* 48:3–12.

Hays, Kirby L. 1956. A synopsis of the Tabanidae (Diptera) of Michigan. *Misc. Publ. Mus. Zool. Univ. Mich.* 98:1–79.

Headley, A. E. 1943. The ants of Ashtabula Co., Ohio. *Ohio J. Sci.* 43 (1):22–31.

Hebard, Morgan. 1932. The Orthoptera of Minnesota. *Univ. Minn. Agric. Exp. Sta. Tech. Bull.* 85:1–65.

———. 1934. The Dermaptera and Orthoptera of Illinois. *Bull. Ill. Nat. Hist. Surv.* 20 (3): 159–62.

Heiss, Elizabeth M. 1938. A classification of the larvae and puparia of the Syrphidae of Illinois exclusive of aquatic forms. *III. Biol. Monogr.* 16 (4): 1–142.

Helfer, Jacques R. 1987. *How to know the grasshoppers, crickets, cockroaches, and their allies.* New York: Dover Publications.

Herring, J. L., and P. D. Ashcock. 1971. A key to nymphs of the families of Hemiptera (Heteroptera) of America north of Mexico. *Fla. Entomol.* 54:207–12.

Hilsenhoff, William L. 1970. Corixidae (water boatmen) of Wisconsin. *Trans. Wisc. Acad. Sci., Arts, Ltrs.* 58:203–35.

———. 1975. Aquatic insects of Wisconsin with generic keys and notes on biology, ecology, and distribution. *Wisc. Dept. Nat. Res. Tech. Bull.* 89:1–53.

———. 1984. Aquatic Hemiptera of Wisconsin. *Great Lakes Entomol.* 17 (1): 29–50.

———. 1985. The Brachycentridae (Trichoptera) of Wisconsin. *Great Lakes Entomol.* 18 (4): 149–54.

———. 1986. Semiaquatic Hemiptera of Wisconsin. *Great Lakes Entomol.* 19 (1): 7–20.

———. 1990. Gyrinidae of Wisconsin, with a key to adults of both sexes and notes on distribution and habitat. *Great Lakes Entomol.* 23 (2): 77–91.

———. 1992. Dytiscidae and Noteridae of Wisconsin (Coleoptera), I. Introduction, key to genera of adults, and distribution, habitat, life cycle, and identification of species of Agabetinae, Laccophilinae, and Noteridae. *Great Lakes Entomol.* 25 (2): 57–69.

———. 1993a. Dytiscidae and Noteridae of Wisconsin (Coleoptera), II. Distribution, habitat, life cycle, and identification of species of Dytiscinae. *Great Lakes Entomol.* 26 (1): 35–54.

———. 1993b. Dytiscidae and Noteridae of Wisconsin (Coleoptera), III. Distribution, habitat, life cycle, and identification of species of Colymbetinae. *Great Lakes Entomol.* 26 (2): 121–36.

———. 1993c. Dytiscidae and Noteridae of Wisconsin (Coleoptera), IV. Distribution, habitat, life cycle, and identification of species of Agabini. *Great Lakes Entomol.* 26 (3): 173–98.

Hilsenhoff, William L., and Steven J. Billmyer. 1973. Perlodidae (Plecoptera) of Wisconsin. *Great Lakes Entomol.* 6 (1): 1–14.

Hilsenhoff, William L., and Warren U. Brigham. 1978. Crawling water beetles of Wisconsin (Coleoptera: Haliplidae). *Great Lakes Entomol.* 11 (1): 11–22.

Hilsenhoff, William L., and Kurt L. Schmude. 1992. Riffle beetles of Wisconsin (Coleoptera: Dryopidae, Elmidae, Lutrochidae, Psephenidae) with notes on distribution, habitat, and identification. *Great Lakes Entomol.* 25 (3): 191–213.

Hine, James S. 1901. A review of the Panorpidae of America north of Mexico. *Bull. Sci. Lab. Dennison Univ.* 11:242–64.

———. 1903. Tabanidae of Ohio and bibliography of species north of Mexico. *Ohio Acad. Sci. Spec. Rept.* 5:1–63.

Hobeke, E. R., and A. G. Wheeler, Jr. 1982. *Rhopalus (Brachycarenus) tigrinus,* recently established in North America, with key to the genera and species of Rhopalidae in eastern North America (Hemiptera: Heteroptera). *Proc. Entomol. Soc. Wash.* 84:213–24.

Holland, W. J. 1968. *The moth book.* New York: Dover Publications.

Holmes, A., et al. 1992. *The Ontario butterfly atlas.* Toronto: Toronto Entomologists' Assoc.

Hottes, Frederick C., and Theodore H. Frison. 1931. The plant lice, or Aphiidae, of Illinois. *Bull. Ill. Nat. Hist. Surv.* 19 (3): 121–447.

Howden, Henry. 1969. Effects of the Pleistocene on North American insects. *Ann. Rev. Entomol.* 14:39–55.

Howden, Henry, and Oscar L. Cartwright. 1963. Scarab beetles of the genus *Onthophagus* Latreille north of Mexico (Coleoptera: Scarabaeidae). *Proc. U.S. Nat. Mus.* 114:1–135.

Howell, J. O., and M. L. Williams. 1976. An annotated key to the families of scale insects (Homoptera: Coccoidea) of America, north of Mexico, based on characteristics of adult females. *Annals Entomol. Soc. Amer.* 69:181–89.

Hughes, J. H. 1944. List of Chrysomelidae in Ohio. *Ohio J. Sci.* 44 (3): 129–42.

Hungerford, H. B. 1922. The Nepidae of North America. *Univ. Kansas Sci. Bull.* 14:425–69.

———. 1936. The Mantispidae of the Douglas Lake, Michigan, region, with some biological observations (Neurop.). *Entomol. News* 47 (2):69–72, 85–88.

Huryn, A. D., and B. A. Foote. 1983. An annotated list of the caddisflies (Trichoptera) of Ohio. *Proc. Entomol. Soc. Wash.* 85:783–96.

Husband, Robert W., et al. 1980. Distribution and biology of bumblebees (Hymenoptera: Apidae) in Michigan. *Great Lakes Entomol.* 13 (4): 225–40.

Hussey, R. F. 1919. The waterbugs (Hemiptera) of the Douglas Lake region, Michigan. *Occ. Pap. Mus. Zool. Univ. Mich.* 75:1–23.

———. 1922. Hemiptera from Berrien County, Michigan. *Occ. Pap. Mus. Zool. Univ. Mich.* 118:1–39.

Imes, Rick. 1992. *The Practical Entomologist.* New York: Fireside Books.

Irwin, Roderick R., and John C. Downey. 1973. Annotated checklist of the butterflies of Illinois. *Ill. Nat. Hist. Surv. Biol. Notes* 81.

Irwin, W. H. 1941. A preliminary list of the Culicidae of Michigan. Part I. Culicinae (Diptera). *Entomol. News* 52 (3): 101–5.

Jackson, C. F. 1906. Key to the families and genera of the order Thysanura. *Ohio Nat.* 6:545–49.

James, H. G. 1949. The distribution in Ontario of the European mantis, *Mantis religiosa* L. *Rept. Entomol. Soc. Ont.* 79:41–44.

———. 1969. Immature stages of five diving beetles (Coleoptera: Dytiscidae), notes on their habits and life history, and a key to aquatic beetles of vernal pools in southern Ontario. *Proc. Entomol. Soc. Ont.* 100:52–96.

Jaques, H. E. 1951. *How to know the beetles.* Dubuque, Ia.: W. C. Brown Co.

Johannsen, O. A. 1905. Mayflies and midges of New York. *New York State Mus. Bull.* 86.

Johnson, Dorothy S. 1935. Leafhoppers of Ohio, subfamily Typhlocybinae (Homoptera: Cicadellidae). *Bull. Ohio Biol. Surv.* 31, 6 (2): 1–122.

Judd, W. W. 1947. The praying mantis (*Mantis religiosa* L.) at Hamilton, Ontario. *Can. Field Nat.* 61:197–98.

———. 1949a. Insects collected in the Dundas Marsh, Hamilton, Ontario, 1946–47. with observations on their periods of emergence. *Can. Entomol.* 81 (1): 1–10.

———. 1949b. Insects collected in the Dundas Marsh, Hamilton, Ontario, 1947–48. *J. New York Entomol. Soc.* 87:225–31.

———. 1950a. Further records of the occurrence of the European mantis, *Mantis religiosa* L., in southern Ontario (Orthoptera). *Entomol. News* 61:205–7.

———. 1950b. Mosquitoes collected in the vicinity of Hamilton, Ontario, during the summer of 1948. *Mosquito News* 10:57–59.

———. 1954. Results of a survey of mosquitoes conducted at London, Ontario, in 1952 with observations on the biology of species collected. *Can. Entomol.* 86:101–8.

———. 1958. Studies on the Byron Bog in southwestern Ontario. V. Seasonal distribution of horse flies and deer flies (Tabanidae). *Can. Entomol.* 90:255–56.

———. 1959. Studies of the Byron Bog in southwestern Ontario. X. Inquilines and victims of the pitcher plant, *Sarracenia purpurea* L. *Can. Entomol.* 91:171–80.

———. 1964. Insects associated with flowering marsh marigold, *Caltha palustris* L., at London, Ontario. *Can. Entomol.* 96:1472–76.

———. 1969. Studies of the Byron Bog in southwestern Ontario. XXXVIII. Insects associated with flowering boneset, *Eupatorium perfoliatum* L. *Proc. Entomol. Soc. Ont.* 99:65–69.

———. 1970. Studies of the Byron Bog in southwestern Ontario. XLV. Insects associated with flowering dandelion, *Taraxacum officinale* Weber. *Proc. Entomol. Soc. Ont.* 101:59–61.

Kathirithamby, J. 1989. Review of the Order Strepsiptera. *System. Entomol.* 14:41–92.

Kellicott, D. S. 1899. Odonata of Ohio. *Ohio Acad. Sci. Spec. Pap.* 3 (2): 1–116.

Kelton, Leonard A. 1978. The insects and arachnids of Canada. Part 4. The Anthocoridae of Canada and Alaska (Heteroptera: Anthocoridae). *Agric. Can. Publ.* No. 1639.

Kim, K. C., and H. W. Ludwig. 1978. The family classification of Anoplura. *System. Entomol.* 3:249–84.

Kinzelbach, R. 1991. The systematic position of the Strepsiptera (Insecta). *Amer. Entomol.* 36:292–303.

Kirk, H. B., and J. N. Knull. 1926. Annotated list of the Cerambycidae of Pennsylvania. *Can. Entomol.* 58:21–26, 39–46.

Klages, H. G. 1901. Supplement to Dr. John Hamilton's list of Coleoptera of southwestern Pennsylvania. *Annals Carnegie Mus.* 1:265–94.

Knapp, Virgil R. 1972. Preliminary annotated list of Indiana Aphididae. *Proc. Ind. Acad. Sci.* 82:242–63.

Knerer, G., and C. E. Atwood. 1961. An annotated list of the non-parasitic Halictidae of Ontario. *Proc. Entomol. Soc. Ont.* 92:160–75.

Knight, Harry H. 1917. A revision of the genus *Lygus* as it occurs in America north of Mexico, with biological data on the species from New York. *Cornell Univ. Agric. Exp. Sta. Mem.* 391:555–645.

———. 1921. Monograph of the North American species of *Deraeocoris* (Heteroptera, Miridae). *Univ. Minn. Agric. Exp. Sta. Tech. Bull.* 1:76–210.

———. 1941. The plant bugs, or Miridae, of Illinois. *Bull. Ill. Nat. Hist. Surv.* 22 (1): 1–234.

Knipping, P. A., et al. 1950. Preliminary list of some fleas from Wisconsin. *Trans. Wisc. Acad. Sci., Arts, Ltrs.* 40:199–200.

Knisley, C. Barry, et al. 1987. Distribution, checklist, and key to adult tiger beetles (Coleoptera: Cicindelidae) of Indiana. *Proc. Ind. Acad. Sci.* 97:279–94.

Knull, Josef N. 1946. The long-horned beetles of Ohio (Coleoptera: Cerambycidae). *Bull. Ohio Biol. Surv.* 39, 7 (4): 133–354.

———. 1951. The checkered beetles of Ohio (Coleoptera: Cleridae). *Bull. Ohio Biol. Surv.* 42, 8 (2).

Knutson, Herbert. 1944. Minnesota Phalaenidae (Noctuidae). *Univ. Minn. Agric. Exp. Sta. Tech. Bull.* 165:1–128.

Kormondy, E. J. 1958. A catalogue of the Odonata of Michigan. *Misc. Publ. Mus. Zool. Univ. Mich.* 104:1–43.

Kosztarab, Michael. 1963. The armored scale insects of Ohio (Homoptera: Coccoidea: Diaspidae). *Bull. Ohio Biol. Surv.* (N.S.) 2 (2): 1–120.

Kurczewski, Frank E., and Robert E. Aciavatti. 1990. Late summer-fall solitary wasp fauna of central New York (Hymenoptera: Tiphiidae, Pompilidae, Sphecidae). *Great Lakes Entomol.* 23 (2): 57–64.

Kurczewski, Frank E., and E. J. Kurczewski. 1963. An annotated list of digger wasps from Presque Isle State Park, Pennsylvania. *Proc. Entomol. Soc. Wash.* 65:141–49.

Kurta, Allen. 1995. *Mammals of the Great Lakes region.* Rev. ed. Ann Arbor: Univ. Michigan Press.

Lager, T. M., et al. 1979. Preliminary report on the Plecoptera and Trichoptera of northeastern Minnesota. *Great Lakes Entomol.* 12 (3): 109–14.

Lauck, D. R. 1964. A monograph of the genus *Belostoma* (Hemiptera). Part III. *B. triangulum, bergi, minor, bifoveolatum,* and *flumineum* groups. *Bull. Chic. Acad. Sci.* 11:102–54.

Lawrence, Vinnedge M. 1967. The distribution of Odonata in Indiana and Ohio. *Proc. North Central Branch Entomol. Soc. Amer.* 22:117–20.

Lawrence, W. H., et al. 1965. Arthropodous ectoparasites from some northern Michigan mammals. *Occ. Pap. Mus. Zool. Univ. Mich.* 639:1–7.

Lawson, H. R., and W. P. McCafferty. 1984. A checklist of Megaloptera and Neuroptera (Planipennia) of Indiana. *Great Lakes Entomol.* 17 (3): 129–32.

Lawton, F. A. 1959. Identification of nymphs of common families of Hemiptera. *J. Kans. Entomol. Soc.* 32:88–92.

Layne, J. N. 1958. Records of fleas (Siphonaptera) from Illinois mammals. *Chic. Acad. Sci. Nat. Hist. Misc.* 162:1–7.

Lehmkuhl, D. M. 1979. *How to know the aquatic insects.* Dubuque, Ia.: William C. Brown Co.

Leonard, J. W., and F. A. Leonard. 1962. Mayflies of Michigan trout streams. *Cranbrook Inst. Sci. Bull.* 43:1–139.

Leonard, M. D. 1928. A list of insects of New York with a list of the spiders and certain allied groups. *Cornell Univ. Agric. Exp. Sta. Mem.* 101:1–1121.

Lindquist, O. H., and W. J. Miller. 1972. A key to sawfly larvae feeding on the foliage of spruce and balsam fir in Ontario. *Proc. Entomol. Soc. Ont.* 102:118–22.

Lindroth, Carl H. 1961. The ground-beetles of Canada and Alaska. Parts 1–6. Entomologiska Sasllskapet, Lund. Pp. 1–1192.

Longridge, J. L., and W. L. Hilsenhoff. 1973. Annotated list of Trichoptera (caddisflies) in Wisconsin. *Trans. Wisc. Acad. Sci., Arts, Ltrs.* 61:173–83.

Lugger, Otto. 1897. The Orthoptera of Minnesota. *Univ. Minn. Agric. Exp. Sta. Tech. Bull.* 55:1–386.

———. 1899. Coleoptera of Minnesota. *Univ. Minn. Agric. Exp. Sta. Tech. Bull.* 66:85–331.

MacArthur, Kenneth. 1948. The louse flies of Wisconsin and adjacent states (Diptera: Hippoboscidae). *Bull. Publ. Mus. Milwaukee* 8 (4): 367–440.

MacDonald, J. F., and Mark Deyrup. 1989. The social wasps (Hymenoptera: Vespidae) of Indiana. *Great Lakes Entomol.* 22 (3): 155–75.

Mallis, Arnold. 1971. *American entomologists.* New Brunswick, NJ: Rutgers Univ. Press.

Martin, H. M. 1952. *Outline of the geological history of Michigan.* Lansing: Mich. Dept. Conserv. Geol. Surv. Div.

Martin, J. E. H. 1977. The insects and arachnids of Canada. Part 1. Collecting, preparing, and preserving insects, mites, and spiders. *Agric. Can. Publ.* No. 1.

Marvin, Daniel E. 1965. A list of fireflies known or likely to occur in Ohio, with special notes on species of *Ellychnia* (Lampyridae: Coleoptera). *Ohio J. Sci.* 65: (1) 37–42.

Masters, C. O. 1949. A study of the adult mosquitoes of a northern Ohio woods. *Ohio J. Sci.* 49 (1): 12–14.

Matheson, R. 1924. The Culicidae of the Douglas Lake region (Michigan). *Can. Entomol.* 56:289–90.

Matsuda, Ryuichi. 1977. The insects and arachnids of Canada. Part 3. The Aradidae of Canada (Hemiptera: Aradidae). *Agric. Can. Publ.* No. 1634.

McAtee, W. L. 1919. Key to the nearctic species of Piesmidae (Heteroptera). *Bull. Brooklyn Entomol. Soc.* 14:80–93.

McCafferty, W. P., and M. C. Minno. 1979. The aquatic and semiaquatic Lepidoptera of Indiana and adjacent areas. *Great Lakes Entomol.* 12 (4): 179–88.

McCafferty, W. P., and J. L. Stein. 1976. Indiana Ensifera (Orthoptera). *Great Lakes Entomol.* 9 (1): 23–56.

McPherson, J. E. 1970. A key and annotated list of the Scutelleroidea of Michigan (Hemiptera). *Mich. Entomol.* 3 (2): 34–63.

———. 1978. A list of the Scutelleroidea (Hemiptera) of southern Illinois. *Great Lakes Entomol.* 11 (3): 159–62.

———. 1979a. Additions and corrections to the list of Michigan Pentatomoidea (Hemiptera). *Great Lakes Entomol.* 12 (1): 27–30.

———. 1979b. A revised list of Pentatomoidea of Illinois (Hemiptera). *Great Lakes Entomol.* 12 (3): 91–98.

———. 1980. The distribution of the Pentatomoidea in the northeastern quarter of the United States (Hemiptera). *Great Lakes Entomol.* 13 (1): 1–16.

———. 1989. An overview of the Heteroptera of Illinois. *Great Lakes Entomol.* 22 (4): 177–98.

———. 1992. The assassin bugs of Michigan. *Great Lakes Entomol.* 25 (1): 25–32.

McPherson, J. E., et al. 1990. Eastern range extension of *Leptoglossus occidentalis* with a key to *Leptoglossus* species of America north of Mexico (Heteroptera: Coreidae). *Great Lakes Entomol.* 23 (2): 99–104.

Medler, J. T. 1942. The leafhoppers of Minnesota (Homoptera: Cicadellidae). *Univ. Minn. Agric. Exp. Sta. Tech. Bull.* 155:1–196.

Medler, J. T., and D. W. Carney. 1963. Bumblebees of Wisconsin (Hymenoptera: Apidae). *Univ. Wisc. Res. Bull.* 240.

Menke, A. S. 1963. A review of the genus *Lethocerus* in North and Central America, including the West Indies (Hemiptera: Belostomatidae). *Annals Entomol. Soc. Amer.* 56:261–67.

Merritt, Richard W., et al. 1978. Larval ecology of some Lower Michigan black flies (Diptera: Simuliidae) with keys to the immature stages. *Great Lakes Entomol.* 11 (4): 177–208.

Metcalf, C. E. 1913. The Syrphidae of Ohio. *Bull. Ohio Biol. Surv.* 1 (1): 1–123.

Metzler, Eric H. 1980. Annotated checklist and distribution maps of the royal moths and giant silkworm moths (Lepidoptera: Saturniidae). *Ohio Biol. Surv. Biol. Notes* No. 14.

———. 1988. Report of Beaver Creek State Park field trip May 13–15, 1988. *Ohio Lepid. Newslet.* 10 (2): 13.

Mickel, Clarence E. 1923. Preliminary notes on the Mutillidae of Minnesota. *Rept. Minn. State Entomol.* 19:97–113.

———. 1928. Biological and taxonomic investigations on the mutilid wasps. *U.S. Nat. Mus. Bull.* 143:1–338.

Middlekauf, Woodrow W. 1958. The North American sawflies of the genera *Acantholyda, Cephalcia,* and *Neurotoma* (Hymenoptera: Pamphilidae). *Univ. Calif. Publ. Entomol.* 14 (2): 51–174.

———. 1964. The North American sawflies of the genus *Pamphilius* (Hymenoptera: Pamphilidae). *Univ. Calif. Publ. Entomol.* 38:1–83.

Miller, William E. 1987. Guide to the Olethreutine moths of midland North America (Tortricidae). *U.S.D.A. Forest Serv. Agric. Handbook* 660.

Milliron, H. E. 1939. The taxonomy and distribution of Michigan Bombidae, with keys. *Pap. Mich. Acad. Sci., Arts, Ltrs.* 24:167–82.

Miskimen, G. W. 1956. A faunal list of Cantharidae of Ohio. *Ohio J. Sci.* 56:129–34.

Mockford, E. L. 1951. The Psocoptera of Indiana. *Proc. Ind. Acad. Sci.* 60:192–204.

Montgomery, B. Elwood. 1956. Anthophilous insects of Indiana. Part 1. A preliminary annotated list of the Apoidea. *Proc. Ind. Acad. Sci.* 66:125–40.

Montgomery, B. Elwood, and A. W. Trippel. 1933. A preliminary list of Indiana Neuroptera. *Entomol. News* 44 (5): 258–61.

Montgomery, R. W. 1931. Preliminary list of the butterflies of Indiana. *Proc. Ind. Acad. Sci.* 40:357–59.

Moore, S. 1922. A list of northern Michigan Lepidoptera. *Occ. Pap. Mus. Zool. Univ. Mich.* 114:1–28.

———. 1930. Lepidoptera of the Beaver Islands. *Occ. Pap. Mus. Zool. Univ. Mich.* 214:1–28.

———. 1939. A list of butterflies of Michigan. *Occ. Pap. Mus. Zool. Univ. Mich.* 411:1–23.

———. 1955. An annotated list of the moths of Michigan (exclusive of Tineoidea) (Lepidoptera). *Misc. Publ. Mus. Zool. Univ. Mich.* 88.

———. 1960. A revised list of the butterflies of Michigan. *Occ. Pap. Mus. Zool. Univ. Mich.* 617:1–39.

Moore, T. M. 1955. Important Illinois spittlebugs. *Ill. Nat. Hist. Surv. Ident. Notes* 2:1–4.

———. 1966. The cicadas of Michigan (Homoptera: Cicadidae). *Pap. Mich. Acad. Sci., Arts, Ltrs.* 51:75–94.

Morofsky, W. F. 1933. Distribution of May beetles (*Phyllophaga*) in Michigan. *J. Econ. Entomol.* 26:831–34.

Morris, C. D., and G. R. DeFoliart. 1971. New records of Tabanidae in Wisconsin. *J. Med. Entomol.* 8:28.

Mumford, Russell E., and John O. Whitacker, Jr. 1982. *The mammals of Indiana.* Bloomington: Indiana Univ. Press.

Munroe, Eugene. 1956. Canada as an environment for insects. *Can. Entomol.* 88:372–476.

Munsee, Jack R., et al. 1985. Revision of the checklist of Indiana ants with additions of five new species (Hymenoptera: Formicidae). *Proc. Ind. Acad. Sci.* 95:265–74.

Myers, R. W., and D. P. Sanders. 1975. New locality records in the genus *Chrysops* (Diptera: Tabanidae) in Indiana. *Proc. Ind. Acad. Sci.* 85:271–73.

Needham, J. G. 1900. Insect drift on the shore of Lake Michigan. *Occ. Mem. Chic. Entomol. Soc.* 1 (1): 19–26.

———. 1904. Beetle drift on Lake Michigan. *Can. Entomol.* 36:294–96.

———. 1917. The insect drift of lake shores. *Can. Entomol.* 49:129–37.

Needham, J. E., and Cornelius Betten. 1901. Aquatic insects in the Adirondacks. *New York State Mus. Bull.* 47:383–612.

Needham, J. E., and Charles A. Hart. 1901. Dragonflies (Odonata) of Illinois. Part 1.

Petaluridae, Aeschnidae, and Gomphidae. *Bull. Ill. State Lab. Nat. Hist.* 6 (1): 1–94.

Needham, J. E., and M. J. Westfall. 1955. A manual of the dragonflies of North America (Anisoptera). Berkeley: Univ. Calif. Press.

Neiswander, C. R. 1963. The distribution and abundance of May beetles in Ohio. *Ohio Agric. Exp. Sta. Res. Bull.* 951:1–35.

New, T. R., and C. C. Loan. 1971. Psocoptera collected near Belleville, Ontario. *Proc. Entomol. Soc. Ont.* 102:16–22.

Nicholson, H. P., and C. E. Mickel. 1950. The black flies of Minnesota (Simuliidae). *Univ. Minn. Agric. Exp. Sta. Tech. Bull.* 192:1–64.

Nicolay, A. S. 1914. Mordellidae of New York. *Bull. Brooklyn Entomol. Soc.* 9:29–32.

Nielsen, M. C. 1992. Preliminary checklist of Michigan butterflies and skippers. *Newslet. Mich. Entomol. Soc.* 37 (1): 5–7.

O'Brien, Mark F. 1988. Records of Ampulicidae in Michigan (Hymenoptera: Sphecoidea). *Great Lakes Entomol.* 21 (2): 81–82.

———. 1989a. New state records of bembecine sand wasps in Michigan (Hymenoptera: Bembicinae). *Great Lakes Entomol.* 22 (2): 103–4.

———. 1989b. Distribution and biology of the sphecine wasps of Michigan (Hymenoptera: Sphecidae: Sphecinae). *Great Lakes Entomol.* 22 (4): 199–217.

Oestlund, O. W. 1887. Aphididae of Minnesota. *Geol. and Nat. Hist. Surv. Minn. Bull.* 4:1–100.

———. 1918. Contribution to the knowledge of the tribes and higher groups of the family Aphididae (Homoptera). *Rept. Minn. State Entomol.* 17:46–74.

Oliver, Donald R., and Mary E. Roussel. 1983. The genera of larval midges of Canada (Diptera: Chironomidae). *Agric. Can., Publ.* No. 1746.

Opler, Paul, and Vichai Malikul. 1992. *A field guide to eastern butterflies.* Boston: Houghton-Mifflin.

Osborn, Herbert. 1900. Remarks on the hemipterous fauna of Ohio with a preliminary record of species. *Ann. Rept. Ohio State Acad. Sci.* 8:60–79.

———. 1903. The Aradidae of Ohio. *Ohio Nat.* 4 (1): 36–42.

———. 1928. The leafhoppers of Ohio. *Bull. Ohio Biol. Surv.* 14, 3 (4) (1): 199–374.

———. 1938. The Fulgoridae of Ohio. *Bull. Ohio Biol. Surv.* 35, 6 (6): 283–357.

———. 1940. The Membracidae of Ohio. *Bull. Ohio Biol. Surv.* 37 ,7 (2): 53–101.

Osborn, Herbert, and Carl J. Drake. 1916. The Tingitoidea of Ohio. *Bull. Ohio Biol. Surv.* 8, 2 (4): 217–51.

Otte, Daniel. 1984a. *North American grasshoppers. Vol. 1. Acrididae: Cyrtacanthacridinae, Gomphocerinae, and Acridinae.* Cambridge: Harvard Univ. Press.

———. 1984b. *North American grasshoppers. Vol. 2. Acrididae: Oedipodinae.* Cambridge: Harvard Univ. Press.

Owen, W. B. 1937. The mosquitoes of Minnesota, with special reference to their biologies. *Univ. Minn. Agric. Exp. Sta. Tech. Bull.* 126:1–75.

Parfin, Sophy I. 1952. The Megaloptera and Neuroptera of Minnesota. *Amer. Midl. Nat.* 47:421–34.

Parfin, Sophy I., and Ashley B. Gurney. 1956. The spongilla-flies, with special reference to those of the western hemisphere (Sisyridae, Neuroptera). *Proc. U.S. Nat. Mus.* 105:421–529.

Pechuman, L. L. 1931. The Tabanidae (horse flies) of Minnesota with special reference to their biology and taxonomy. *Univ. Minn. Agric. Exp. Sta. Tech. Bull.* 80:1–132.

———. 1957. The Tabanidae of New York. *Proc. Rochester Acad. Sci.* 10:121–79.

———. 1972. The horse flies and deer flies of New York (Diptera: Tabanidae). *Search: Agric.* 2:1–72.

Pechumen, L. L., et al. 1961. The Tabanidae (Diptera) of Ontario. *Proc. Entomol. Soc. Ont.* 91:77–121.

———. 1983. The Diptera, or true flies, of Illinois. Part I. Tabanidae. *Bull. Ill. Nat. Hist. Surv.* 33:1–121.

Perry, T. E. 1983. Additions to state and local lists of dragonflies and damselflies (Odonata). *Ohio J. Sci.* 83 (3): 141.

Peters, Harold S. 1928. Mallophaga from Ohio birds. *Ohio J. Sci.* 28:215–28.

Pettit, J. 1869. List of Coleoptera taken at Grimsby, Ontario. *Can. Entomol.* 1:106–7.

———. 1870. List of Coleoptera taken at Grimsby, Ontario. *Can. Entomol.* 2:7, 17–18, 53–54, 65–66, 84–86, 102–3, 117–18, 131–33, 151.

———. 1871. List of Coleoptera taken at Grimsby, Ontario. *Can. Entomol.* 3:105–7.

———. 1872. List of Coleoptera taken at Grimsby, Ontario. *Can. Entomol.* 4:12–14, 98–99.

Pettit, R. H. 1903. Mosquitoes and other insects of the year 1902. *Mich. State Agric. Coll. Spec. Bull.* 17:1–12.

Pettit, R. H., and E. McDaniel. 1918. Key to Orthoptera of Michigan with annotations. *Mich. State Agric. Coll. Spec. Bull.* 83:1–48.

———. 1920. The *Lecania* of Michigan. *Mich. State Agric. Coll. Agr. Exp. Sta. Tech. Bull.* 48:1–35.

Philip, C. B. 1931. The Tabanidae (horseflies) of Minnesota, with special reference to their biologies and taxonomy. *Univ. Minn. Agric. Exp. Sta. Tech. Bull.* 80.

Price, R. D. 1960. Identification of first-instar aedine mosquito larvae of Minnesota (Diptera: Culicidae). *Can. Entomol.* 92:544–60.

Profant, Dennis. 1991. An annotated checklist of the Lepidoptera of the Beaver Island archipelago, Lake Michigan. *Great Lakes Entomol.* 24 (2): 85–97.

Pyle, Robert. 1981. *The Audubon Society field guide to North American butterflies.* New York: A. A. Knopf.

Rahn, Russell A. 1973. Moths taken in Berrien County, Michigan (with 102 new county records). *Great Lakes Entomol.* 6 (1): 23–29.

Rehn, J. W. H. 1950. A key to the genera of North American Blatteria, including established adventives. *Entomol. News* 61 (2): 64–67.

Restifo, Robert A. 1982. Illustrated key to the mosquitoes of Ohio. *Ohio Biol. Surv. Biol. Notes* No. 17.

Ricker, W. E. 1945. A first list of Indiana stoneflies (Plecoptera). *Proc. Ind. Acad. Sci.* 54:225–30.

Ries, Mary Davis. 1967. Present state of knowledge of the distribution of Odonata in Wisconsin. *Proc. North Central Branch Entomol. Soc. Amer.* 22:113–15.

Rings, Roy W., et al. 1987. A nine year study of the Lepidoptera of the Wilderness Center, Stark Co., Ohio. *Ohio J. Sci.* 87 (2): 55–61.

———. 1992. The owlet moths of Ohio, Order Lepidoptera, Family Noctuidae. *Bull. Ohio Biol. Surv.* (N.S.) 9 (2): 1–120.

Rings, Roy W., and Eric H. Metzler. 1988. A preliminary annotated check list of the Lepidoptera of Atwood Lake Park, Ohio. *Ohio J. Sci.* 88:159–69.

———. 1989. A preliminary check list of the Lepidoptera of Mohican State Forest and Mohican State Park, Ashland Co., Ohio. *Ohio J. Sci.* 89 (4): 78–88.

———. 1990. The Lepidoptera of Fowler Woods State Nature Preserve, Richland Co., Ohio. *Great Lakes Entomol.* 23 (1): 43–56.

———. 1991. A check list of the Lepidoptera of Fulton Co., Ohio with special reference to the moths of Goll Woods State Nature Preserve. *Great Lakes Entomol.* 24 (4): 265–80.

———. 1992. A check list of the Lepidoptera of Beaver Creek State Park, Columbiana Co., Ohio. *Great Lakes Entomol.* 25 (2): 115–31.

Rivard, I. 1964. Carabid beetles (Coleoptera: Carabidae) from agricultural lands near Belleville, Ontario. *Can. Entomol.* 96:517–20.

———. 1965. Additions to the list of carabid beetles (Coleoptera: Carabidae) from agricultural lands near Belleville, Ontario. *Can. Entomol.* 97:332–33.

Roberts, R. H., and R. J. Dicke. 1958. Wisconsin Tabanidae. *Trans. Wisc. Acad. Sci., Arts, Ltrs.* 47:23–42.

Robertson, D. J. 1984. The aquatic insect community in Penitentiary Glen, a Portage Escarpment stream in northeastern Ohio. *Ohio J. Sci.* 84 (3): 113–19.

Rockburne, E. W., and J. D. LaFontaine. 1976. The cutworm moths of Ontario and Quebec. Ottawa: Can. Dept. of Agric.

Rogers, J. S. 1942. The crane flies (Tipulidae) of the George Reserve, Michigan. *Misc. Publ. Mus. Zool. Univ. Mich.* 53:1–128.

Ross, Herbert H. 1937. Studies on Nearctic aquatic insects. I. Nearctic alder flies of the genus *Sialis* (Megaloptera, Sialidae). *Bull. Ill. Nat. Hist. Surv.* 21 (3): 57–77.

———. 1944. The caddis flies, or Trichoptera, of Illinois. *Bull. Ill. Nat. Hist. Surv.* 23 (1): 1–326.

———. 1947. The mosquitoes of Illinois. *Bull. Ill. Nat. Hist. Surv.* 24 (1): 1–96.

Roth, J. C. 1967. Notes on *Chaoborus* species from the Douglas Lake region, Michigan, with a key to their larvae (Diptera: Chaoboridae). *Pap. Mich. Acad. Sci., Arts, Ltrs.* 52:63–68.

Rutherford, Jane E. 1985. An illustrated key to the pupae of six species of *Hydropsyche* (Trichoptera: Hydropsychidae) common in southern Ontario streams. *Great Lakes Entomol.* 18 (3): 123–32.

Sabrosky, C. W. 1944. A mosquito survey of southern Michigan. *J. Econ. Entomol.* 37:312–13.

Sanders, Darryl P., and John L. Petersen. 1974. The occurrence of the pigeon fly, *Pseudolynchia canariensis* (Macquart) in Indiana. *Proc. Ind. Acad. Sci.* 84:287–88.

Sanders, J. G. 1904. The Coccidae of Ohio. *Proc. Ohio Acad. Sci.* 4 (2): 25–92.

Sanders, J. G., and D. M. DeLong. 1917. The Cicadellidae (Jassoidea—Fam. Homoptera) of Wisconsin, with descriptions of new species. *Annals Entomol. Soc. Amer.* 10 (1): 79–95.

Schmid, F. 1980. Genera Trichopteres du Canada et des Etats adjacents. *Agric. Can. Publ.* No. 1692.

Schmunde, Kurt L., and William L. Hilsenhoff. 1986. Biology, ecology, larval taxonomy and distribution of Hydroscaphidae (Trichoptera) in Wisconsin. *Great Lakes Entomol.* 19 (3): 123–46.

Schrock, J. R. 1985. Checklist of adult carabid beetles from Indiana. *Proc. Ind. Acad. Sci.* 94:341–56.

Schuh, T. 1967. The shore bugs (Hemiptera: Saldidae) of the Great Lakes region. *Contrib. Amer. Entomol. Inst.* 2:1–35.

Scott, H. G. 1961. Collembola: Pictorial key to Nearctic genera. *Annals Entomol. Soc. Amer.* 54 (1): 104–13.

Selman, Charles L. 1975. A pictorial key to the hawkmoths (Lepidoptera: Sphingidae) of eastern United States (except Florida). *Ohio Biol. Surv. Biol. Notes* No. 9:1–31.

Shapiro, Arthur M. 1974. Butterflies and skippers of New York State. *Search: Agric.* 4 (3). Agric. Exp. Sta. Cornell Univ., Ithaca.

Shull, A. F. 1911. Thysanoptera and Orthoptera. Pp. 117–231 *in* A. G. Ruthven ed., A biological survey of the sand dune region on the south shore of Saginaw Bay, Michigan. *Mich. Geol. Biol. Surv.*

Shull, Ernest M. 1987. *The butterflies of Indiana*. Bloomington: Indiana Univ. Press.

Shull, Ernest M., and F. Sidney Badger. 1972. Annotated list of the butterflies of Indiana. *J. Lepid. Soc.* 26:13–24.

Sites, R. W., and J. E. McPherson. 1980. A key to the butterflies of Illinois (Lepidoptera: Papilionoidea). *Great Lakes Entomol.* 13 (3): 97–114.

Slater, J. A., and R. M. Baranowski. 1978. *How to know the true bugs (Hemiptera-Heteroptera)*. Dubuque, Ia.: William C. Brown Co.

Smith, C. L., and J. T. Polhemus. 1978. The Veliidae (Heteroptera) of America north of Mexico—keys and checklists. *Proc. Entomol. Soc. Wash.* 80:56–68.

Smith, C. W. 1958. New records of the European mantis, *Mantis religiosa* L. (Orthoptera: Mantidae) in Ontario. *Proc. Entomol. Soc. Ont.* 89:70.

Smith, Roger C. 1932. The Chrysopidae of Canada. *Annals Entomol. Soc. Amer.* 25:579–601.

Snider, Richard J. 1967. An annotated list of the Collembola (springtails) of Michigan. *Mich. Entomol.* 1 (6): 179–234.

Snow, L. N. 1902. The microcosm of the drift line. *Amer. Nat.* 36:855–64.

Somes, M. P. 1914. The Acrididae of Minnesota. *Univ. Minn. Agric. Exp. Sta. Tech. Bull.* 141:1–100.

Sommers, Lawrence M., ed. 1977. *Atlas of Michigan*. East Lansing: Michigan State Univ. Press.

Spillman, T. J. 1973. A list of the Tenebrionidae of Michigan (Coleoptera). *Great Lakes Entomol.* 6 (3): 85–91.

Stannard, Lewis J. 1963. Post-Wisconsin biogeography in eastern North America, based, in part, on evidence from thrips (Insecta). *Proc. North Central Branch Entomol. Soc. Amer.* 18:31–35.

———. 1968. The thrips, or Thysanoptera, of Illinois. *Bull. Ill. Nat. Hist.* 29 (4): 215–552.

Stehr, W. C. 1930. The Coccinellidae of Minnesota. *Univ. Minn. Agric. Exp. Sta. Tech. Bull.* 75.

Stein, J. L., and W. P. McCafferty. 1975. Diagnostic tables to the long-horned grasshoppers and crickets in Indiana. *Purdue Univ. Agric. Exp. Sta. Res. Bull.* 921:1–20.

Steinly, B. A., et al. 1987. The distribution of shore flies (Diptera: Ephydridae) in Illinois. *Entomol. News* 98 (4): 165–70.

Steward, C. C., and J. W. McWade. 1961. The mosquitoes of Ontario (Diptera: Culicidae), with keys to species and notes on distribution. *Proc. Entomol. Soc. Ont.* 91:121–88.

Stone, A., and H. A. Jamnback. 1955. The black flies of New York State. *New York State Mus. Bull.* 349:1–144.

Strickler, Jeffrey D., and Edward D. Walker. 1993. Seasonal abundance and species diversity of adult Tabanidae (Diptera) at Lake Lansing Park-North, Michigan. *Great Lakes Entomol.* 26 (2): 107–12.

Surdick, Rebecca F. 1985. Nearctic genera of Chloroperlinae (Plecoptera: Chloroperlidae). *Ill. Biol. Monogr.* 54:1–146.

Surdick, R. F., and K. C. Kim. 1976. Stoneflies (Plecoptera) of Pennsylvania. *Penn. State Coll. Agric. Exp. Sta. Bull.* 808:1–73.

Syme, P. D., and D. M. Davies. 1958. Three new Ontario black flies of the genus

Prosimulium (Simuliidae). Part 1. Descriptions, morphological comparisons with related species, and distribution. *Can. Entomol.* 90:697–719.

Taboada, Oscar. 1964. An annotated list of the Cicadellidae of Michigan. *Quart. Bull. Mich. Agric. Sta.* 47:113–22.

———. 1979. New records of leafhoppers (Homoptera: Cicadellidae) for Michigan, including vector of X-disease. *Great Lakes Entomol.* 12 (3): 99–100.

Talbot, Mary. 1934. Distribution of ant species in the Chicago region with reference to ecological factors and physiological tolerance. *Ecology* 15:416–39.

———. 1975. A list of the ants (Hymenoptera: Formicidae) of the Edwin S. George Reserve, Livingston Co., Michigan. *Great Lakes Entomol.* 8 (4): 245–46.

Telford, Horace S. 1939. The Syrphidae of Minnesota. *Univ. Minn. Agric. Exp. Sta. Tech. Bull.* 140:1–76.

Teskey, H. J. 1960. Survey of insects affecting livestock in southwestern Ontario. *Can. Entomol.* 92:531–41.

Thomas, C. A. 1941. Elateridae of Pennsylvania. *J. New York Entomol. Soc.* 49:233–63.

Thomas, M. K. 1968. Some notes on the climatic history of the Great Lakes. *Proc. Entomol. Soc. Ont.* 99:21–30.

Thornhill, Albert R., and James B. Johnson. 1974. The Mecoptera of Michigan. *Great Lakes Entomol.* 7 (2): 33–53.

Throne, Alvin L. 1971a. The Neuroptera—suborder Planipennia of Wisconsin. Part I. Introduction and Chrysopidae. *Mich. Entomol.* 4 (3): 65–78.

———. 1971b. The Neuroptera—suborder Planipennia of Wisconsin. Part II. Hemerobiidae, Polystoechotidae, and Sisyridae. *Mich. Entomol.* 4 (3): 79–88.

———. 1972. The Neuroptera—suborder Planipennia of Wisconsin. Part III. Mantispidae, Ascalaphidae, Myrmeleontidae, and Coniopterygidae. *Great Lakes Entomol.* 5 (4): 119–28.

Tietz, Harrison M. 1936. The Noctuidae of Pennsylvania. *Penn. State Coll. Agric. Exp. Sta. Bull.* No. 335.

———. 1952. The Lepidoptera of Pennsylvania. State College: Penn. State Univ.

Tomlin, A. D. 1978. Protura of Canada *in* H. V. Danks, ed., Canada and its insect fauna. *Mem. Entomol. Soc. Can.* 108:303–4.

Tufford, Sarah, and Rudolph Hogberg. 1965. Guide to fossil collecting in Minnesota. *Minn. Geol. Surv. Educ. Publ.* No. 1.

Urquhart, F. A. 1941a. The Blattaria and Orthoptera of Essex County, Ontario. *Contrib. R. Entomol. Mus. Zool.* 20:1–32.

———. 1941b. A faunal investigation of Prince Edward County, Ontario. An annotated list of the crickets and grasshoppers (Orthoptera: Saltatoria) of Prince Edward County, Ontario. *Univ. Tor. Stud. Biol. Ser.* 48:116–19.

———. 1942. The Dermaptera of Ontario. *Can. Field Nat.* 56:3.

Urquhart, F. A., and C. E. Corfe. 1940. The European praying mantis (*Mantis religiosa* L.) in Ontario. *Can. Field Nat.* 54:130–32.

Venard, C. E., and F. W. Mead. 1953. An annotated list of Ohio mosquitoes. *Ohio J. Sci.* 53 (6): 327–31.

Vickery, Vernon R., and D. Keith McE. Kevan. 1985. The insects and arachnids of Canada. Part 14. The grasshoppers, crickets, and related insects of Canada and adjacent regions (Ulonata: Dermaptera, Cheleutoptera, Notoptera, Dictuoptera, Grylloptera, and Orthoptera). *Agric. Can. Publ.* No. 1777.

Voss, Edward G. 1969. Moths of the Douglas Lake region (Emmet and Cheboygan Counties), Michigan: I. Sphingidae-Ctenuchidae (Lepidoptera). *Mich. Entomol.* 2 (3–4): 48–54.

————. 1981. Moths of the Douglas Lake region (Emmet and Cheboygan Counties), Michigan: II. Noctuidae (Lepidoptera). *Great Lakes Entomol.* 14 (2): 88–102.

————. 1983. Moths of the Douglas Lake region (Emmet and Cheboygan Counties), Michigan: III. Thyatiridae, Drepanidae, Lasiocampidae, Notodontidae, Lymantriidae (Lepidoptera). *Great Lakes Entomol.* 16 (4): 131–38.

————. 1992. Moths of the Douglas Lake region (Emmet and Cheboygan Counties), Michigan: IV. Geometridae (Lepidoptera). *Great Lakes Entomol.* 24 (3): 187–201.

Walker, Edmund M. 1902. A preliminary list of Acrididae of Ontario. *Can. Entomol.* 34:251–58.

————. 1904. Crickets of Ontario. *Can. Entomol.* 36:142–44, 181–88, 249–55.

————. 1909. On the Orthoptera of northern Ontario. *Can. Entomol.* 41:173–78.

————. 1912. The Blattidae of Ontario. *Can. Entomol.* 44:171–72.

————. 1915. The occurrence of *Mantis religiosa* L. in Canada. *Can. Entomol.* 47:135.

————. 1953. The Odonata of Canada and Alaska. Vol. 1. Toronto: Univ. Toronto Press.

————. 1958. The Odonata of Canada and Alaska. Vol. 2. Anisoptera. Toronto: Univ. Toronto Press.

Walker, Edmund M., and P. S. Corbet. 1975. *The Odonata of Canada and Alaska.* Vol. 3. Toronto: Univ. Toronto Press.

Wallis, J. B. 1961. *The Cicindelidae of Canada.* Toronto: Univ. of Toronto Press.

Wallis, J. B., and D. J. Larson. 1973. An annotated list of the Hydradephaga (Coleoptera: Insecta) of Manitoba and Minnesota. *Quaest. Entomol.* 9:99–114.

Washburn, F. L. 1918. The Hymenoptera of Minnesota. *Rept. Minn. State Entomol.* 17:145–240.

Watson, S. A. 1928. The Miridae of Ohio. *Bull. Ohio Biol. Surv.* 16, 4 (1): 3–44.

Webb, Donald W., et al. 1975. The Mecoptera, or scorpionflies, of Illinois. *Bull. Ill. Nat. Hist. Surv.* 31 (1).

Wegner, G. S., and H. D. Niemczyk. 1979. The *Ataenius* of Ohio. *Ohio J. Sci.* 79 (6): 249–55.

Wellso, Stanley G., et al. 1976. Keys and notes on the Buprestidae (Coleoptera) of Michigan. *Great Lakes Entomol.* 9 (1): 1–22.

Whedon, A. D. 1914. Preliminary notes on the Odonata of southern Minnesota. *Rept. Minn. State Entomol.* 13:77–102.

White, Charles E., et al. 1984. Checklist of the aquatic Coleoptera of Indiana. *Proc. Ind. Acad. Sci.* 94:357–64.

White, R. E. 1962. The Anobiidae of Ohio (Coleoptera). *Bull. Ohio Biol. Surv.* (N.S.) 1 (4): 1–58.

————. 1983. *A field guide to the beetles of North America.* Boston: Houghton-Mifflin.

Wickham, H. F. 1897. The Coleoptera of Canada. *Can. Entomol.* 29:81–88, 105–11, 148–53, 169–73, 187–93, 201–8.

————. 1898. The Cerambycidae of Ontario and Quebec. *Can. Entomol.* 30:37–44.

Wilcox, John A. 1954. The leaf beetles of Ohio. *Bull. Ohio Biol. Surv.* No. 43:353–506.

Williamson, E. B. 1900. The dragonflies of Indiana. *Rept. Ind. State Geol.*, 1003–10.

Wilmot, Thomas R., et al. 1992. A key to container-breeding mosquitoes of Michigan (Diptera: Culicidae), with notes on their biology. *Great Lakes Entomol.* 25 (3): 137–48.

Wilson, Nixon. 1957. Some ectoparasites from Indiana mammals. *J. Mammal.* 38:281–82.

Wilson, Nixon, and Wendel J. Johnson. 1971. Ectoparasites of Isle Royale, Michigan. *Mich. Entomol.* 4 (4): 109–16.

Wilson, S. W., and J. E. McPherson. 1980. A list of the Fulgoroidea (Homoptera) of southern Illinois. *Great Lakes Entomol.* 13 (1): 25–30.

Wood, D. M., et al. 1963. The black flies of Ontario. Part II. Larval identification with descriptions and illustrations. *Proc. Entomol. Soc. Ont.* 93:99–129.

———. 1979. The insects and arachnids of Canada. Part 6. The mosquitoes of Canada (Diptera: Culicidae). *Agric. Can. Publ.* No. 1686.

Woodworth, Charles W. 1887. Jassidae of Illinois. *Bull. Ill. State Lab. Nat. Hist.* 3 (2): 9–37.

Wright, J. F., and J. Whitehouse. 1941. Additions to the list of Cincinnati Coleoptera. *Bull. Brooklyn Entomol. Soc.* 36: 69–73.

Wright, Mike. 1939. *Argia fumipennis* in Ohio. *Ohio J. Sci.* 39 (3): 155–56.

Wright, Mike, and Alvah Peterson. 1944. A key to the genera of Anisopterous dragonfly nymphs of the United States and Canada (Odonata, Suborder Anisoptera). *Ohio J. Sci.* 44 (3): 151–66.

Zesch, Frank H., and Ottomar Reinecke. 1881. List of Coleoptera observed and taken in the vicinity of Buffalo. *Bull. Buffalo Soc. Nat. Sci.* 4:2–15.

Zoltai, S. C. 1968. Glacial history of the Great Lakes. *Proc. Entomol. Soc. Ont.* 99:15–20.

Index